Varieties of Scientific Experience

Lewis S. Feuer

Varieties of Scientific Experience

EMOTIVE AIMS IN SCIENTIFIC HYPOTHESES

TRANSACTION PUBLISHERS
NEW BRUNSWICK (U.S.A.) AND LONDON (U.K.)

Copyright © 1995 by Transaction Publishers,
New Brunswick, New Jersey 08903

All rights reserved under International and Pan-American Copyright Conventions. No part of this book may be reproduced or transmitted in any form or by any means, electronic or mechanical, including photocopy, recording, or any information storage and retrieval system, without prior permission in writing from the publisher. All inquiries should be addressed to Transaction Publishers, Rutgers—The State University, New Brunswick, New Jersey 08903.

This book is printed on acid-free paper that meets the American National Standard for Permanence of Paper for Printed Library Materials.

Library of Congress Catalog Number: 95–1470
ISBN: 1–56000–223–9
Printed in the United States of America

Library of Congress Cataloging-in-Publication Data

Feuer, Lewis S., 1912–
    Varieties of scientific experience : emotive aims in scientific hypotheses / Lewis S. Feuer.
       p.   cm.
    Includes bibliographical references and index.
    ISBN 1–56000–223–9 (alk. paper)
    1. Science—Philosophy. 2. Religion and science. I. Title.
Q175.F418   1995
501—dc20   95–1470
                                                                                          CIP

To the Memory

of

Harry Austryn Wolfson

Contents

	Introduction	ix
	Acknowledgements	xix
1.	Noumenalism and Einstein's Argument for the Existence of God	1
2.	Teleological Principles in Science	41
3.	God, Guilt, and Logic: The Psychological Basis of the Ontological Argument	73
4.	Sociological Aspects of the Relation Between Language and Philosophy	99
5.	The Principle of Simplicity	121
6.	The Genetic Fallacy Re-examined	141
7.	The Reasoning of Holocaust Theology	165
8.	Confronting Evil and Its Unreason	181
9.	The Philosophical Method of Arthur O. Lovejoy: Critical Realism and Psychoanalytical Realism	191
10.	Lawless Sensations and Categorial Defenses: The Unconscious Sources of Kant's Philosophy	211
11.	The Dream of Benedict de Spinoza	253
12.	The Dreams of Descartes	269
13.	Anxiety and Philosophy: The Case of Descartes	289

14. Spinoza's Thought and Modern Perplexities:
     Its American Career   321
15. John Stuart Mill as a Sociologist:
     The Unwritten Ethology   355
16. The Sociobiological Theory of Jewish Intellectual
     Achievement: A Sociological Critique   385
17. Causality in the Social Sciences   421
    Index   437

Introduction

A collection of essays is literally a collection of efforts, and mine spanning forty years have ranged from the logical to the metaphysical, from the sociological to the psychoanalytical. Actually, however, apart from the dates in which they were written, there was a certain continuity in their problems and ideas.

To various readers, it seemed strange that I wrote of "teleological" principles in the thinking of scientists in the twentieth century. My usage, however, was intended to emphasize the fact, sometimes forgotten, that the gestation of the hypotheses of original-minded scientists, such as Darwin, Einstein, or Bohr, is, in large part, an unconscious process; during their brooding upon problems, such scientists try to project upon the world structural laws that besides fitting the given physical realities will also realize their own emotional longings among the alternative world views. Even the problems as the given individual scientist defines them will express emoticnal choices and directions peculiar to himself.

To be sure, the philosopher-scientist, Immanuel Kant, maintained that certain a priori principles were common to all thinkers. More significant, however, for explaining the fruitful conflicts and divergences among scientists has been the gradual recognition that diverse varieties of scientific character are apt to seek and discover corresponding kinds of scientific truth. If such an experimental biologist, for instance, as the radical immigrant, Jacques Loeb, hoped to take pleasure in finding physico-chemical laws under which he could deterministically subsume the activities of German militarists, there was, on the other hand, the sensitive young Belgian priest, the Abbé Georges Lemaître, who after enduring the ordeal of the trenches during the First World War, conceived the hypothesis of an originating "display of fireworks" that was a milestone in the theorizing of an expanding astronomical universe. It seemed likewise a psychological corollary that the young physicist, J. Robert Oppenheimer, the only young American scientist reliably reported to have read all three volumes of Karl Marx's *Capital* aiming to prove the inevitability of capitalist collapse, was also moved to study the likely patterns of catastrophic gravitational collapse in the macro-physical world. In such isomorphic convergences, the scientist's character, with its underlying purposive, sometimes unconscious aims, may exert a crucial bearing on the scientist's work. It is told, for instance, that when the great Austrian physicist, Erwin Schrödinger, listened in Copenhagen to the arguments of the indefatigable Danish physicist-philosopher, Niels Bohr, in favor of the quantum approach as against that of the wave mechanics, Schrödinger commented bitterly that if such were indeed the case, he would rather have done with physics. If the scientific unconscious of one given inquirer may find its intuitive aims and instinctive predilections most at home with some particular methodological approach or procedural guide, then that of another may feel itself enclosed by the formulae of an emotionally inconsonant alternative.

When I first began studying these questions, the dominant aim of the younger philosophers in the universities was the study of linguistic usage and distinctions. Although I had read Wittgenstein's *Tractatus* in my senior undergraduate year, having been misled perhaps by its title's resemblance to Spinoza's *Tractatus*, I was not drawn to an outlook and style so much out of keeping with my own sociological and political interests. Among the articles I thus wrote shortly after the Second World War was one on "Sociological Aspects of the Relation between Language and Philosophy." It aimed to show that, as a matter of sociological fact, the structural grammar of a society's language had little to do with the formation of its basic philosophical ideas. The essay, cordially received by social scientists, was then reprinted in an anthropological textbook, where it was regarded as a reply to the then fashionable sociological relativism. For I had found that not only had the most diverse philosophies arisen among the users of the same given language but, by contrast, the same metaphysics had frequently flourished among the users of grammatically diverse languages. Then, too, if in some language the words for "table" or "thought" were feminine in gender, the contrary might be the case in others, so that the distinguished philologist, Otto Jespersen, concluded that it was "impossible to find any single governing principle in the chaos." If the subjunctive mood were associated by many with feelings of doubt or disappointment, it was also true that Kierkegaard's anxieties seem to have originated and communicated themselves well enough in a language that was discarding that mood.

Then too, sociological antipathies have a way of projecting themselves into scientific divergences. Although the noted experimental physicist, P.W. Bridgman, owed much to Einstein for his own "operational" theory of meaning, Bridgman warned that "the Jew is particular prone to mystical modes of thought." On the other hand, A.N. Whitehead, who was something of a mystic himself, exhorted the Aristotelian Society in 1916 not to allow themselves "to be bludgeoned into accepting the rather bizarre doctrine of relativity." Though it is hard to picture Einstein, a pacifist throughout the First World War, as wielding an intellectual bludgeon, it is obvious that passionate feelings and preferences had been evoked, or were underlying in such scientific controversies. Einstein did indeed enjoy reading Spinoza for whom scientific knowledge merged with the intellectual love of God. On the other hand, Einstein also regarded himself as a follower of the positivist physicist, Ernst Mach, who regarded such concepts as space and time as required to satisfy experimental or observational criteria rendering them serviceable for scientific work. Bertrand Russell, on the other hand, holding fast to the intuitionist privilege of G.E. Moore and the circle of Cambridge Apostles, had rejected the Machian relativist approach.

If the scientific unconscious in its formation or adherence to hypotheses is affected by underlying emotional aims in world-seeking, it is also true that abstract arguments in philosophy such as the oft-invoked "ontological argument" for the existence of God have evidently rested basically on nonlogical, psychological fixations. This most celebrated "logical" argument was first introduced in the latter part of the eleventh century by the celebrated St. Anselm, an Italian monk who rose to the archbishopric of Canterbury, and is still eloquently advocated by logical Wittgensteinians. The "argument," remarkably simple, affirms that from the very definition of a Perfect Being, it follows that such a Being necessarily exists, for if it didn't exist, the Perfect Being would not be perfect but imperfect; a Perfect Being, by definition, is thus regarded as necessarily existent. To which the common sense of ordinary working people responds by saying that if a Perfect Being were to include necessary existence among its properties, then it might well be the case that whatever exists is imperfect, that is, that nothing exists necessarily or, as a popular song of Ira and George Gershwin once had it, "it ain't necessarily so."

But then one notices too a curious sociological uniformity, that in the twentieth century as in the eleventh, those philosophers who have adopted the "ontological argument" for the existence of God have usually conjoined it with a commentary on the human guilt for human sinfulness. Somehow it is suggested that only those persons overwhelmingly sensitive to their personal guilt will be sufficiently logically sensitive to grasp the overwhelming validity of the ontological argument. A frame of mind that we might call "logical masochism" seems to be a necessary condition for perceiving the alleged validity of the ontological argument. The alleged Necessary Perfect Being, which many philosophers find neither necessary nor perfect, is taken to be such only when a sense of personal guilt has been so overwhelming, as to humiliate or browbeat our intellectual processes into a total surrender.

Such considerations have led me to reopen from a scientific standpoint the status of the so-called "genetic fallacy." For from John Locke onward, the advances in our understanding of such basic concepts as causation and sundry "intuitions" have been based on advances in psychogenetic understanding. Curiously, even the experimental physicist or astronomer, concerned with either the precision of his measuring equipment or the utter cleanliness of his telescopic lenses and mirrors, is engaged in obviating "technogenetic" fallacies, akin logically to those arising from "psychogenetic" interferences. When Eddington in 1919 led his little expedition to Principe Island, a lonely strip of ten miles off the coast of West Central Africa, to take a series of photographs during the sun's total eclipse, and measure the successive displacements that occurred to the light emitted from a distant star as it traversed the sun's gravitational field, he was much concerned with the clean readiness of his photographic plates and the angle of their mounting; for the experimental testing of Einstein's theory of relativity depended on the meticulous, unbiased reportage from the expedition's physical equipment. Possibly, as a Quaker, Eddington may even have prayed for a cloudless sky.

Probably neither Einstein nor Eddington knew of the unspeakable human torments that had beclouded this Principe Island. In 1492 the Portuguese Inquisition had torn thousands of small Jewish children between the ages of two and ten from their parents, and delivered them to Principe Island. Most of them died soon, though six hundred still survived into the sixteenth century. Tse-tse flies in the nineteenth century reduced the total native population to 550 altogether; a few may have had some Jewish blood.[1] By some unfathomable cosmic irony, the crucial astronomical verification of a profound human theoretical achievement was being made on an island redolent with unspeakable deeds of a cruel company of humankind that professed an intimacy with Divine Intent.

Genetic analyses indeed can well be helpful, and sometimes indeed constitute a precondition for restoring intellectual clarity and firmness. In their dreams, indeed, the philosopher-scientists of the seventeenth century coped with their anxieties as the bearers of the new revolutionary ideas in searching for truths. Descartes, the most sympathetic personality among them, sincerely held that God's endorsement for his scientific enterprise was conveyed to him during the night of November 10, 1619, while he, a young man twenty-three years old, serving as a soldier in the army of the Holy Roman Emperor, and resting close to a stove in a house near the Danube River, was dozing and meditating. Orphaned of his mother when he was a little child, and rather ignored by his father, young Descartes was finding his joy in scientific thinking, and naturally appreciated that in his dreams, God seemed to be reassuring him that the external world was real, not a mirage. A Perfect Father replaced the biological one who seemed to affect Descartes with an "amputative" pain. And Descartes' disciple, the excommunicate lens-grinder in Amsterdam, Benedictus de Spinoza, told too of a dream he had had evidently in the winter of 1663-64, when "a certain black and scabby Brazilian" came to menace him. The symbolism had a historic timeliness, for a few years earlier the Dutch Jewish colony in Brazil had been attacked by a Portuguese and Negro force that aimed to exterminate them; at the last moment, a Dutch fleet of rescue appeared as if miraculously. The spiritual leader of the Brazilian Jewish colony was the Rabbi Isaac de Aboab, who was none other than the same rabbi who in 1656 read the decree to the Amsterdam Jewish synagogue that Spinoza was excommunicated. Spinoza's dream was laden with the lonely anxieties he experienced as a young scientist-philosopher, trying to make his living as a grinder of telescopes, but also as a political philosopher, casting his lot with the remarkable republican leader and advocate of science, John de Witt. When a few years later, a Dutch mob, enraged by a military defeat, lynched John de Witt and his brother Cornelius, Spinoza wanted to rush out to confront the mob, and was forcibly restrained by his landlord who refused to see him going to a similar death. The scientist-philosopher perforce asked the question: how was a liberal democracy to survive if the citizenry oscillated at the edge of savagery? The imagery of the threatening savage, which the rabbi and the Brazilian Jewish colony had indeed experienced, was a reminder of the irrational passions that could threaten the liberal democratic thinker himself.

If it was Spinoza's determinism, together with his identifying God with Nature, that most affected Einstein's philosophy as a scientist, it was also the case that Einstein, together with such elder colleagues of his as Max Planck, was deeply impressed by Kant's doctrine of a priori, intuitive principles; Kant was long respected as the foremost German scientific philosopher. Kant all his life, however, was troubled by curious anxieties with respect to his sensory mechanisms—sight, smell, hearing—and by dramatic anxiety-laden dreams as well. Probably these anxieties led to Kant's countermeasures, his strict categorial control of the mind. Though Einstein dispensed with such a Kantian rigidity as Euclidean geometry, Einstein held fast to the importance of determinist theory for the understanding of reality. I have used the psychoanalytical approach to explore how Kant's categorial principles were largely imposed as measures to constrain the otherwise lawless provocations in sensations themselves, provocations evidently sexual in character.

On the other hand, it was also true that among contemporary philosophers, the presumable use of Occam's Razor could also lead to a kind of methodological solipsism in which one dispensed with the assumption of others' experiences, and aggressively enlarged Occam's Razor toward a status akin to Occam's Axe or Guillotine. When methodological rigor then clearly culminated in solipsism, one could take pleasure, as Russell once suggested (perhaps humorously) in omitting the assumption of others' existence, a kind of methodological sadism. The use of Freud's psychoanalytical method to enquire into the origins of epistemologies seems to be the natural extension of John Locke's psychological inquiry into the origins of both scientific knowledge and dogmatic, antiscientific error.

Although Western philosophy has, on the whole, since Descartes preserved an optimist outlook, a certain measure of gloom affected during the nineteenth century such thinkers as Herbert Spencer, John Stuart Mill, and Charles Darwin. The fresh American pragmatic standpoint of William James and John Dewey, by contrast, evinced a zest for the future and the democratic liberation of individuals that was new in philosophy.

Nonetheless, the current "Holocaust Theology" now transmits a traumatic foreboding of evil which is unprecedented in the history of Western philosophical thought. The "Holocaust," the mass eradication of European Jewry, was evidently the first instance in the world's history where one advanced people was trying to obliterate altogether another advanced one. All of Freud's pessimistic prognoses for the future of European civilization seemed validated. Central European, German civilization rebelled against the containment and repression of underlying primitive aggressive drives. Could one sincerely hope for a new Enlightenment, led by a new generation of Voltaires, Lessings, and Humes when a collective, massive trauma impugned the whole sincerity and vestigial sympathy of the dominant European psyche? None of Schopenhauer's homespun remedies of musical concerts for the melancholy were of serious avail, and a fresh generation of theologians dared look for God's presence even in the loci for human sacrifice.

The plain fact, moreover, was that the Holocaust had overwhelmingly made disbelievers of its survivors, and this fact, by itself was perhaps an insuperable obstacle to a Holocaust theology, as it was called. The ancient prophets tried to justify a God who admonished errant adults, but when He permitted stunned children to be thrust into poisonous chambers and crematoria, the natural inference was that God was either extinct, or non-existent to begin with, or irrevocably bewildered through repeated defeats; the creative evolutionary zest that Henri Bergson had once inspired was now dissipated. When Bergson bravely insisted on standing in line to sign his name on the Nazi lists for potential Jewish executees, he probably realized that his *Creative Evolution* lacked indeed the chapter on Retrogressive Devolution that he was now too tired or even disillusioned to try to write. The wisest of the American scientist-philosophers, Benjamin Franklin, had spoken, on the other hand, forthrightly against the Holocaust-makers of his own time, (massacrists of a harmless Indian tribe), in words more straightforward than we now customarily use. With no qualification, hesitation, or scholarly hesitation, Franklin wrote thunderously: "But the wickedness cannot be covered, the Guilt will be on the whole Land, . . . The Blood of the Innocent will Cry to Heaven for Vengeance."[2] Holocaust Theology seems indeed a sorry intellectual anodyne for the macro-political failure on the part of the world's peoples and their politicians; they had refused to deal early and thoroughly with the well-advertised plans of the Nazi agents for destruction. One can well understand that the doubt never left Franklin, as he wrote in 1782 to his friend, the scientist Joseph Priestley, discoverer of oxygen, "whether the Species were really worth producing or preserving."[3]

The scientist-philosophers of the future will, in the measure of their sincerity, have the responsibility as the permanent skeptics as to the claims of Western civilization; for they made Western civilization possible through their advances in science and technology, and theirs therefore is the chief responsibility for its wise or unwise basic evaluation. Theirs is the primary duty for challenging the world's peoples to a higher vocation and more profound communion than that which perished with their collapsed God.

Probably a new mode of atheism will challenge that it would be far better if no God existed rather than One capable of permitting such massive collective evils as the Holocaust. Such thinkers, however, as Thomas Hardy, Bernard Shaw, and Samuel Alexander have dared to envisage the evolutionary emergence of a more comprehending World Deity, who will assist the foundering human enclaves to cooperate indeed with His efforts; in that case, the tired human species will have then done well to endure its struggle, not always victorious, with evil intent and malicious ignorance.

The guiding philosopher-sociologist to have emerged from the nineteenth century with this standpoint, was in my view, John Stuart Mill. In Parliament itself, he raised, for instance, squarely the basic problem of the depletion of Britain's coal reserves. Whereas Marx and Engels were simply confident that a socialist society would somehow ensure the indefinite development of the productive resources, Mill cited the evidence assembled by the patient economist-logician, W. Stanley Jevons, of the diminishing finitude of Britain's mineral reserves. Mill respected logic too much to allow himself the ease of projecting our wishes into alleged laws of sociology, and unlike Marx, he refused to flatter the workingmen by telling them that history had assigned to them the role as its most progressive agency. Instead, Mill advised the

International Workingmen's Association to work for socialistic reforms but without using the language of violence and revolution. Yet he would defend the legally elected Paris Commune and its proposed reforms against a hostile British press. And if Karl Marx couldn't restrain himself from trying to find fault with Charles Darwin's scientific method, Mill pleased Darwin greatly by reassuring him that the *Origin of Species* was a work "in the most exact accordance with the strict principles of logic."[4] If, as a militant youngster, Mill had cherished a revolutionary project for eliminating every British person with an annual income of £500 a year or more, in his maturity he concluded that the "revolutionary socialists" with their aim to direct by one "central authority" all the productive operations were actually being impelled by "malevolent" motives. Mill was indeed the first psychologist of socialism, an inquiry which Marx and Engels carefully avoided, though occasionally, in controversy with more militant advocates, they would caution that many years would be required before the "proletariat" would master the arts of social administration, and might in that interim even be fighting colonial wars.

Of all America's philosophers during the twentieth century, Arthur O. Lovejoy was generally regarded as probably the most learned; he was also among the few philosophers who regarded themselves, like Freud and Einstein, as "psycho-physical dualists"; mental and physical events or entities were for him both real and irreducible one to another. Moreover, Lovejoy was always ready to debate with all manner of philosophers, and had argued publicly (and perhaps cogently) with Royce, William James, and John Dewey. Probably his most historic encounter, however, was that in Baltimore in 1935 when he was chosen by the American Philosophical Association to argue with the newly arrived logical positivist refugee from Nazi Germany, Rudolf Carnap. With all his rigor and vigor, Carnap ridiculed the outmoded, dualistic realist who when he affirmed, "This animal—say a worm—has consciousness," believed he was saying something more than that this worm reacted physically in certain ways to certain stimuli. Lovejoy replied that a mother does indeed infer that her child is feeling pain when it cries, and that if children were regarded as "insensible automata" when they cried, we would not respond to appeals from relief agencies for funds. Carnap replied that such statements lacked "theoretic content." Thus, during the onset of the Hitlerite era, the Vienna Circle, with its socialists and liberals, professed themselves as solipsistic behaviorists. Perhaps redolent of that era was a cultural sado-masochism that misdirected philosophers, especially Central European ones, into affirming the alleged scientific meaninglessness of statements concerning the "consciousness" of worms, together logically, with that of persons. The despair that mothers and children later voiced in Hitler's death-installations had thus unwittingly received a preparatory logical analysis as devoid of "theoretic content." Probably the unusual thesis as to the meaning or meaninglessness of propositions about persons' feelings was a sociological corollary deriving from the varieties of aggressive feelings to which philosophers were not immune, from those founded on self-aggression to diverse kinds of other-aggression. But such analyses were frowned upon in those days as tainted with "genetic fallacy." Thales' reminder at the beginning of philosophy to "Know Thyself" had been forgotten.

Finally, I should like to express my deep appreciation to Professor Irving L. Horowitz, the guiding spirit of Transaction Publishers, who invited me to assemble this collection of essays written during the decades after the Second World War. Would that they might contribute to the more rational society that was the aim of our intellectual generations.

Notes

1. Tony Hodges and Malyn Newitt, *São Tomé and Principe, From Plantation Colony to Microstate,* Boulder, Col., 1988, pp. 18,55.
2. Benjamin Franklin, "A Narrative of the Late Massacres in Lancaster County," in Benjamin Franklin, *Representative Selections,* ed. Frank Luther Mott and Chester E. Jorgenson, New York, 1948, p. 310.
3. Ibid., p. 444.
4. Also, see *The Letters of John Stuart Mill,* ed. Hugh S.R. Elliot, London, 1910, Vol. I, p. 236; Vol. II, p. 181. Darwin wrote to Henry Fawcett: "You could not have told me anything which would have given me more satisfaction than what you say about Mr. Mill's opinion." See Letter of Henry Fawcett to C. Darwin, July 16, 1861, in *More Letters of Charles Darwin,* ed. Francis Darwin, Vol. I, London, 1903, p. 189.

Acknowledgements

1. "Noumenalism and Einstein's Argument for the Existence of God," *Inquiry,* Vol. 26 (1983), pp. 251–285.

2. "Teleological Principles in Science," *Inquiry,* Vol. 21, Winter, 1978, pp. 377–407.

3. "God, Guilt, and Logic: The Psychological Basis of the Ontological Argument, *Inquiry,* Vol. 11, (1968), pp. 257–281.

4. "Sociological Aspects of the Relation Between Language and Philosophy," *Philosophy of Science,* Vol. 20, No. 2, April. 1953, pp. 85–100.

5. "The Principle of Simplicity," *Philosophy of Science,* Vol. 24, (1957), pp. 109–122.

6. "The Genetic Fallacy Re-examined," in *Sidney Hook: Philosopher of Democracy and Humanism,* ed. Paul Kurtz, Buffalo, New York, 1983, pp. 227–246.

7. "The Reasoning of Holocaust Theology," in *Judaism: A Quarterly Journal of Jewish Life and Thought,* Vol. 35, (1986), pp. 198–210.

8. "Confronting Evil and Its Unreason," *Encounter,* Vol. LXX, No. 5, May, 1988, pp. 67–70.

9. "The Philosophical Method of Arthur O. Lovejoy: Critical Realism and Psychoanalytical Realism," *Philosophy and Phenomenological Research,* Vol. XXIII, June 1963, pp. 493–510.

10. "Lawless Sensations and Categorial Defenses: The Unconscious Sources of Kant's Philosophy," in *Psychoanalysis and Philosophy,* ed. Charles H. Hanly and Morris Lazerowitz, New York: International Universities Press, 1970, pp. 76–125.

11. "The Dream of Benedict de Spinoza," *American Imago,* Vol. 14. No. 3, (1957), PP. 76–242.

12. "The Dreams of Descartes," *American Imago,* Vol. 20, Spring, No. 1, (1963), pp. 3–26.

13. "Anxiety and Philosophy: The Case of Descartes," *American Imago,* Vol. 20, No. 4, Winter (1963), pp. 411–449.

14. "Spinoza's Thought and Modern Perplexities: Its American Career," in *Spinoza: A Tercentenary Perspective,* ed. Barry S. Kogan, Hebrew Union College–Jewish Institute of Religion, Cincinnati, 1978, pp. 36–79.

15. "John Stuart Mill as a Sociologist: The Unwritten Ethology," in *James and John Stuart Mill: Papers of the Centenary Conference,* ed. John M. Robson and Michael Laine, University of Toronto Press, (1976), pp. 86–110.

16. "The Sociological Theory of Jewish Intellectual Achievement: A Sociological Critique," in *Ethnicity, Identity, and History,* ed. Joseph B. Maier and Chaim I. Waxman, New Brunswick, 1983, pp. 93–123.

17. "Causality in the Social Sciences," *The Journal of Philosophy,* Vol. LI (1954) pp. 681–695.

1

Noumenalism and Einstein's Argument for the Existence of God

I

During most of his latter life, Einstein used to say that he held to Spinoza's standpoint toward the universe; he believed that all events were instances of the operation of determinist laws, the grandeur of which filled one with a cosmic, religious emotion, with the sense of an impersonal intelligence manifested in nature. Toward the end of his life, furthermore, Einstein's view approached a conviction that a personal intelligence was underlying the cosmic order, that otherwise the universe that existed would far more likely have been a chaos than a cosmos. Einstein explained himself in a letter written in March, 1952, to an old friend of his Bern years, Maurice Solovine:

> You find it curious that I regard the intelligibility of the world (in the measure that we are authorized to speak of such an intelligibility) as a miracle or an eternal mystery. Well, *a priori* one should expect that the world would be rendered lawful only to the extent that we intervene with our ordering intelligence. It would be a kind of order like the alphabetical order of the words of a language. The kind of order, on the contrary, created, for example, by Newton's theory of gravitation, is of an altogether different character. Even if the axioms of the theory are set by men, the success of such an endeavor presupposes in the objective world a high degree of order that we were *a priori* in no way authorized to expect. This is the 'miracle' that is strengthened more and more with the development of our knowledge.
> Here is where the weak point is found in the positivists and dedicated atheists who regard themselves fortunate because they feel that not only have they, with entire success, deprived the world of gods, but also 'despoiled' it of miracles The curious thing is that we have to content ourselves with recognizing the 'miracle' without having a legitimate way of going beyond it . . .[1]

According to Einstein, it would have been *a priori* far more probable that the world should have been chaotic and unintelligible than orderly and intelligible. The fact that it is intelligible is 'a miracle or an eternal mystery'. There is no 'legitimate way', that is, a scientific way, for explaining this 'miracle or eternal mystery'. But he thinks it is a fact that contravenes the positivist and atheist standpoints; in other words, the miracle of an intelligible world points to the other alternative, a realistic theism of some kind.

In a chaotic world, events, in Einstein's view, might at most be arranged like the words in a dictionary or the names in a telephone book, but the analogy goes no further; there would be no philological or linguistic laws to explain the attribution of gender in nouns, or the cases or the conjugations of verbs, or explanations as to how certain sounds came to be used for certain meanings. An 'empirical formula' for the world's events might, in principle, be contrived, but it would be as complex as the chaos itself, for its numbers of both coefficients and variables would equal that of the events.[2]

A miracle in ordinary parlance is an inexplicable departure from law or regularity. Einstein's argument, in one sense, employs the world 'miracle' with another significance, for he means by it not a departure from law but the existence of law itself. On the other hand, in another sense, he has extended by analogy the meaning of law or regularity. For in the most abstract sense of possibility, an infinite number of worlds might possibly be conceived, embracing all sorts of chaotic universes as well as others governed by law, of varying degrees of complexity and simplicity. Each such universe would, in the utter absence of any anterior law, be as equipossible *a priori* as any of the others. The fact that of all these possible universes the one that does exist is characterized by simple, beautiful mathematical laws is for Einstein, by analogy, the ultimate scientific miracle. This miracle, moreover, makes human knowledge possible, for if the physical universe were not governed by simple, beautiful mathematical laws, a creature such as man would not have been able to trace its workings. Thus, Einstein reverts to the conception of Descartes that clear and distinct ideas are man's avenue to truths because God indeed created a world that expressed clear and distinct thoughts.[3]

For all Einstein's veneration of Spinoza, and admiration for his conception of determinism and the intellectual love of God, Einstein's argument for the scientific miracle constituted an abandonment of

Spinoza's pantheism, which rendered Nature equivalent with God. According to Spinoza, whatever was possible as an idea of God would be actualized; there are no pure possibilities, according to Spinoza's metaphysics, that remain unfulfilled; for God's power is infinite, and expresses itself in all the infinitely possible ways. Furthermore, from Spinoza's standpoint, such notions as 'order', 'simplicity', 'beauty' are but aids to the human imagination, and it would be anthropomorphic to limit God's power to the perspective of such arbitrary human preferences. Indeed, Einstein's argument has moved him from a Spinozist pantheism toward a Leibnizian theism, the belief that God chose to actualize that one among the infinite possible universes which had the most beautiful simple laws combined with the maximum of architectonic detail.

Einstein usually averred that he rejected the notion of a personal God; but he defined a personal God as one 'interfering with natural events'; such a doctrine, in Einstein's view, took 'refuge in those domains in which scientific knowledge has not yet been able to set foot';[4] therefore, he called upon teachers of religion and ethics to relinquish it. As for himself, he wrote:

> I believe in Spinoza's God who reveals himself in the orderly harmony of what exists, not in a God who concerns himself with fates and actions of human beings ... I believe that intelligence is manifested throughout all nature ... .[5]

Einstein's actual argument for the existence of God, however, is one on behalf of a personal God, an Intelligence that manifested Himself in a cosmos rather than chaos. The notion of a personal God does not necessarily include the idea of what we might call an 'Intervening Deity', that is, one who chooses from time to time to intervene in biological or human processes so that their outcomes are altered. The notion of an 'Intervening Deity' is independent of that of a 'Rational Designer'. Einstein's conception of God did evolve from the Impersonal Natural Order to that of a 'Rational Designer'. The steadfast repudiation of an 'Intervening Deity' was related to his Schopenhauerian pessimism, to his experience of human cruelty and human misery. He enthroned a Divine Intelligence that some would say had a coefficient of cruelty, or intellectual sadism.

Now Einstein's philosophy was not a speculative addendum to his work as a scientist; rather it both guided and was affected by his scientific researches. His theory of the gravitational field law, said

Einstein, emerged amazingly in a 'purely formal investigation'; 'it is mathematically determined by the principle of general relativity', by the requirement that the laws of nature be covariant with respect to all continuous transformations of the coordinates.[6] This same standpoint, which we might call Einstein's *noumenalism,* guided the efforts of his latter years to formulate a Unitary Field Theory; thus Einstein stated his general problem as one of deriving the empirical laws of nature from a theory formulated primarily in accordance with aesthetic-mathematical requirements: 'Which are the simplest formal structures that can be attributed to a four-dimensional continuum, and which are the simplest laws which may be conceived to govern these structures?'[7]

At this juncture, Einstein acknowledged that physicists who rejected his philosophical standpoint, and who inclined 'towards realism or positivism' would regard the method of the General Theory as a 'weakness'; they would cavil at the 'degree of formal speculation, the slender empirical basis, the boldness in theoretical construction . . .'. In partial reply, Einstein invoked the studies of Emile Meyerson, the historical epistemologist of science, who had long argued that scientific theory demanded a metaphysical, rationalist conception of causality, that the positivist notion of scientific law as bare uniformities omitted the condition of identity which our reason finally required between the antecedent and the consequent of every scientific causal proposition:

> Meyerson, in his brilliant studies on the theory of knowledge, justly draws a comparison of the intellectual attitude of the relativity theoretician with that of Descartes, or even of Hegel, without thereby implying the censure which a physicist would naturally read into this.[8]

Einstein well realized that 'the sceptic will say that this is a "miracle creed" ', but, he responded, 'it is a miracle creed which has been borne out to an amazing extent by the development of science'.[9] Many years earlier, around 1912, the mystical religious philosopher, Martin Buber, who repudiated scientific knowledge as a path to a knowledge of God, pressed Einstein concerning his faith: 'Finally he [Einstein] burst forth. "What we strive for," he cried, "is just to draw his lines after *Him*".'[10] That man could draw such lines after Him, as a child traces the lines of a picture book, was possible only because He had fashioned a world whose laws indeed were simple and mathematically elegant. God, for Einstein, became a heuristic principle. 'Many a time, when a new theory appeared to him arbitrary or forced, he remarked: "God doesn't do anything like that" ', recalled the eminent physicist, Arnold Sommerfeld. Einstein, seeking the law underlying two classes of phenomena, asked himself: 'How would God have solved the problem?'[11]

Now, from the logical positivist point of view, the theistic cast in Einstein's methodological maxims was simply a humorous turn of speech, a way of talking both reverentially and ironically about the scientist's search for the simplest verifiable hypothesis in accordance with the facts. Therefore, according to his appreciative empirical biographer, Philipp Frank, Einstein 'never encouraged the religious interpretation of recent physics which became current by the popular books of such scientists as Jeans and Eddington'.[12] The truth of the matter, however, is that Einstein shared, at least during his last thirty-five years, the same philosophy of science as the mystical rationalist Quaker, Eddington, and his countryman Sir James Jeans; in their respective ways all were noumenalists. When Einstein's eminent physicist friend, Max Born, wrote an essay sharply criticizing Eddington's mathematical mystical apriorism from an empiricist standpoint, Einstein, on the contrary, defended the role of this 'quixotic', 'Hegelian' propensity; when the latter is lacking, he wrote, 'the inveterate philistine rules'. Accepting sardonically the role of a mathematical Jewish intuitionist that the Nazi ideologists derided, Einstein declared: 'I am therefore confident that "Jewish physics" is not to be killed.' To his old friend, Michele Angelo Besso, who had encouraged him when he was composing his first paper on the theory of relativity, Einstein wrote in 1953 that 'there is perhaps some truth in Eddington's reasonings. The latter has always appeared to me as a man truly filled with ideas . . .'.[13] Eddington indeed claimed to be able to derive the basic laws and constants of physics as Necessities of Mind, as the consequences of essentially epistemological premises. Einstein conceded that Eddington was 'lacking in a critical spirit', and that Eddington's philosophy reminded him of 'a *prima ballerina* who doesn't herself quite believe in the correctness of her elegant leaps', but he nonetheless recognized their community in spirit.

The student of the general theory of relativity finds himself, indeed, enveloped in a noumenalist setting. Arbitrary constants, brute facts, irrational ones, certified purely on empirical grounds are replaced by a rational deductivism. A former director of mathematical studies at

Trinity College, Cambridge, E. W. Barnes,[14] has vividly described the experience of its study:

> The astonishing thing about Einstein's equations is that they appear to have come out of nothing. In deducing them we have, first of all, assumed that the interval-element is such that its square is a quadratic function of the differentials of the space-time coordinates. We have further assumed that the laws of nature must be capable of expression in a form which is invariant for all possible transformations of the spacetime coordinates. But, so far as can be seen, no other assumption has been made; and yet *from this exiguous basis we deduce formulae of gravitation more accurate than those given by Newton's law of the inverse square.*
>
> The conclusion seems to be irresistible that such laws of nature as the principle of conservation of energy, the principle of conservation of momentum and the law of gravitation *are necessary consequences of our modes of measurement.* They are, in fact, elaborately disguised identities which could have been predicted *a priori* by a being of sufficiently powerful analytical insight . . .[15]

Now Einstein would not agree that his formulae are simply consequences of our modes of measurement. For he would acknowledge that a certain empirical content is introduced into his theory by the requirement that the laws of nature be invariant for all possible transformations of the space-time coordinates. If such experiments as that by Michelson and Morley had adduced positive results, the requirements of the Lorentz transformations would have been obviated, and the constancy of the velocity of light for all observers moving uniformly relatively to each other would not have been postulated. The strong limiting condition of invariance that Einstein, however, imposed on the structure of the laws of nature has a rationality, a simplicity about itself, that seems to constitute a claim to ultimate truth.

What then is noumenalism? It stands in contrast to the phenomenalist standpoint that eschews any commitment to or guidance from, any metaphysical idea. The French scientists who originated phenomenalism, Ampère and Fourier, wished to avoid the materialist ideology associated with the French Revolution;[16] they rejected not only atomistic hypotheses, but all the speculative idealistic verbalisms that were emanating from the German universities and theological seminaries. The phenomenalist limits himself to exploring the laws and relations of observable phenomena. The noumenalist, on the other hand, believes he has an *a priori* sense, intuition, or divination of an underlying reason in things, an underlying reality that expresses itself in phenomena that conform astonishingly to simple, beautiful, and elegant laws. Noumenalism too is a heuristic standpoint, leading its practitioner, such as Einstein, with a feeling of prescience as to the structure of scientific theories. When Hermann Bondi, an originator of the 'steady-state' cosmology, was investigating the problem of unified field theories, he proposed to Einstein, as he recalls, 'a suggestion that seemed to me cogent and reasonable', but Einstein only said:

> 'Oh, how ugly'. As soon as an equation seemed to him to be ugly, he really rather lost interest in it and could not understand why somebody else was willing to spend much time on it. He was quite convinced that beauty was a guiding principle in the search for important results in theoretical physics.[17]

A collaborator of Einstein at Princeton observed that the phrase

> 'God created the world' . . . so often repeated by Einstein with variations, was his peculiar religious feeling that laws of nature can be formulated simply and beautifully. When he had a new idea he asked himself: 'Could God have created the world in this way?' or 'Is this mathematical structure worthy of God?'

To all of which his materialist collaborator, Infeld, appended the comment: 'Translated into ordinary language, this sentence means: "Is the theory logically simple enough?"'[18] But is this an adequate translation of Einstein's creed? It omits precisely the noumenal component, the belief that God's Intelligence expressed itself in the structure of the laws of nature; the tenet of the metaphysical simplicity of laws would be maintained even if the phenomena, remaining recalcitrant, seemed to require complex, inelegant structures. The simplicity of natural law was not one hypothesis among others that could be tested, and even discounted. Perhaps it was an 'illusion', Einstein conceded, to believe that 'man is able to comprehend the objective world rationally, by pure thought, without any empirical foundations, in short, by metaphysics'. Nevertheless, he held 'that every true theorist is a kind of tamed metaphysician, no matter how pure a "positivist" he may fancy himself'. He was 'tamed' in so far as he recognized that 'from the standpoint of physics there is nothing to warrant the assumption that a theory which is "logically simple" should also be true'; for the inter-metaphysical standpoint of physics is maintained by the test of experiment. Nevertheless, as a metaphysician, Einstein affirmed intuitively that 'premises of great simplicity' would suffice to comprehend the totality of all sensory experience, his conviction that the limit of truth would coincide with the limit of mathematical beauty.[19]

In such a friend of Einstein as the eminent physicist, Max Born, this noumenalist standpoint evoked a deep anxiety. When Born in 1913 first studied Einstein's papers on general relativity, he found them 'fascinating but . . . almost frightening'. The theory always remained for him

> the greatest feat of human thinking about Nature, the most amazing combination of philosophical penetration, physical intuition and mathematical skill. But its connections with experience were slender. It appealed to me like a great work of art, to be enjoyed and admired from a distance.[20]

Born 'decided never to work in this field'. Born's word 'frightening' tells much concerning the phenomenalist (and materialist) resistance to Einstein's noumenalist standpoint. Einstein might reply that the ancient Hebrews too were 'frightened' to behold the face of God, and that his deiform laws and noumenalist method evoked the ancient anxiety. Born reiterated to Einstein more than thirty years later in 1944 that it was his 'honest opinion that when average people try to get hold of the laws of nature by thinking alone, the result is pure rubbish. Schroedinger might be able to do it'.[21] But even such powerful mathematicians as Eddington and Milne, in his view, were in all likelihood misled into aprioristic fantasy; 'there is no philosophical highroad in science, with epistemological signposts. No, we are in a jungle, and find our way by trial and error . . .'.[22] An old adage has it, 'purus mathematicus, purus asinus', and that indeed was Born's view of scientific noumenalism [23]

Nonetheless, Max Born, critic though he was of noumenalism, found himself using its language spontaneously when he spoke of the greatest moments of scientific insight or intuition. Clerk Maxwell's powerful formulation of the electromagnetic field 'was not mathematics pure and simple, but an amazing act of divination . . . without proper empirical foundation' but guided mainly 'by reasons of mathematical perfection or beauty'. Similarly, physicists before Einstein had failed to grasp the basic significance of the fact that inertial mass was equivalent to the gravitational; Einstein's general theory of relativity sprang from his 'divining' its importance. Born doubted whether one could analyze 'the idea of beauty or perfection or simplicity of a natural law which has often guided the correct divination'.[24] Is this a way of saying that precisely the divine ingredient in divination eludes analysis?

The noumenalist philosophy was one that Einstein shared with several of his great scientific contemporaries. Sir James Jeans outraged many militant atheists when he wrote: 'we may think of . . . the laws of nature, as the laws of thought of a universal mind', yet he expressed Einstein's methodology in his assertion that 'nature and our conscious mathematical minds work according to the same laws'.[25] Like Einstein, Jeans saw evidence in the universe of a 'designing' power that had something in common with our intelligence rather than with our emotion or morality;[26] for Einstein in his frequent Schopenhauerian moods, doubted that the Deity would much bewail the end of the human race.

Jeans, however, did not use the noumenalist philosophy in the heuristic fashion that was characteristic of Einstein; indeed he later dissented from Eddington's effort to derive from epistemological grounds the laws and constants of theoretical physics, and his account of scientific method did not invoke the theistic standpoint.[27] The noumenalist outlook has characteristically been that of scientists who have consecrated themselves to relativity and unitary field theory—Einstein, Eddington, E. A. Milne, Dirac, Schrödinger. Thus, Eddington's efforts to effect a rational deduction of the basic laws of nature and physical constants, drew the critical approbation of Dirac; although Eddington's arguments, in his view, were not always rigorous, 'they give me the feeling', he wrote, 'that they are probably substantially correct in the case of the smaller numbers', such as the ratio of the mass of the proton to that of the electron.[28] To Erwin Schrödinger it seemed that as the knowledge of a given class of facts accumulated, it evolved from the empirical to the rational; that which was first discovered on an experimental level evolved into propositions 'more and more tautological'.[29] Bertrand Russell said much the same thing, though with a negative overtone, when he wrote that the laws of nature, as a philosophical consequence of the theory of relativity, have taken on the character of 'truisms'.[30] Even the anti-mystical P. W. Bridgman, the proponent of the operationalist standpoint, wanted to see the arbitrary universal physical constants, such as that of gravitation, the velocity of light, the quantum, and the charge and mass of the electron, derived in a rationalist way, 'instead of always having to carry them in our equations as elements imposed from without'.[31] The philosopher William James used to revel in the arbitrary, contingent, irrational, stubborn, irreducible character of facts and laws; not so the noumenalist. From Leibniz, who conceived of God as the Mathematical Architect who achieved the maximum of effects with the minimum of means, there first issued that interest in the principle of least action which, as Max Planck wrote, 'was fundamentally based upon the metaphysical idea that the rule of the Deity reveals itself in Nature'.[32] This has been the underlying metaphysical intuition of its formulators from the Leibnizian Maupertuis to the Coleridgean idealist, Sir William Rowan Hamilton. Even the evolution of thermodynamics has likewise been in the noumenal direction. To Boltzmann, in the latter part of the nineteenth century, the presence of such words of human agency as 'sink' or 'reservoir' in formulations of the second law, or of the concept of entropy, marred its logic and language by their anthropomorphism. Consequently he undertook to achieve a more noumenal foundation for the second law of thermodynamics that thereupon virtually approximated to the proposition of pure logic, or reason, that all physical and chemical changes proceed from improbable to more probable states.[33] The noumenalist standpoint, with its tenet that the laws of nature will take a rational form, and that the residuum of irrational facts, or arbitrary constants, will tend to vanish, has been mainly either guided or co-involved with a belief in the Rational Designer of the laws of nature.[34] To be sure, there are various combinations of the phenomenalist and noumenalist standpoints in different proportions and situations, and then, too, there are scientists who revel in the sheer diversity of empirical facts to the point of being skeptical concerning the existence of any underlying law. Thus, the great experimenter, Albert A. Michelson, queried one day as he exhibited to the theorist Max Born the irregularities in the bands of various atomic spectra: 'Do you really think that there is a simple law behind it if horrible things like that happen?'[35]

II

Is the noumenalist standpoint, however, therefore confined to the handful of human beings such as Einstein, Eddington, Planck, Jeans, Milne, who have had the powers to discover, re-trace, or re-experience in highly mathematical theory the thoughts of an Originative Intelligence? Actually every novice in scientific study experiences a sense of wonderment at the simplicity of basic scientific laws; they seem so rational that he wonders, once he has understood them, whether any other pattern could have been as appropriate.

The law of conservation of energy, for instance, to take the most striking example, seems even to the beginning student a virtually *a priori* law. Can we reconstruct his logical processes, what goes on in his mind? Probably the following intellectual sequence:

'Two alternatives are possible: either the total energy in a closed system is conserved or it changes. If it is conserved, its changes are equal to zero; but if the total energy varies, it will increase or decrease by some arbitrary increment or decrement. *A priori,* the probability of its staying the same or changing is one-half for each of the two alternatives. But the second alternative, that of a varying total energy, embraces an infinite class of possible increments or decrements and their combinations; an infinite number of arbitrary constants for the increase or decrease of energy-units in the universe would be candidates for the possible laws. The alternative of permanence, of a conservative system, with its probability of one-half, is a more rational one than any of the class of an infinite number of arbitrary increments or decrements, with its total probability of one-half apportioned to each member with its infinitesimal probability. There seems thus an *a priori* more rational character to the law of the conservation of energy; it is less arbitrary than any of the infinite possible values to an increase or decrease of energy.'

Such an argument, though it expresses the sense of the rationality of the universe, contravenes, however, the logic of Einstein's argument for the miracle of intelligibility. Though the rational universe, unsullied by arbitrary constants, is more rational than the infinite number of possible universes characterized by arbitrary increments or decrements, it is not therefore more probable; if the language of probability were to be used strictly, that universe would still be no more probable than any of the other infinite alternatives. To segment all the possible universes into two alternative classes, the unchanging one, which is a class with only one member, that of zero alteration in energy, and the one of a changing sum of energy, with its infinite possibilities, is to introduce our own 'rational' preferences into the classification of universes. For, if we enumerate all the varieties of universes in a neutral spirit, without regard to meanings of 'chaos' or 'cosmos', then the cosmos, with its rational order, appears with no privileged status as a favored alternative among an infinite number. To make a special class out of a zero permanence would be as illicit as to favor, let us say, some multiple of three units. Our language, experience, and basic preferences do prompt us to weight the probability-classes in favor of the zero alternative, but to do so is to beg the question by assuming beforehand the higher probability of the simpler universe. But if we were to eliminate not only all post-foundational statistical experience but our favoring intuitions as well, then the sheer occurrence of the simplest alternative among an infinite class is a 'miracle'; that the most rational alternative among the infinite number of possibles has been actualized is indeed a miracle.

A sense of the miraculous rationality of the universe characterizes scientists of noumenalist mind. Like Schrödinger, their feeling about the principles of energy is that 'though they are not tautological, it would seem a rather hypocritical broadmindedness to maintain that they are empirical regularities only and could be contradicted by experiment at any time'.[36] But where the noumenalist tends to feel intuitively the antecedent, greater likelihood that the simplest alternative is true, the empirical critic responds that such intuitive reasoning concerning antecedent probabilities actually congeals, or summates, a long statistical experience with hypotheses; such experience, he maintains, has confirmed that the simplest hypotheses have a higher frequency of success, that is, a higher probability of truth. In the view of the empirical critic, Einstein and the noumenalists err when they affirm as an independent intuition that the laws of nature are simple, beautiful, and deiform; the accumulated class of the frequencies of favorable outcomes of the simplest hypotheses as compared to the total number of hypotheses is like a conditioning psychological mechanism that has generated in the noumenalist the sense that he enjoys an independent intuition.[37]

Yet the fact of the matter is that a noumenalist intuition does seem to underlie the judgments made in accordance with the frequency theory of probability itself. Let us examine, for instance, a typical inductive, frequentist argument such as 'the probability of getting a heads in the toss of a coin is one-half'; the probability of a class of events is herein defined as the limit approached by the ratio of the favorable class of events (heads) to the total number of events (tosses of the coin) as the events are indefinitely increased toward infinity. Let us suppose, indeed, that in the course of a hundred million tosses, the ratio of the heads to the total tosses does indeed approach one-half. The notion, however, cannot be excluded that henceforth heads will predominate, and that the ratio will begin to approach two-thirds. Nevertheless, after the first hundred million tosses, we obviously would have felt justified in drawing our curve for future tosses to approximate the limiting ratio of one-half for the heads, and we would have rejected suggestions that our curve might undergo complex involutions and devolutions. We would exclude the indefinite number of such needlessly complicated curves not on the basis of the frequency definition itself but only through invoking an intuitive belief in the truth of the simplest hypothesis. We adopt the simplest curve consonant with the facts. The assignment of the limiting ratio within the frequency theory of probability cannot itself be done without an *a priori* intuitive belief in the maximal simplicity of the truthful alternative.

Though we can verify inductively our belief that the laws of nature are simple by using the history of scientific laws discovered, yet each inductive inference we have made, to begin with, has been founded on an intuitive postulate of simplicity. What the psychological basis of this intuitive postulate of simplicity may be remains unknown. Perhaps it is part of our genetic endowment; perhaps, in a world whose natural laws are simple, those of our distant forebears in whom a genetic mutation took place toward 'least action' thinking, or toward an austere economy in reasoning, or perhaps an aesthetic attraction toward lines of simple beauty, survived as against their slovenly-minded competitors. Perhaps our sense of beauty is appalled by arbitrary constants the way a blemish on a maiden appalls her lover; perhaps our pleasure in elegantly conceived, comprehensively conceived theories is itself a special case among the psychological laws of thought of the cosmic universality of the principle of least action. Or finally, such an intuition may be implanted in human minds as part of what they share with the Divine Mind.

The sense that because the law of conservation of energy is a more rational state of affairs, a simpler one than any of the infinite number of conceivable possibilities, it is therefore probably true, rests indeed on the intuition of a world fashioned by intelligence. All judgments of probability, even those assigning equiprobabilities to alternatives whose causal conditions we cannot analyze, assume indeed some background of constraining physical law or order. We know that the laws of mechanics will continue to preside over the tosses of coins, though the relevant physical conditions are always obscure. We take for granted that when coins are tossed, they will land, and come to rest in accordance with laws of gravitation, momentum, and reaction. We take for granted that when we mix bags of beads, they won't go flying off spontaneously, or disappear. Only against the background of physical law are probabilities observable and calculable. According to the 'Ergodic Theorem' (we are told), the positions of the particles in any given system at a given time will, after an unknown, gigantesque number of seconds have elapsed, repeat themselves; the Nietzschean Recurrence takes on a probabilitistic guise.[38] But such a theorem is founded as well on conditions of transformation, or laws of physics, that are taken as continuously operative.

Einstein's argument for the existence of God is actually based not only on the *a priori* improbability of the existence of simple, beautiful mathematical laws (the 'high degree of order that we were *a priori* in no way authorized to expect') as opposed to ungainly, unaesthetic ones, characterized by arbitrary constants; it involves as well the contrast between the existence of law itself in contrast to the infinite possibilities of lawless chaos. By 'chaos' is meant a universe in which all events are uncaused; only in a tautological sense could there be a 'law' to describe a chaos, for the terms of its equation would in number equal that of the chaotic occurrences. The meaning of 'chaos', moreover, is to be distinguished from that of 'chance' and 'disorder'. Chance events are compatible with the existence of an underlying causal order; chance events, for instance those described by the principle of indeterminacy, are traits or characteristics that occur within carefully circumscribed limits, set forth in a law that presumes a setting of causal laws; chance events are a partially uncaused segment within a predominantly causal world; chance operates relative to underlying law. But a chaotic universe is one devoid of all underlying law; and the infinite varieties of chaos are, from Einstein's standpoint, each as equiprobable as any of the worlds of law, of cosmos itself.

Furthermore, a 'chaotic' world differs in meaning from one of 'disorder'. A world of 'disorder' is associated by scientists especially with those conceptions of the law of entropy that were advanced by Clausius and Boltzmann. The second law of thermodynamics, that heat would not by itself flow from a cooler place to a warmer one, is replaced by the generalization that in all physical processes, the movement of molecules in organized systems tends toward a more disorganized, random aggregation; physical change is thus from order to disorder, though the phenomenon itself, in so far as it is subsumed under the law of entropy, conforms to an 'order' in its direction of change, namely, toward that state characterized by an increase in the logarithm of the probability of its molecular distribution. Physical movement is in the direction from the less probable to the more probable pattern, but not in a tautological sense, since the more probable is linked to the physical outcomes of 'randomization' through collisions. But the evolution toward a 'disorderly' world itself rests on physical laws; Newton's laws of motion, the law of the conservation of momentum, the law of action and reaction in collisions, are taken for granted; and the demonstration that the absolute temperature of a given system is the measure of the kinetic energy of its molecular motions is based on the Newtonian laws. A world of 'disorder' in the thermodynamic sense is thus not the kind of disorder that Einstein has in mind when he envisages the possible worlds of chaos; the latter are lawless, subsumable under no generalization, 'without form', as the cosmologist in *Genesis* speculated.

From the empiricist standpoint, however, Einstein's argument for the existence of God is based on a misuse of the concept of probability. According to the empiricist, the universe, as a whole, is simply given; it is contingent; it is what it happens to be. A contingent universe, or system of objects and their laws, is a member of no probability-distribution of a class of universes; it is neither probable nor improbable; it is non-probable. This universe eludes all probabilistic reasoning because it is necessarily, as the universe-class, a class with only one member, and thus, no member of any probability-distribution. According to the empiricist, we cannot speak with logical right, as Einstein would, of the 'miracle' of the scientific knowledge of an 'objective world' founded on 'a high degree of order' that was improbable, that is, 'that we are *a priori* in no way authorized to expect'.

From the empiricist standpoint, Einstein's argument is founded on a misuse of the 'ignorance', subjectivist theory of probability. Einstein argues that the world of simple, beautiful laws was infinitely improbable, because an infinite number of alternative, less beautiful worlds were equipossible; and that these equipossible worlds were equiprobable. To the empiricist, the successes of the 'ignorance' conception in everyday life are due to the fact that in ordinary experience there is a high correlation between alternatives called equipossible, as the two sides of a coin, and the actual statistical experience in tossing them. Behind every subjectivist decision as to an event's probability lies an inarticulate, concealed experience of statistical frequency. And Einstein, in presuming the antecedent improbability of the universe of simple, beautiful laws, is allegedly guilty of an extrapolation that is illicit. If

16 Varieties of Scientific Experience

our world were merely contingent, neither probable nor improbable, Einstein could not coherently speak of its existence with its simple laws as a 'miracle', for a miracle contravenes either law or the probable inference.

III

In effect, Einstein's argument for the existence of God thus constitutes his culminating break with the philosophy of the Scottish sceptic, David Hume. Together with his friends in the Berne Olympia Academy during the formative youthful years that led to the classic papers of 1905, Einstein had read, argued, and accepted Hume's standpoint.[39] And Hume was the most penetrating critic that the argument from design has ever had. What then would Einstein reply to the criticisms that Hume had made, which most empiricist philosophers have accepted as convincing?

Hume aimed to show that we can trace the causes of only particular individual things, but as he added: we cannot unite them together, and ask for 'the cause of the whole'. Curiously, Hume introduces in the course of his reasonings, a sociological argument, indeed an *argumentum ad hominem,* against the mathematical type of mind for its propensity toward an *a priori* argument for the existence of God. Through his protagonist Philo, Hume writes:

> [T]he argument *a priori* has seldom been found very convincing, except to people of a metaphysical head who have accustomed themselves to abstract reasoning, and who, finding from mathematics that the understanding leads to truth through obscurity and contrary to first appearances, have transferred the same habit of thinking to subjects where it ought not to have place.[40]

To which Einstein can indeed reply: we place our trust in the method of analysis that has brought the human mind its greatest achievements, rather than appeal to 'people, even of good sense', as the arbiters.[41] Many such persons would have projected their 'good sense' to veto both the theories of relativity and the quantum. Is the mathematical type of mind to be regarded as genetically handicapped, or does experience itself show it to be genetically emprivileged?

Hume, moreover, denies that the notion of 'design' can be extended from the parts of nature to nature as a whole. He acknowledges that 'thought, design, intelligence', is 'one of the springs' of 'particular parts of nature'. 'But', he challenges, 'can a conclusion, with any propriety, be transferred from parts to the whole?'[42] Einstein would reply: reasoning from a segment of the universe to its entirety is of the essence of scientific thought: the relativistic law of gravitation, the law of the conservation of mass-energy, the law of entropy, are cosmological laws, that is, laws valid for the universe as a whole (or with qualification, for the universe as a closed system).

Hume reiterates: 'But is a part of nature a rule for another part very wide of the former? . . . Is a very small part a rule for a whole?' According to the Humean, in the scientific cases, an extrapolating takes place from the operations of one observed gravitational field to those of distant, and even unobserved gravitational fields; but the terms of the equation remain homogeneous, and the relations unchanged. One could similarly infer, on the basis of past experience, that the presence of a house indicates that an originative intellect has been at work; but we cannot assume 'the universe bears such a resemblance to a house' as to enable us to infer that a Divine Architect exists; we cannot reason from a part to a whole, except he might add where the 'whole' merely signifies a class of homogeneous entities.

To which the noumenalist, Einstein, or Eddington, or Jeans replies:

> Assuredly, the universe is not a house, but it is a cosmos, characterized by beautiful simple mathematical laws, rather than any one of the infinite conglomerations of chaos, and to this extent it is analogous to all the works of human art that have been created through the ordering of some chance distribution, whether of paints, stones, metals or words, into a work expressive of meanings. And upon that analogy, proceeding from a class of instances similar in one respect. that of their having been created in accordance with patterns of simple beauty, we found our inference of the existence of a rational designer.

At which point Hume invokes what we might call his 'observationist restriction': no inference to a Designer is valid which leaps beyond those classes of designers and designed works that have actually been observed. We can infer from a footprint-shaped indentation in the sand that it has been caused by a man's walking, for we have seen many men walking, and making footprints; any such inference is based on the correlation of two 'collectives', classes with many observed members; but no collectives have been experienced of Designers and Designed Universes. As Hume states it, an observed order cannot constitute 'any proof of design; but only so far as it is experienced to proceed from that principle';[43] therefore, we cannot transfer a conclusion 'from parts to the whole'. 'When two species of objects have always been observed to be conjoined together, I can infer . . . the existence of one whenever I see the existence of the other . . .'.[44] But when the universe as a whole is in question, an object 'single, individual, without parallel, or specific resemblance . . .', no such inference is possible.[45]

What grounds, however, does Hume advance for the 'observationist restriction' on the domain of cause and effect? At this point, he seems to rely on the cultural social-intellectual postulate of his time; the cultivated man of the eighteenth century tended to be a scientific sceptic; in France among the *philosophes,* he would be an anticlerical with Voltaire and Diderot; in Scotland, among the scientists and philosophers of the Scottish Renaissance, he would rejoice in disenthrallment from Calvinist domination. Together with Hume, such thinkers would, in accordance with an emotional *a priori,* tend to disapprove of reasonings that went beyond the observationally verifiable. By contrast, the Quaker scientist, John Dalton, the scientific discoverer of the atomic theory, felt constrained by no 'observationist restriction'. A pure observationist might have been content to state the observed uniformities of the laws of multiple and simple proportions in chemical compounds without advancing any hypothesis concerning the existence of underlying atoms. For although the existence of atoms, indivisible, indestructible, and of the appropriate relative weights, would explain the chemical uniformities, it was highly doubtful in Dalton's time that they themselves would ever be experimentally observed. Yet, like every hypothesis, Dalton's too was based on an inductive analogy, in his case, perhaps, that of the combinations of various kinds of unit-objects, as, for example, different balls with diverse, constant weights.

To which a Humean might well respond that the sheer difference between a hypothesis concerning the existence of atoms and one concerning the existence of God cannot be obliterated; the atomic hypothesis proved to be a heuristic one, that is, it helped guide the discovery of countless new compounds and a whole series of atoms; in principle, moreover, one might have envisaged from its first formulation, the possible experimental conditions under which atoms might perhaps be observed. The theistic hypothesis, on the other hand, according to the Humean, is an isolated speculative extension of the causal principle that is utterly sterile, unheuristic, fruitless, a blind alley, not an opening path, and leading to no novel experimental search or discovery.

Einstein's reply, however, would be clear. He rejects the 'observationist restriction'. The process of extending scientific knowledge, particularly in its most theoretical reaches, is indeed akin to one of divination: we literally ask ourselves, sometimes unconsciously, how would a highly rational and mathematical God have designed, for instance, the law of gravitation? Hume denigrates the human reason when he denies that its imaginations can be in any way analogous to a divine creative reason: 'why select so minute, so weak, so bounded a principle as the reason and design of animals' as 'the foundation of our judgment concerning the *origin* of the whole . . .?', asks Hume; '[w]hat peculiar privilege has this little agitation of the brain which we call *thought,* that we must make it the model of the whole universe?' Yet the human intelligence does retrace gradually the beautiful comprehensive underlying laws of nature, and it finds its own longing for beauty and simplicity satisfied with the congruent structures of nature that emerge into view. Hume, sceptical about any inference from part to whole, bids us mistrust our reason; but he has never had the experience of awaiting the crucial test of a cosmological law of gravitation by observations made during a night in a small segment of nature, an island off the West African coast, and then receiving word that his prediction of the bending of light rays in the neighborhood of the sun during its eclipse was confirmed.

And, if Einstein speculates on the miracle that a cosmos exists rather than one of the infinite varieties of chaos, Hume, notwithstanding his formal 'observationist restriction', speculates informally upon the same miracle of intelligibility. For Hume did wonder if in a world of primal chaos, bereft of any guiding force, matter would remain 'forever in disorder and continue an immense chaos, without any proportion or activity'; he conjectured in a way later pursued by Charles Peirce and Samuel Alexander, that 'a perpetual restlessness', sustained by 'an actuating force' might 'at last' be superseded by some 'uniformity of appearance', emerging 'from the eternal revolutions of unguided matter'.[46] Hume, in other words, for all his observationist self-limitation, did wonder, as Einstein did, why a cosmos rather than a chaos existed, and pondered whether some 'actuating force', that is, a Deity with an aim other than chaos, had somehow transmuted the latter into a world of physical law.

The cautious Hume warned appropriately (through Demea) against those who assumed too easily that they could read God's Mind in Nature's work. When he read an ordinary volume, wrote Hume, he entered 'into the mind and intention of the author; I become him, in a manner, for the instant . . .'. But, added Hume, 'so near an approach we never surely can make to the Deity. His ways are not our ways'. Contrariwise, Einstein affirms that in our highest scientific thought, we do recapitulate or retrace God's Mind; herein was the significance for him of Spinoza's doctrine of the intellectual love of God, the experience of God's Laws of Nature as the necessary expressions of the Highest Rationality. Thomas Henry Huxley, loyal Humean though he was, and a comparative zoological anatomist, paid tribute to Spinoza's *Ethics,* for its 'wonderful piece of psychological anatomy'.[47] Spinoza had indeed described the ethos of the noumenalistic scientist.

Hume's principle of 'observationist restriction' asserts that probable inferences can be made only if the antecedent-consequent clauses of the proposed causal laws refer to classes of events, members of which have been previously experienced. This principle, as we have seen, rests on non-logical emotions; it is the premise of a culture-circle sharing certain common values and disvalues. Indeed, when Einstein was a young Humean, he experienced aggressive feelings and resentments similar to those that had affected Hume. Toward the German military state, he felt much as did the Hume who wrote how he was 'delighted to see the daily and hourly progress of madness and folly and wickedness of England'. Awaiting the crushing of Berlin militarism, Einstein might have spoken as the Hume who hoped for 'the third of London in ruins, and the rascally mob subdued!'.[48] The 'observationist restriction principle' expresses a resentment against the pretensions of human beings, almost bestial in nature, to noumenal knowledge. It pronounces a new punishment for the sinfulness of man, a decree that the fruits of the Tree of Knowledge shall be phenomenal only. But Einstein, as he explored general relativity and field theory, felt that the Tree, perhaps by some mutation, was at last yielding its noumenal fruit.

Curiously, Charles Peirce, the logician of probability theory, though endorsing Hume's rejection of the argument from design, yet conceded that if the material universe were of limited extent and finite age, 'it is conceivable that a general law or design embracing the whole universe should be discovered . . .'.[49] Einstein's theism would satisfy Peirce's condition; his theory of a Riemannian finite universe would have done so plainly, and indeed, theories of an expanding universe do so as well, for, according to them, the 'age' of the universe is calculable with the use of Hubble's law of the recession of galaxies, from the value of the radius of the universe at the given moment.

Einstein's argument for the existence of God, a form of the argument from design, is based solely, however, on a considering of the laws of the physical universe; he found them beautiful, elegant, rational; he never claimed to find evidence of God's Intelligence in the biological universe or in the social and political history of the human races. The biological argument from design received its overwhelming criticism at the hands of Charles Darwin in his *The Origin of Species*. Darwin had emerged from Cambridge University 'charmed and convinced' by the teleological arguments for God's existence that he found set forth in Paley's *Natural Theology*.[50] But his years on the cruises of the H.M.S. *Beagle* in distant lands and oceans, and his subsequent years of reflection, led him to substitute for almost every case of purported biological design an equivalent cumulative series of variations selected within the given species by the struggle for existence. If the biological argument from design thus entered upon its declining days, the sociological argument from design never flourished to begin with. No outstanding sociologist in modern times has seen evidence for God's design in the nature of men's societies, neither Mill, Durkheim, Marx, Pareto, nor Weber. Even in late medieval times, the Jewish philosopher Don Hasdai Crescas was so repelled by the details of human hatred and misery that he decided God did not know about them since his knowledge is confined to universals, and excludes particulars; hence, 'the positing of Divine science of the world's affairs is untenable as the disorder in the natural sphere and the existence of evil in human affairs testifies'.[51] Regrettably, even this argument did not safeguard the Deity from a knowledge of human evil, for if God had been spared a knowledge of the particularities of human genocides, for instance, He would nonetheless have known and presumably rejoiced in the universal laws of genocide, those generalizations of which the particular incidents are cases.

Einstein's basic argument for design, however, is not open to the kind of refutation that Darwin directed against its biological form. Darwin could show how various traits and organs had evolved among the various species so that they seemed so wonderfully adapted, as he remarked at the end of a later edition of his *The Origin of Species,* 'endless forms most beautiful and most wonderful have been, and are being evolved'.[52] But the basic laws of physics, those that Einstein helped formulate, are not themselves the outcome of an evolutionary process. Their invariant structural forms are coeval with the physical universe itself. No trial and error with chance variations, no cosmic selection took place to determine what kinds of laws might be perpetuated rather than others. Darwin's refutation of the biological argument from design is altogether irrelevant to the physical argument. One philosopher-scientist, Charles S. Peirce, did, in a daring speculation, affirm that the physical laws of nature were themselves the outcome of cumulative variations that had diffused through the continuum of world-feelings; thus, through a basic tendency to take on habits, the reign of law, according to Peirce, had gradually replaced that of chance.[53] But such a projection of Darwinian variation and selection into an isomorphic origin of physical laws, though Peirce claimed it was 'really inseparable' from the idea of a personal creator, involved a huge bio-panpsychism; such a multiplied number of auxiliary hypotheses concerning a continuum of feelings diffusing themselves, and an originative tendency for habits among physical feelings to arise spontaneously, would contravene the scientific canons of methodological simplicity and verification.

Einstein's noumenal argument for the existence of God is thus unaffected by the Darwinian critique of biological arguments from design. The elegant, beautiful simplicity of universal physical laws, does stand indeed in pleasant contrast to the wastefulness and cruelty of natural selection, the carnage of lives, the extravagant losses in reproduction, and the destruction of species. Einstein stirred to exhilaration by his discoveries, published them quickly, and thanked God for the wonderful moment when in 1919 a postcard brought him news of the verifying observations during the sun's eclipse off the coast of Africa. Darwin, on the contrary, grew gloomier with the years as he accumulated facts confirming his theory, and he kept postponing publication. He tried to feel exhilarated by his great generalization but rather failed.

IV

Curiously, among scientists, it has been observed that physicists tend more frequently to believe in the existence of God than do biologists, and that the least believing groups are the sociologists and psychologists. In other words, to the extent that a given discipline becomes less amenable to simple elegant laws, to the extent, in other words, that it fails to satisfy the noumenal sense, its seeking into the underlying reason of things, to that extent its practitioners become disbelievers.

The religious beliefs of American scientists were most fully investigated by James H. Leuba in two studies carried out in 1914 and 1933. Professor Leuba polled one-tenth of the scientists listed in the directory of *American Men of Science,* which in 1933 included about 23,000 names, and about half the membership of the Sociological and Psychological Associations; the sample was selected on a random basis. A high proportion of the scientists, at least seventy-five percent in each class, responded. It transpired that the disbelievers and doubters in the existence of God (defined in a personal sense, as in the Christian religion) were the majority. But among the various classes of scientists, the physicists had the highest percentage of believers, actually thirty-eight percent of them; on the other hand, only twenty-seven percent of the biologists believed in God's existence, and the percentage diminished further to twenty-four percent of the sociologists, and reached its nadir with ten percent of the psychologists. Furthermore, the greater scientists (marked by asterisks in the directory) were generally even more disbelieving than their less distinguished colleagues; only seventeen percent of the greater physicists were believers in contrast to the forty-three percent of the 'lesser physicists', while among the 'greater psychologists', the percentage of believers was extraordinarily low, two percent, with the 'lesser psychologists' not too far behind at thirteen percent. During the two decades between 1914 and 1933, moreover, 'a marked increase in unbelief' had taken place; whereas in 1914, the believers constituted forty-two percent of all the responders; that number was diminished in 1933 to thirty percent.[54]

Why, however, was the percentage of physicist-theists almost four times that of the psychologist-theists? The psychologist Leuba asked whether 'psychological learning' made belief in God almost impossible. Was it because the psychological laws discovered were so materialistic and deterministic as to weaken the basis for belief in a supra-physical God? That scarcely seemed the case when one considered the hodge-podge of loosely-grounded correlations that were the staple of psychology. Was it perhaps because those who undertook the study of psychology were to begin with frequently anti-theists, rebelling against religion, and hoping to find in their science grounds with which to substantiate their leanings? Were they desirous, as the votaries of a new and borderline science, to claim to be at least as, if not more scientific, than their physicist colleagues? Certainly, the psychologist could not share the experiences of the relativity physicist in great theoretical generalizations that enunciated cosmological laws, profound and original.

As for the sociologists, who among them could point sincerely to generalizations concerning the advancement of societies that gave one a noumenal joy? Bertrand Russell, turning in the post-World War I years to the study of the Soviet society and the Chinese revolution, seeking a new social hope for freedom in the radical sociological theories, wandered one night along the Volga River among famine-fleeing nomads:

> And at last I began to feel that all politics are inspired by a grinning devil, teaching the energetic and quick-witted to torture submissive populations for the profit of pocket or power or theory.[55]

If the noumenalistic physicist asks: 'How would God have decreed the law for this cosmic reality?', the noumenalistic sociologist might adopt a Satanic heuristic principle: 'How would Satan have contrived the laws of social reality to delude men with mirages, and mislead them into quagmires?' Max Weber saw men as pursuing the rationalization of industry to solve their social problems, but achieving instead an irremediable disenchantment; Freud foresaw the advent of a peaceful order making for men even more tormented by self-aggression; Engels saw history ruled as by a cruel goddess who demanded the sacrifice of millions of lives for every installment of progress; Malthus tried to justify a God who had created a world in which progress could be achieved only through famine, war, disease, vice, massacre; Durkheim envisaged a rising rate of suicide that was the inevitable concomitant of industrial advance and the decline of religion; Herbert Spencer's faith in his law of evolution foundered as he foretold, in his latter years, an era of re-barbarization. John Stuart Mill dropped the utilitarian's axiom of progress, and hoped at best for a sociologically stationary state.[56] Both Bernard Shaw and H. G. Wells, nurtured alike in the nineteenth century's confidence in progress, nonetheless wondered whether a diabolical vector was perhaps the central agent for advance.

God has been portrayed as a Great Mathematician, Great Mechanic, Great Watchmaker, Great Architect, Great Physicist, Great Artist; but no philosopher has had the effrontery to depict God as a Great Sociologist. Laws propounded by sociological masters do indeed have a certain simplicity, but they scarcely impart a feeling of beauty. Grotesque evils pervade the sociological universe, with its cruelties, massacres, genocides, tyrannies. Max Weber's law of bureaucratic inevitability, Robert Michels's law of oligarchy, Vilfredo Pareto's law of the circulation of elites, Karl Marx's law of class struggle, Durkheim's law of an increasing rate of suicide in advanced society, all delineate a sociological universe bereft of inspiring grandeur. Theoretical sociologists have little of the sense of retracing God's thoughts that theoretical physicists enjoy. The latter can take pleasure even in the so-called impotence principles into which all physical theory can be axiomatized, for instance, that no perpetual motion machine can be constructed, that no velocity can be contrived to exceed that of light;[57] sociological theory too has its own counterpart impotence principles, such as that no Utopian society is constructible, or that no society can satisfy men's incompatible drives, that no cultural reconstruction can obliterate biogenetic differences, that a classless society is a sociological impossibility, that no advanced technological society can maintain its equilibrium upon a diminishing base of natural resources. But such sociological impotence principles entail the end of dreams, not their fruitful assimilation into the formation of emotionally satisfying structures of sociological theory.

And if God is not a Great Sociologist, neither does He seem to be a Great Biologist. The wasteful processes in natural selection, the random proliferations, and remorseless extinctions, the slow struggles, the occasional advancement in evolution often being entrapped in degeneration, all were superadded in a spectacle that destroyed Charles Darwin's belief in a Rational Designer, and confirmed him in a lifelong melancholy. What then happens to Einstein's argument that the beautiful, rational simplicity of the laws of nature are evidence of a Divine Intelligence? Is this a Physicist's Fallacy of Selection? Or did

God falter when His Intelligence was confronted with the phenomena of life and society? Einstein himself got no intellectual joy in contemplating the sociological phenomena of Nazi Germany, except for a retaliative pleasure in subsuming them under determinist patterns of human irrationality. Certainly, a Lamarckian law of evolution, with a creature's needs determining the direction of its structural changes would have been far more beautiful, and pleasing to our aesthetic and human sensibilities, than the Darwinian scheme of natural selection. Then how could Einstein salvage his belief in a Divine Intelligence?

It may well be, however, that the rationality of the physical universe itself precludes the existence of such more beautiful laws as the proposed pattern of Lamarckian evolution. In order for emotional, physiological, or intellectual desire to have had the degree of causal efficacy that a Lamarckian pattern of evolution would presuppose, a domain of physical indeterminacy would have been required, so large in its dimensions for ordinary persons as to have been inconsistent with the ordinary laws of mechanics. An efficacious striving of elementary organisms to be other than they are might well be incompatible with the operations of the physical law of least action; for the latter's corollary in psychology, the law of least effort, would be abrogated in the perpetual search for novelties. The rational laws of physics may thus pose strict constraints upon the possible phenomena of biology; no pattern of evolution, other than that of variation, natural selection, and the struggle for existence would then be consistent with the laws of physics. The law of least action, that governs all causal world-lines, including quantum phenomena, would contravene the Lamarckian alternative.

Perhaps, too, the discovered laws of biology are beginning slowly to take the form of the elegant, mathematical law that to Einstein signified the handiwork of God. As the foremost British mathematical geneticist, R. A. Fisher, wrote in an oft-cited passage on the latent power of *a priori* reasoning in the biological domain:

> It is a remarkable fact that had any thinker in the middle of the nineteenth century undertaken, as a piece of abstract and theoretical analysis, the task of constructing a particulate theory of inheritance, he would have been led, on the basis of a few very simple assumptions, to produce a system identical with the modern scheme of Mendelian or factorial inheritance . . . It thus appears that, apart from dominance and linkage . . . all the main characteristics of the Mendelian system flow from assumptions of particulate inheritance of the simplest character, and could have

been deduced *a priori* had any one conceived it possible that the laws of inheritance could really be simple and definite.[58]

The researches into the physico-chemical structure of genes in the last generation have given likelihood to the possibility that a science of evolutionary laws will arise whose logical structure will be akin to that of the advanced physical sciences. Already the evolutionary kinships of species are being traced most fruitfully through the kinships of their genetic material; and the processes of natural selection become isomorphic with laws of environmental or genetic disturbances of systemic equilibria in which the relative populations of species and their variants are altered to reinstate novel equilibria with the least degree of change. The research that culminated in the discovery of DNA, the physico-chemical agent for self-reproducing states, realized biology's 'perfect principle', that of replication. With the next approximation to a 'perfect principle' of evolution, such imprecise terms as Darwin's 'tendency' to variation together with the conjectural histories of species will be replaced by laws of the differential effects of modifications in genetic material that can be empirically validated. A rational system of biological law begins to take shape: one consequence may be the derivation of a set of evolutionary-genetic laws that will be trans-galactic and trans-planetary. It should be possible to predict what kind of living organisms would have survived under different physical environmental conditions. Such invariant, trans-galactic biological laws would be the only ones compatible with the underlying system of physical law. Given the system of relativist-quantum physical law, just as the number of chemical elements and their periodicities in the Mendeleev Table were predictable, together presumably with their relative abundance in the physical universe following a hypothetical disequilibrating event, so likewise might be derivable the approximate number and type of the biological species in the course of a planetary evolution. The cruelties and destruction in biological evolution may be the cosmic cost of our system of beautiful mathematical physical laws. Divine Wisdom itself would have no alternative; though a non-Euclidean geometry was conceivable in physics, a non-Darwinian world of biological species would contravene the physical laws and limitations of the environment. Can a set of physical laws of a non-arbitrary kind be imagined that would preclude any struggle for existence, and define a physical environment where the available energy of closed systems would not diminish, and resources would remain unlimited? A Rational Designer might thus be constrained by his own rationality. Possibly He might have designed more complex laws with a different mode of life. But in that case, the miracle of an intelligible universe would not have existed; for only the existence of *simple* laws makes possible the understanding of nature.

That the noumenalist standpoint has its pitfalls the history of science abundantly shows; Newton thought an absolute space and time were most consonant with his theological conviction, whereas Leibniz deemed a relational theory most in keeping with God's nature. Einstein held steadfast to the determinist standpoint as the only one consistent with reason, and regarded the probabilities in quantum theory as a 'dice-playing' altogether unworthy of God's sense of mathematical beauty. Max Born, on the contrary, like his assistant, Werner Heisenberg, found that a physical indeterminacy fulfilled his noumenal principle of the freedom of the human will. 'For I am unable to separate the two', wrote Born to Einstein in October, 1944,

> and I cannot understand how you can combine an entirely mechanistic universe with the freedom of the ethical individual.... To me a deterministic world is quite abhorrent—this is a primary feeling.[59]

Arthur Eddington had a noumenal sense concerning the evolution of stars; the notion therefore that a large star might radiate its energy away, and keep contracting until, having diminished to a radius of a few kilometers, its gravitational forces would imprison all further radiation, this notion of what is now called a 'black hole', seemed to him a *reductio ad absurdum*, contravening some underlying law of nature: 'I think there should be a law of nature to prevent a star from behaving in this absurd way', he said. To the theory's author, Subrahmanyan Chandrasekhar, however, such thinkers as Eddington and Einstein seemed to have been misled by their successes into a certain arrogance toward nature, in their presuming that a standpoint they had found fruitful in solving their scientific problems at a given time was canonical for all problems henceforth.[60]

V

Is the noumenal sense a matter, then, of a cultural postulate or a generational axiom, or a trait of individual psychology that derives from either biogenetic variation or the influence of personal circumstances? The history of the formation of Einstein's character suggests indeed that he had repressed his own voluntarist sense, and adopted a strict determinism as a congenial defensive retaliatory mechanism against the irrational tyrants and tendencies of his time. His determinist postulate rather than being the outcome of his noumenalist sense, his own original apperception of reason and beauty, was perhaps a culturally conditioned response to trauma. The noumenal sense needs itself to be investigated genetically to clarify the source of noumenalist insight. But that, common to the men of all cultures, and above all to its men of scientific temper, there is a trans-cultural sense of a universal mathematical simplicity and beauty does seem to be a foundational tenet for the society of scientists.

Doubtless an element of sociological fashion obtains in the view prevalent among noted physical scientists that not only living organisms but 'advanced civilizations' are to be found elsewhere, scattered through our galaxy. As Carl Sagan has written:

> It is now OK to talk about life elsewhere or intelligent life elsewhere, whereas a decade ago it was not OK: It was considered too speculative to be worth any investment in time. . . . The most optimistic estimates, in the view of many, about the number of civilizations that might be in the galaxy is on the order of a million, which means that only one in roughly 200,000 stars has such a civilization.[61]

A half-century ago some scientists still inclined to the view of Lawrence J. Henderson that the peculiar collection of elements on the Earth, especially its proportion of such elements as carbon, hydrogen, and oxygen, with their unusual life-sustaining properties, was an occurrence of infinitesimal probability, whose actual existence suggested a purposive, miraculous intervention.[62] To which the logician Bertrand Russell rebutted that in an infinite, or exceedingly vast universe, the highly improbable could occur

> if chance had brought the right components together in the right structure. If it be said that this is very improbable, the answer is that only an infinitesimal proportion of the matter known to science is alive . . .[63]

Oddly enough, however, Einstein's own cosmology and the theory of the 'expanding universe' likewise would weaken Russell's argument against the 'miraculous' character of life's existence. For Einstein's Riemannian model of a universe, 'finite but unbounded', was nonetheless finite. And all arguments that took the form that in an infinite universe, infinitesimal probabilities are realized, would lose much of their force; for a finite universe might well be too small to actualize infinitesimal probabilities. Moreover, a universe that 'began' with a 'big bang' several billion years ago, and which at any given time is finite, though expanding its bounds, is no friendlier to the infinitely improbable; there is always a radical gap between the finite and infinite classes.

The advance of astronomical science, however, and the knowledge of vast expanding galaxies and the stellar systems in various stages of evolution, made Henderson's argument seem parochial. That many planets elsewhere in the universe satisfy what we might call the 'Henderson conditions' for 'the fitness of the environment' does not seem a far-fetched idea. Though the empirical evidence for such 'Henderson worlds' is lacking, the new scientific expeditions into the solar system have brought a sense that such worlds might yet be discovered Among biologists, however, especially those of the elder generation, a sense persists of the overwhelming improbability of an evolution such as that to a human civilization.[64] Even the biologists, however, though emphasizing the improbabilities of the evolutionary process, would usually have agreed that the details of that process are physico-chemically determined; it is a question of a highly improbable combination of physico-chemical world-lines. Possibly future discoveries in the mechanisms for genetic variation and the origin of novel traits will better explain the directionality in evolution, and the rise of human civilization.

Nonetheless, the advent of consciousness and intelligence into the lives of biological creatures still remains inexplicable and irreducible to physical law. For if Einstein speaks of the miracle that an intelligible world exists founded on laws mathematically simple and beautiful, there is the 'co-miracle' (as we might call it) that life, human consciousness, and intelligence are supervenient upon that physical world, making possible the fact of its knowledge. If Einstein's world was utterly 'unexpected' in contrast to any one among the infinite varieties of chaos, so likewise only the infinite forms of non-consciousness would have been potential existents at the beginning of the physical universe. For life, consciousness, and intelligence were not to have been expected at all from the physical universe with its purely physical dimensions. In this sense, the divergence between 'mono-sapientists' and 'pluri-sapientists', between believers that our planet alone has seen the rise of an advanced civilization and those who believe that this evolution has been repeated elsewhere, is a secondary one. For no one could possibly show from the laws of physics alone, together with some set of initial physical conditions, that any probability whatsoever existed, in a scientific sense, that life, consciousness, and intelligence would arise. Every instance of a physical probability, if it is to be part of a theory, must be grounded in principle in a set of physical laws; thus, the probability of one-half that a coin tossed upwards will fall on the tails side is grounded in the laws of rotation and translation, momentum, both linear and angular, and gravitational acceleration; the number of spins and turns of a coin, when tossed with a certain force and direction, is thus in principle calculable. But the advents of feeling, consciousness, and thought are grounded not at all in any physical theory; the dimensions of all physical concepts are definable in terms of mass, length, and time, but a feeling eludes such dimensional analysis. If some Maxwellian demon, either diabolical or divine, had been observing the originative 'big bang' of the physical universe, he could not have defined from the terms of his equations any term such as 'feeling' or 'conscious' or 'thinking', despite his own selfhood, nor would any law containing them have been deducible. Presumably if consciousness and feeling originated from some physico-chemical reaction, then presumably there ensued at some time disequilibria in physico-chemical systems that were restored to equilibria only through the emergence of mental entities or events. But introducing a mental term as a condition for the reinstatement of a chemical equilibrium that has been disturbed is an entirely miraculous extension of Le Chatelier's Law, with the multiplicities of such miracles presumed to have occurred in physico-chemical systems during the origination of living organisms.

The probability of life or thought occurring in association with a given physical functioning body remains an isolated statistical correlation unless it can be absorbed in principle within a body of explanatory theory. As Emile Meyerson, the analyst of scientific method whom Einstein most admired, wrote: 'the extreme limits' of law are reached when the irrationals are reached that defy causal explanation.

> Thus, that which is irrational—transitive action, sensation—always, and no matter what the future development of science is, may be conceived of as teleological, as due to divine intervention . . .[65]

In the aftermath of the World War I, as Western civilization tried to regain its sense of balance, a number of philosophers sought for a noumenal explanation for evolutionary advance; they took up a theme that had been adumbrated before the War, that of 'emergent evolution'. It had been conjectured that novel, inexplicable properties 'emerged' when given entities were conjoined, that a 'whole' in this sense could bring into existence properties that were not derivable from a knowledge of its parts; thus, the properties of water, it was said, were unpredictable from a knowledge of its constituents, oxygen and hydrogen. The notion of emergent evolution brought hope in the possibly limitless progress of man and sentient being; among its theorists in Britain and America were C. Lloyd Morgan, Alfred N. Whitehead, C. D. Broad, Arthur O. Lovejoy, Samuel Alexander, L. T. Hobhouse. All had experienced strong theistic longings together with a sense of the inexplicable, irreducible miraculousness of life.[66] Yet the emergentist 'laws', however, stood altogether outside the structure of causal laws. The laws of physical science, as Meyerson emphasized, conform to a condition of identity; its if-then statements, the relations between antecedent and consequent, conform to such 'identities' as for instance, the conservation of mass-energy, or the conservation of charge. An emergentist 'law', on the contrary, always has a surd term in the consequent standing wholly outside any relationship of identity with the terms of the antecedent. A physical collocation of a body's molecular and sub-atomic reactions is followed not only by consequences that satisfy the conservational requirement, but by the appearance of an unlocalizable, hence partially unmeasurable term such as 'thought' or 'feeling'. One might argue that the concept of emergence has only a provisional validity, that a given phenomenon is taken as an 'emergent' when, relative to the terms of some given theory, it cannot as yet be reduced. From such a standpoint, the working hypothesis remains that, in due course, such an 'emergence' theory will be superseded by a more explanatory one that defines its concepts without introducing any novel term irreducible to the base of physico-chemical ideas. But 'consciousness' and 'thinking' are more than provisional examples of 'emergence'; for no possible physical theory can overcome the dualistic separation of entities exempt from location in spatial dimensions from those enjoying such attributes. All 'emergence' theories involve a rupture with the causalist 'identity' principle.

Moreover, the optimist social philosophy of the emergent evolutionists was scarcely sustained by their weak theological premise. The brief known history of the human race gave little warrant to the cosmically-scaled optimism of the emergentists.[67] On the one hand, they were believers in the progress of the human races. On the other hand, they envisaged a finite God, with limited powers and capacities, but struggling against Nature's odds to achieve the human advance. But if God were finite, his impotence, as Anatole France once observed, would be infinite.[68] For measured against the infinitude of the universe, a finite God would seem destined to defeat; obstacles toward the advancement of rationality would always arise inexhaustibly from a limitless non-rational or irrational environment. David Hume, in a daring speculation upon the attributes of a finite Deity, susceptible to 'error, mistake, or incoherence, in his undertakings', wondered whether 'many worlds might have been botched and bungled, throughout an eternity', and 'many fruitless trials made', as the Deity endeavored to improve itself 'in the art of world-making'.[69] Is there any assurance that a finite Deity would be secure in a trajectory of improvement? John Stuart Mill, continuing in this intellectual direction, adhered in his latter years to the view that the order in Nature pointed toward its origin in the Design of a Deity working, however, within the constraints of a given Matter and Force; at the same time, while believing that 'benevolence is one of the attributes of the Creator', Mill thought that human history was no record of high success on the part of the Deity:

> If the motive of the Deity for creating sentient beings was the happiness of the beings he created, his purpose, in our corner of the universe at least, must be pronounced, taking past ages and all countries and races into account, to have been thus far an ignominious failure.

Was man's own capacity for improving himself, Mill wondered, 'worth purchasing by the sufferings and wasted lives of entire geological periods?'[70]

In short, the emergent evolutionist standpoint, apart from failing to satisfy the causal principle of identity between both sides of a scientific equation, fails to find a permanent place to sustain its sense of the abiding worth of the human venture. If the latter were terminated, and a bleak physical universe ceased to provide the setting for an even partially rational human race, if the lines of evolution became degenerative and extinctual, would the conviction of a noumenal universe with a permanent abode for man's spiritual achievement be declared an illusion?

From the standpoint of the causal principle, the 'emergence' of consciousness and thought is a phenomenal given that must then be explained. To conform to the causal principle of identity, however, we must find some conscious and intellectual equivalent in the antecedents of 'emergence'. Unless we were to embark on the vast auxiliary hypotheses of a panpsychism, the simplest hypothesis available is one affirming that from some meta-causal source of an Infinite Consciousness and Intelligence, there sprang the 'co-miracle' of an imparted consciousness and thought. And that meta-causal agent is to be identified most naturally with the Intelligence that Einstein saw objectified in the miracle of the world of intelligible laws.

The biological and social worlds remain, however, for all their triumphs, a scene of tragic defeats, sufferings, and annihilations. And, as such, is the meta-causal agent a collaborator in the unredeemed worldlines of the histories of sentient beings? Curiously, the development of Einstein's general theory of relativity points toward a non-temporal counterpart to our own perceived temporal world; perhaps the sentient Jobs of existence may find their Creator's redemption. The logician, Kurt Gödel, in his investigations of the cosmological solutions consistent with the gravitational equations of the general theory of relativity, found there were some in which the temporal aspect of things was radically altered; the past, for instance, ceased to be irrevocable; it became possible, for instance, according to the equations, 'to travel into any region of the past, present, and future, and back again, exactly as it is possible in other worlds to travel to distant parts of space'.[71] Einstein, while generously acknowledging Gödel's analysis as an 'important contribution to the general theory of relativity', thought that his cosmological solutions might prove to be 'excluded on physical grounds';[72] in that case, their status would be akin to the imaginary solutions that every high school boy learns to discard when he solves for the roots of simultaneous equations. Gödel himself noted that, to travel into one's near past,

> the velocities which would be necessary in order to complete the voyage in a reasonable length of time are far beyond everything that can be expected ever to become a practical possibility;

Nonetheless, it could not be excluded *a priori* that the world's spacetime structure is of the kind described.[73] The common axiom of mankind that the past is irrevocable, will, of course, not be surrendered unless an overwhelming scientific evidence is forthcoming; nevertheless, Gödel's work is a bewildering sign-post. Gödel felt, indeed, that his researches tended to confirm the insight of such philosophical idealists as Kant and McTaggart that time was self-contradictory, and the world of change unreal. There seems to be no logical relationship, however, between McTaggart's philosophical argument for the unreality of time and Gödel's relativist analysis; both men shared, however, an atemporal temperament.[74] Einstein too, though rejecting the idealistic theory of knowledge, and its propensity to adjudge realities as self-contradictory, still felt increasingly that the world of past, present, and future was supplanted by a timeless reality. His last words to his old friend, Michele Besso, written on March 21, 1955 less than a month before he died, were: 'For us, believing physicists, the separation between past, present and future has the significance only of an illusion, however tenacious it may be.'[75] Has the noumenalist standpoint, from Spinoza to Einstein, from Descartes and Leibniz to Eddington and Gödel, discovered a reality in the equations of mathematical science more profound than that vouchsafed to common sense? Rather than denying physical reality to unimaginable formulae, such thinkers have been prepared to follow their noumenalist faith, seeking novel, hitherto unknown realities signified in the logical forms. Perhaps in the noumenal timeless domain is found that part of us which as Spinoza wrote, we 'feel and know by experience' to be eternal.[76]

The mathematical realities that find no exemplification in physical terms may point to, and converge with, a meta-physical reality.

Perhaps indeed, in the timeless domain, those spiritual realities in ourselves that are most in keeping with the Infinite Intelligence are actualized without the dross of the evolutionary struggle for existence. For if, as Einstein believed, Rational Design is objectified in the rationality of the laws of nature, one might think that the rationality of the timeless domain would provide too for the fulfillment of the spirit. The properties of physical objects, their inertial masses, for instance, are altered when their velocities approach that of light; possibly, the properties of our spiritual selves are trans-realized in that timeless domain whose existence the laws of nature themselves may seem to suggest.

Notes

1. Albert Einstein, *Lettres à Maurice Solovine*, Gauthier-Villars, Paris 1956, pp. 114–15.
2. Frederic William Westaway, *Scientific Method, Its Philosophical Bases and Its Modes of Application*, Hillman-Curl, New York 1937, p. 277.
3. *The Philosophical Works of Descartes*, trans. by Elizabeth S. Haldane and G. R. T. Ross, Cambridge University Press, Cambridge 1911, 1931, Dover Publications, reprinted New York 1955, pp. 158–9.
4. Albert Einstein, *Out of My Later Years*, Philosophical Library, New York 1950, new ed. Littlefield, Adams & Co., Totowa, New Jersey 1967, p. 32.
5. Albert Einstein, *The New York Times*, April 25, 1929, p. 60. Cited in Virgil G. Hinshaw, Jr., 'Einstein's Social Philosophy', in Paul Arthur Schilpp (Ed.), *Albert Einstein: Philosopher-Scientist*, Open Court, La Salle 1945, reprinted Harper & Row, New York 1959, Vol. II, p. 659.
6. Albert Einstein, *Ideas and Opinions*, trans. by Sonja Bargmann, Crown Publishers, New York 1954, reprinted Dell Publishing Co., New York n.d., pp. 342, 338.
7. Albert Einstein, 'The New Field Theory, II—Structure of Space-Time', *The Times of London*, 5 Feb. 1929.
8. Loc. cit.
9. Albert Einstein, 'On the Generalized Theory of Gravitation', in *Ideas and Opinions*, op. cit., p. 333.
10. Martin Buber, *The Knowledge of Man: Selected Essays*, trans. by Maurice Friedman and Ronald Gregor Smith, Harper & Row, New York 1965, p. 156.
11. Arnold Sommerfeld, 'To Einstein's Seventieth Birthday', in Paul Arthur Schilpp (Ed.), *Albert Einstein: Philosopher-Scientist*, op. cit. (1959), Vol. I, p. 103.
12. Philipp Frank, *Einstein: His Life and Times*, trans. by George Rosen, Alfred A. Knopf, New York 1947, p. 284.
13. Albert Einstein, Michele Besso, *Correspondance 1903–1955*, trans. by Pierre Speziali, Hermann, Paris 1972, p. 502.
14. W. H. McCrea, 'Edward Arthur Milne', *Obituary Notices of Fellows of the Royal Society*, London (n.p.), Vol. 7, Nov. 1951, p. 421.
15. Ernest William Barnes, *Scientific Theory and Religion: The World Described by Science and its Spiritual Interpretation*, Macmillan Co., New York/Cambridge University Press, Cambridge 1934, p. 178.
16. Lewis S. Feuer, *Einstein and the Generations of Science*, Basic Books, New York 1974, p. 327.
17. G. J. Whitrow (Ed.), *Einstein: The Man and His Achievement*, 1967, reprinted Dover Publications, New York 1973, p. 82.
18. Leopold Infeld, *Quest: The Evolution of a Scientist*, Doubleday, Doran & Co., New York 1941, p. 267.
19. Einstein, *Ideas and Opinions*, op. cit., pp. 333, 339.
20. Max Born, *Physics in My Generation*, Longman, New York 1969, pp. 109, 156.
21. *The Born-Einstein Letters, Correspondence between Albert Einstein and Max and Hedwig Born, 1916–1955*, trans. by Irene Born, Walker & Co., New York 1971, p. 156.
22. Max Born, *Experiment and Theory in Physics*, Cambridge University Press, Cambridge 1944, p. 44.

23. *The Collected Mathematical Papers of James Joseph Sylvester,* Vol. II, Cambridge University Press, Cambridge 1908, reprinted Chelsea Publishing Co., New York 1973, p. 100.
24. Max Born, *Experiment and Theory in Physics,* op. cit., pp. 10, 14, 34.
25. James Jeans, *The Mysterious Universe,* rev. ed., Macmillan, New York 1933, pp. 150, 145.
26. Ibid., p. 159.
27. Sir James Jeans, *Physics and Philosophy* (1942), University of Michigan Press, Ann Arbor 1958, pp. 72–80. E. A. Milne, *Sir James Jeans: A Biography,* Cambridge University Press, Cambridge 1952, p. 160.
28. P. A. M. Dirac, 'The Cosmological Constants', *Nature,* Vol. 139, 20 Feb. 1937, p. 323.
29. Erwin Schrödinger, 'Sir Arthur Eddington, The Philosophy of Physical Science', *Nature,* Vol. 145, 16 March 1940, p. 402.
30. Bertrand Russell, 'Relativity: Philosophical Consequences', *Encyclopaedia Britannica,* 14th ed., New York 1930, p. 100.
31. Percy Williams Bridgman, *Reflections of a Physicist,* Philosophical Library, New York 1950, p. 112.
32. Max Planck, *A Survey of Physics,* trans. by R. Jones and D. H. Williams, Methuen & Co., London 1925, p. 120.
33. Ibid., p. 66.
34. 'Cosmology assumes the rationality of the universe, but can give no reason for it short of the creator of the laws of nature being a rational creator.' E. A. Milne, *Modern Cosmology and the Christian Idea,* Clarendon Press, Oxford 1952. It is interesting to observe that Milne's director of studies at Cambridge, E. W. Barnes, subsequently became Bishop of Birmingham. W. H. McCrea, op. cit. (see Note 14), p. 421.
35. Max Born, *My Life: Recollections of a Nobel Laureate,* Charles Scribner's Sons, New York 1978, p. 149.
36. Schrödinger, op. cit. (see Note 29), p. 402.
37. 'It is sometimes said . . . that scientists assign higher *a priori* probability to simple laws of nature as against more complex laws, on the basis of an *a priori* probability that the character of nature is simple. It would be as though someone from his own inner consciousness or prejudice began specifying the characteristics of the Deity. There is no *a priori* probability that nature is simple. But inductive logic gives *a posteriori* grounds for preferring a simple law.' Roy Harrod, *Sociology, Morals and Mystery,* Macmillan, London 1971, p. 37.
38. Philipp Frank, *Philosophy of Science: The Link Between Science and Philosophy,* Prentice-Hall, Englewood Cliffs, New Jersey 1957, p. 295.
39. Albert Einstein, 'Autobiographical Notes', in Schilpp, op. cit., Vol. I, pp. 13, 53. M. Solovine, 'Introduction', in Einstein, *Lettres à Maurice Solovine,* op. cit. (see Note 1), p. viii.
40. David Hume, *Dialogues Concerning Natural Religion,* Pt. IX, ed. by Henry D. Aiken, Hafner Publishing Co., New York 1957, p. 6.
41. Loc. cit.
42. Ibid., Pt. II, p. 21
43. Ibid., Pt. II, pp. 20, 21.
44. Ibid., Pt. II, p. 23.
45. Loc. cit.

46. Ibid., Pt. VIII, p. 54.
47. Professor (Thomas Henry) Huxley, *Hume,* Harper & Brothers, New York 1879, p. 64.
48. Ibid., p. 40.
49. Charles S. Peirce, *Chance, Love and Logic: Philosophical Essays,* ed. by Morris R. Cohen, Harcourt, Brace & Co., New York 1923, p. 125.
50. *Autobiography of Charles Darwin,* ed. by Sir Francis Darwin, Watts & Co., London 1929, p. 2.
51. Meyer Waxman, *The Philosophy of Don Hasdai Crescas,* Columbia University Press, New York 1920, p. 99.
52. Charles Darwin, *The Origin of Species,* intro. by Sir Julian Huxley, Mentor ed., New American Library, New York 1958, p. 450.
53. Charles S. Peirce, *Chance, Love and Logic: Philosophical Essays,* op. cit., pp. 200, 260, 264, 234.
54. James H. Leuba, 'Religious Beliefs of American Scientists', *Harper's Magazine,* Vol. 169, August, 1934, pp. 291–300. Also James H. Leuba, *The Belief in God and Immortality, a Psychological, Anthropological and Statistical Study,* Sherman, French & Co., Boston 1916, pp. 230–59, 270–80. Another study by Lehman and Witty found similarly that, among the eminent scientists, the percentage of physicists who stated their church membership for the sketches in *Who's Who in America* exceeded that of the chemists and zoologists, and was more than twice that of the psychologists. Harvey C. Lehman and Paul A. Witty, 'Certain Attitudes of Present-Day Physicists and Psychologists', *The American Journal of Psychology,* Vol. XLIII (1931), p. 675.
55. Bertrand Russell, *The Problem of China,* The Century Co., New York 1922, pp. 18–19.
56. Cf. L. S. Feuer, 'John Stuart Mill as a Sociologist', in *James and John Stuart Mill: Papers of the Centenary Conference,* ed. by John M. Robson and Michael Laine, University of Toronto Press, Toronto 1976, 98–99.
57. Sir Edmund Whittaker, *From Euclid to Eddington: A Study of Conceptions of the External World,* Cambridge University Press, Cambridge 1949, pp. 58 ff. Max Born, *Physics and Politics,* Basic Books, New York 1962, p. 39.
58. R. A. Fisher, *The Genetical Theory of Natural Selection,* Oxford University Press, Oxford 1930, rev. ed. Dover Publications, New York 1958, pp. 7–8.
59. *The Born-Einstein Letters,* op. cit. (see Note 21), pp. 155–6.
60. John Tierney, 'Quest for Order', *Science* 82, Vol. 3 (1982), No. 7, pp. 71–72.
61. Richard Berendzen (Ed.), *Life Beyond Earth and the Mind of Man,* National Aeronautics and Space Administration, Washington, D.C. 1973, pp. 6–7, 11. George Wald, 'Introduction' to Lawrence J. Henderson, *The Fitness of the Environment,* Beacon Press, Boston 1958, p. xxiii. Cf. Fred Hoyle, *The Nature of the Universe,* rev. ed., Harper, New York 1960, pp. 104–5.
62. Lawrence J. Henderson, *The Fitness of the Environment* (1913), Beacon Press, Boston 1958, pp. 276 ff., 292, 307–9. Henderson later evolved into a Paretian positivist. But in his earlier years, he was, according to the chief student of his work, 'obviously influenced' by the views of Bernard Bosanquet, the English philosophical idealist, on the necessity for conjoining teleology with mechanism. John Parascandola, 'Organismic and Holistic Concepts in the Thought of L. J. Henderson', *Journal of the History of Biology,* Vol. 4 (1971), p. 76.
63. Bertrand Russell, 'C.D. Broad, *The Mind and Its Place in Nature*', *Mind,* Vol. XXXV (1926), p. 73. 'It may seem odd that life should occur by accident, but in

such a large universe accidents will happen.' Bertrand Russell, *Religion and Science,* Henry Holt, New York 1935, p. 216.
64. [A]ll the great contemporary experts in the theory of evolution—Francisco Ayala, Theodosius Dobzhansky, Ernst Mayr, Simpson—are unanimous in claiming that the evolution of an intelligent species from simple one-celled organisms is so improbable that we are likely to be the only intelligent species ever to exist.' Frank J. Tipler, 'We Are Alone in Our Galaxy', *New Scientist,* Vol. 96, 7 Oct. 1982, p. 33.
65. Emile Meyerson, *Identity and Reality,* trans. by Kate Loewenberg, London 1930, reprinted Dover Publications, New York 1962, pp. 318–19.
66. 'For my own part I believe that, in the end, a theistic conception of God's deity is demanded by the facts of nature.' Samuel Alexander, *Philosophical and Literary Pieces,* ed. by John Laird, Macmillan, London 1939, p. 275. 'God, as being, is the nisus of the universe pressing onwards to levels as yet unattained.' C. Lloyd Morgan, *Emergent Evolution,* Holt, New York 1923, p. 34. Also, p. 301. L. T. Hobhouse, *Development and Purpose: An Essay Towards a Philosophy of Evolution,* Macmillan, London 1913, pp. 369–72. J. A. Hobson and Morris Ginsberg, *L. T. Hobhouse: His Life and Work,* G. Allen & Unwin, London 1931, p. 256. C. D. Broad, *The Mind and Its Place in Nature,* Harcourt, Brace & Co., New York 1925, p. 647. 'I feel in my bones that the orthodox scientific account of man as an undesigned calculating machine . . . is fantastic nonsense.' C. D. Broad, 'Autobiography', in P. A. Schilpp (Ed.), *The Philosophy of C. D. Broad,* Tudor, New York 1959, p. 58, Also, p. 49. Joseph Needham, 'A Biologist's View of Whitehead', in Paul Arthur Schilpp (Ed.), *The Philosophy of Alfred North Whitehead,* Northwestern University Press, Evanston 1941, p. 257. Joseph Needham, *Moulds of Understanding: A Pattern of Natural Philosophy,* ed. by Gary Werskey, George Allen & Unwin, London 1976, pp. 267–9. Daniel J. Wilson, *Arthur O. Lovejoy and the Quest for Intelligibility,* The University of North Carolina Press, Chapel Hill 1980, pp. 206, 39. 'Advance or Decadence are the only choices offered to mankind. The pure conservative is fighting against the essence of the universe.' Alfred North Whitehead, *Adventures of Ideas,* Macmillan, New York 1933, p. 354.
67. Thus Lovejoy, though he adopted the tenet of emergent evolution, wrote: 'The history of a single planet or solar system, during the brief span of a few hundreds of millions of years, can scarcely yield a cogent inductive inference respecting the disposition of so large a thing as a universe, or the tendency running through all time.' Arthur O. Lovejoy, 'L. T. Hobhouse, Development and Purpose', *The Nation,* Vol. 97 (1913), p. 164.
68. Anatole France, *Under the Rose,* trans. by J. Lewis May, Dodd, Mead & Company, New York 1926, p. 64.
69. Hume, *Dialogues Concerning Natural Religion,* op. cit. (see Note 40), Pt. V, pp. 38–39.
70. John Stuart Mill, *Three Essays on Religion,* Henry Holt, New York 1874, reprinted Greenwood Press, New York 1970, pp. 192–3, 178. Alfred Russel Wallace recalled how, at a dinner with Mill in 1870, the latter vehemently rejected the notion that an omnipotent, benevolent God could have made a mankind afflicted with so much misery and folly. *My Life: A Record of Events and Opinions,* Dodd, Mead & Co., New York 1905, Vol. II, pp. 254–5.
71. Kurt Gödel, 'A Remark about the Relationship between Relativity Theory and Idealistic Philosophy', in P. A. Schilpp (Ed.), *Albert Einstein: Philosopher-Scientist,* op. cit., Vol, II, p. 560. Cf. Rudy Rucker, 'Master of the Incomplete', Science 82, Vol. 3, No. 8 (Oct. 1982), pp. 56–60.

Varieties of Scientific Experience

72. Albert Einstein, 'Reply to Criticisms', in Schilpp, op. cit., pp. 687–8.
73. Kurt Gödel, op. cit., p. 561.
74. Cf. G. J. Whitrow, *The Nature of Time,* 1972, reprinted Thames & Hudson, Harmondsworth 1975, p. 137.
75. Einstein, Besso, *Correspondance 1903–1955,* op. cit. (see Note 13), pp. 538–9.
76. 'At nihilominus sentimus, experimurque, nos aeternos esse.' Spinoza, *Opera,* ed. by Carl Gebhardt, Heidelberg 1925, Vol. II, p. 296. Spinoza, *Ethics,* Bk. V, Prop. XXIII, Scholium, trans. by W. H. White, in John Wild (Ed.), *Spinoza Selections,* Charles Scribner's Sons, New York 1930, p. 385.

2

Teleological Principles in Science

I

From the standpoint of intellectual history, the most unusual fact about contemporary scientific theory is that it has seen the revival of 'perfect principles'. Whether in the search for elementary particles, the cosmological structure, or the pattern of the gene, 'perfect principles' have been enlisted as guides. The teleological mode of thought had occasionally been recognized for its heuristic value; after all, Einstein said that when he coped with a problem, he asked himself how would God have arranged the matter. But Einstein was regarded by many as a mystic mutation among scientists. What is noteworthy now is how teleological principles express themselves frankly in the work of scientific thinkers.

A few years ago, for instance, a vigorous search was going on to ascertain whether there exist particles which travel with velocities higher than that of light; it was argued that the existence of 'tachyons', as they were called, is consistent with the basic principles of science, such as the conservation of energy, and Einstein's postulates. Therefore, it was maintained such particles must exist; for if the existence of any given class of entities is compatible with the laws of physics, then that class of entities must exist. As two physicists expressed it:

> There is an unwritten precept in modern physics, often facetiously referred to as Gell-mann's totalitarian principle, which states that in physics, 'anything which is not prohibited is compulsory'. Guided by this sort of argument we have made a number of remarkable discoveries, from neutrinos to radio galaxies.[1]

Leibniz made the whole basis of his philosophy 'the principle of perfection', that whatever is possible, or more accurately, compossible with the laws of nature (or ultimately God), must exist.[2] The scientific theorist today is apt to be an unconscious Leibnizian. In its most general form, the model for this class of 'perfect principles' is the axiom: every ontological niche must be filled.

It is best to call such principles straightforwardly 'teleological'. They are not merely 'paradigms', preferred modes of language, favored forms of laws, or convenient models. For what underlies the choice of these principles is some underlying emotional aim; a particular kind of world is sought which will answer to the scientist's emotional longings. A teleological principle, in its most general sense, is one which affirms that some ethical, extra-logical purpose is fulfilled in the structure of the laws of nature. Such a principle, moreover, serves then as a heuristic agent for discovering those laws of nature. It is not an after-the-fact theological commentary but an active participant in the work of exploration. It was exemplified in the classical case of the formulation of the physical principle of least action, which realized Leibniz's concept of a God who as a Supreme-Economist achieved the maximum of effects with the minimum of means. But it will be found among the most diverse theorists. Whether it is Ernst Mach, formulating the sensory basis for physical theory, or Max Born and Werner Heisenberg striving for indeterminist lacunae in the physical world, or Jacques Loeb, intent on wholly determinist laws in biology, teleological principles are always the underlying directors.

Gell-mann's principle, 'anything which is possible is compulsory', is a scientific form of what Arthur O. Lovejoy called the 'principle of plenitude', the notion that the perfection of Nature expresses itself in the actualization of every possible being. This doctrine, as a metaphysical one, encountered insuperable difficulties during the course of its history. It was hard to credit, for instance, that there was some real world in which live sphinxes and mermaids existed, simply on the ground that we could conceive such entities, and make statuary of them. Or that there were an infinite number of real worlds in which gravitational attraction varied through all the possible powers of the distance. The principle of scientific plenitude does not take 'possible' as signifying 'conceivable' or 'imaginable'. Rather 'possible' is taken as relative to a given system of scientific theory or laws; what is then possible in a scientific sense, that is, relative to a given system of scientific theory, is then held to exist in a scientific, empirical sense. Thus, though objects such as sphinxes and mermaids are conceivable, they are not possible in a scientific theoretical sense, for their structures contravene all the laws of comparative anatomy. We could adapt Leibniz's usage, and assert that all those entities 'compossible' with a given system of scientific entities must exist; the principle of scientific plenitude would then affirm that 'whatever is compossible exists'.

II

It was precisely during the latter part of the nineteenth century that, as the metaphysical principle of plenitude fell into desuetude, disbelieved and discarded, the principle of scientific plenitude came to the fore. Its great embodiment was in the Periodic Table of the Elements, proposed in 1869 by the Russian chemist, Dmitri Mendeleev. To this day Mendeleev's Periodic Table remains the supreme example of how the search for novel entities will be stimulated by a theory which has a place for their existence. The 'symmetry approach' in particle physics is recognized as a lineal descendant of the Periodic Table, as another example of the principle of scientific plenitude: 'It has become customary to compare our tabulation of particles and resonances with the inauguration of the Periodic Chart of the elements by D. I. Mendeleev in 1870', writes Y. Ne'eman.[3] Indeed, in 1923, when Wolfgang Pauli, the later discoverer of the Exclusion Principle, gave his inaugural lecture, it was appropriately enough on the periodic system of the elements, affording him a setting for articulating his dissatisfactions with the problem of 'the closing of the electronic shells', essential to the architecture for the new classification of the elements.[4]

When in 1932 the 'positron' was discovered experimentally, 'the first example in the history of physics where the existence of a new particle was predicted on a purely theoretical basis', the application of the principles of scientific plenitude, first applied by Mendeleev, entered a new domain. Mendeleev, to fulfill the symmetry in his law that the properties of the elements are periodic functions of their atomic weights, had predicted that an element would be found between calcium and titanium; 'eka-boron', he called it, describing its properties. And in the vacant places between zinc and arsenic, he likewise predicted the properties of two as yet undiscovered elements, 'eka-aluminium' and 'eka-silicon'. Then, during the years from 1875 to 1888, chemists discovered the elements scandium, gallium, and germanium, which corresponded to those for which Mendeleev had indicated the waiting, vacant places. Those entities that were composible with the Periodic Law were found to exist. Thus, too, the discovery of the positron was guided by the conviction that all those entities that were mathematically composible with a satisfactory physical theory should be physically real. Wherever a given physical equation was consistent with alternative solutions, these should have all their counterparts realized in physical existence. Such was the principle of scientific plenitude that guided Paul Dirac when he first noted in an early paper that the wave equation yielded, beside the solutions for which the kinetic energy of the electron was positive, 'an equal number of unwarranted solutions with negative kinetic energy for the electron, which appear to have no physical meaning'; as he continued his inquiry, these apparently physically meaningless entities transmuted themselves into 'a new kind of particle, unknown to experimental physics, having the same mass and opposite charge to an electron'. Dirac called these unknown entities 'anti-electrons', but still doubted whether we could 'find any of them in Nature' because their rates of combination would be so rapid.[5] Then on 2 August 1932 the 'positron' was discovered in actual experiment. No more beautiful example could have been provided of Leibniz's 'principle of perfection' that 'all possible things . . . tend by equal right toward existence . . .'.

But then how can we show that the principle of scientific plenitude is more than a 'methodological' one, that it is indeed founded on a teleological mode of thought, and thus constitutes a teleological principle? To do so we must necessarily enter upon biographical and psychological considerations to ascertain what indeed were the basic emotional longings of the scientist, what the kind of world it was that he, on emotional grounds, sought to realize in his scientific theorizing? What, in other words, was the scientist's *emotional a priori*? Naturally, Mendeleev, as the exemplar in the search for unknown elements and particles, is the one into whose scientific standpoint we must first inquire.

Dmitri Ivanovich Mendeleev was, unlike the Russian revolutionists of his time, no devotee of materialistic doctrines. He found his innermost feelings expressed in his favorite poet, Fyodor Ivanovich Tyutchev, Russia's lyric poet of mysticism, and especially in the poem he most admired, *Silentium*:[6]

> How does a heart speak out?
> How explain yourself to another?
> Will he grasp what you live by?
> Pronounced thought is a lie . . .
>
> Know how to live within
> Your soul contains a world
> Of mysterious, magical thoughts;
> The outer tumult stifles
> Hear their song and be silent.[7]

Tyutchev blended his mysticism with a socially conservative outlook, looking upon the ego as a disruptive power, destroying harmony, and undermining unity with its anarchy. The ego for him signified the lawless impulse that was the basis of Revolution; while to escape from the ego by merging it with the all was Tyutchev's aim.[8] But that All, Nature, was itself suffused with an all-pervasive life. So Tyutchev wrote in his poem *Nature*:

> Nature is not what you may fancy:
> No lifeless image, not a copy,
> She has a soul, a voice, inspired
> By love and by her own free will.[9]

Indeed, Tyutchev in his poetry had also transcribed some of Schelling's *Naturphilosophie,* the view that nature is a living organism, that in the course of its development from unconscious to conscious life and spirituality, traverses virtually a Periodic Table of Existence.

Mendeleev indeed, had said that he conceived his Periodic System when, undertaking to write a textbook of chemistry, he was faced with the necessity of systematizing a huge body of diverse facts and information. But evidently something of a pantheistic, mystical longing to see the Chain of Being that ramified through the chemical facts was the teleological *a priori* of his systematizing task.[10] The pantheistic mystic communicated silently with the textbook writer. For Mendeleev himself was indeed explicity aware that the personality of the scientist, his emotional longings and values, shaped the character of his scientific theorizing. 'If statements of fact themselves depend upon the person who observes them', he wrote at the outset of his great book,

The Principles of Chemistry, 'how much more distinct is the reflection of the personality of him who gives an account of methods and of philosophical speculations which form the essence of science'.[11] His own personality had been largely shaped by his mother, a woman of strong character; she had managed their glass factory in Siberia by herself for many years. Dmitri recalled her last words to him in the introduction to his work *Solutions*: 'When dying, she said, "Refrain from illusions, insist on work, and not on words. Patiently search divine and scientific truth". She understood how often dialectical methods deceive . . . Dmitri Mendeleeff regards as sacred a mother's dying words. October 1887.'[12] It was 'divine and scientific truth', in his mother's words, that Mendeleev sought. And 'natural reason, simplicity and plan' (in Mendeleev's words) were of the essence of that divine truth which must be fulfilled in Nature. He used that phrase in a letter on landscape painting that he published in a newspaper, in which he expressed his pantheistic creed. Landscape painting, he observed, only arose with the Renaissance when 'nature became an object of study':

> External nature thus ceased to be merely subservient to man, and became his equal, his friend. Dead and senseless as it had been, it now became alive. Everywhere it presented motion, stores of energy, natural reason, simplicity and plan. Inductive and experimental science became a crown of knowledge. . . . Landscape painting was born simultaneously with this change . . .[13]

The work of science as it approached the divine plan brought to Mendeleev's eyes 'pure enjoyment', 'that pure enjoyment experienced on approaching to the ideal, in that eagerness to draw aside the veil from the hidden truth'. This pure eudaemonism in the pursuit of science was linked with the teleological longing which expressed itself in the principle of scientific plenitude. In Mendeleev's words:

> To conceive, understand, and grasp the whole symmetry of the scientific edifice . . . is equivalent to tasting that enjoyment only conveyed by the highest forms of beauty and truth. Without the material, the plan alone is but a castle in the air—a mere possibility; whilst the material without a plan is but useless matter.[14]

The teleological principle affirmed that the physical universe must be such as to realize the highest form of beauty and truth.

III

What, however, of those scientists who utterly repudiate any form of mysticism or teleological principle? In what sense can one presume to say that a teleological principle nevertheless runs through their work? The teleological motivation in every individual scientist, of course, has its own unique features; the name of Jacques Loeb, the outstanding mechanist and anti-teleologist in the history of American biology, however, comes perhaps most challengingly to mind. And indeed, in the world of art, the character of Jacques Loeb has been projected into the most celebrated literary portrait of a scientist that has ever been written—that of the bacteriologist and biochemist Max Gottlieb in Sinclair Lewis's *Arrowsmith,* ruminating about 'Father Nietzsche and Father Schopenhauer (but damn him, he was teleological-minded!)'.[15]

Discoverer of chemically-induced parthenogenesis,[16] elaborator of the theory of tropisms (which he regarded as virtually a perfect biological principle), Loeb, at first a student of philosophy, had then turned strongly against metaphysics. Above all, it was the problem of the freedom of the will which stirred Loeb, releasing his own life-long tropismic intellectual movement. Loeb had read Schopenhauer and Hartmann, seeking 'in vain for the solution of his problems in the current philosophies of the day: then came his conversion to mechanism. Faith in mechanism became the religion to which he devoted his life . . .'. It followed on his reading of Voltaire's *Candide* when Loeb was just a 'lazy bank clerk'. Loeb became convinced that the problem of the freedom of the will had to be dealt with experimentally: 'if the will be free, it cannot be controlled; this question must be tested experimentally.'[17]

The conviction that 'mechanism will explain the most baffling mysteries' was for Loeb 'a militant faith calculated to move monuments', animating his studies from his first in brain physiology to his last on colloidal systems. It led Loeb to tackle 'apparently impossible problems', sometimes solving them, and always giving to his discourse 'a prophetic fire'.[18]

For Loeb was indeed the prophet of a special kind of teleological principle. If it was not a divine perfection whose outline he sought to delineate in the workings of nature, he had a teleological demand all his own: the world and its objects must be mechanistically explicable because only in a mechanistic world, recognized as such, could the values of human freedom and dignity be realized. Jacques Loeb's teleological demand indeed entailed the perfect principle of mechanism: the world must be such as it would have been designed if its designer had meant to confute the reactionaries, and make for the achievement of humanitarian values. The enemies to be defeated were superstition, religion, nationalism, militarism, all those ideas advocated by those whom Loeb called 'romanticists'. Their phrases had led nations and races to hate one another; German reactionaries followed Nietzschean words, French reactionaries marched to Bergsonian ones. By contrast, Loeb felt himself of the company of Voltaire and Diderot; to the latter he dedicated his book *The Organism as a Whole*. 'What progress humanity has made', wrote Loeb, 'not only in physical welfare but also in the conquest of superstition and hatred, and in the formation of a correct view of life, it owes directly or indirectly to mechanistic science.'[19]

The ugly beings, according to Loeb, were the makers of hatred and romantic metaphysics, who believed in the freedom of the will, and war among the peoples. He detested the 'idealistic' German scientists: 'The head-professors in Germany', he said, 'were such Jew-baiters that the quicker they died out the better.' Yet he rejected Zionism too, for he felt that it was another nationalism, and 'nationalism is the very curse of the world'.[20] Every time he achieved a mechanistic explanation of a biological phenomenon Loeb felt he had struck a blow against the reactionaries of the world. When he showed that plants, plant-lice, and caterpillars' larvae in their movements toward light conformed to the Roscoe-Bunsen light energy law, that the chemical effect of light is equal to the product of its strength by the time during which it is acting, he felt that the world had been just that much further disenthralled from the men of evil. 'Not only is the mechanistic conception of life compatible with ethics: it seems the only conception of life which can lead to an understanding of the source of ethics.'[21] He thought that the search for justice and truth, and the love for work, are physico-chemically implanted in us, whereas the effect of exploitative systems that made 'economic slaves' of us, and the use of 'certain phrases' such as 'racial supremacy' were meant to deprive us of our 'degrees of freedom' in much the same fashion that an admixture of carbonated water in a jar will constrain the movements of crustaceans.[22]

The biophysicist Loeb and the historical economist Thorstein Veblen both cited each other's words approvingly in their writings.[23] For the purposive aim of Loeb's work was to show that the biological world and its laws were such that an egalitarian, socialistic society (of the sort Veblen advocated) was not only possible but perhaps inevitable.

On the face of it, Loeb's work seemed to be inspired wholly, in the poet's fashion, to grasp 'the beauty . . . of the thermodynamic laws which lay beneath the quivering system that is life'.[24] But there are alternative ways, or constraining conditions, that can be imposed on the universal quest for beauty; Loeb's had to be a beauty of law that was completely deterministic. The observed phenomena, in Loeb's view, had to be explicated 'in the form of equations containing no arbitrary constants'.[25] For an arbitrary constant might conceal the guise of a voluntarist reactionary. No chance, no contingency. Others would regard this total determinism as violating one's innermost sense of beauty. Loeb's scientific logic was indeed constrained by an extra-logical, social teleological principle; the ethical-political outlook of Voltaire and Diderot from his standpoint had to be confirmed in biological research.[26]

During Loeb's last years, the total deterministic principle was being undermined by a group of physicists who felt that their perfect world must have an ingredient of indeterminacy sufficient to allow for the existence of free will. A teleological principle was thus the basis, sometimes articulated, for the formulations of Max Born and Werner Heisenberg. But it is noteworthy that the anti-teleologist Loeb was guided by a social teleological principle. Though the language of purposes and values was excluded within the scientific system, purposes and values constituted the extrasystematic conditions that the deterministic system as a whole was expected to satisfy.

IV

No doubt it is in cosmological theories that the workings and conflicts of alternative perfect teleological principles stand out most clearly. For, as P. W. Bridgman once observed, the limitations of the observational facts allow for theoretical projections in diverse models that verge on

> the artistic, the emotional, and the metaphysical. The artistic instinct in the cosmologist finds expression in selecting those formulations or solutions out of the many possible ones, which are most elegant or most simple. The emotional element is well illustrated by the attitude which various cosmologists take toward the extrapolation into the indefinite past or future.[27]

What perhaps is overlooked, however, is that all the principal cosmological theories were evidently guided by social teleological principles.

We must distinguish between social teleological principles and universal teleological ones. A universal teleological principle is a requirement of scientific theories that is based on some universal psychological trait; the demand that a scientific theory should have a generic mathematical simplicity, the preference for 'beauty' and 'elegance' in scientific systems, are shared by the scientists of diverse societies, cultures, and political commitments. A social teleological principle, on the other hand, is based on an emotional preference or longing which is characteristic of a particular social or political group; the scientific theory that it then seeks is one which actualizes the values of its particular group. The meaning which is attached to the 'simplicity' of a given theory is usually therefore not wholly logical; for a theory congenial to one's social teleological standpoint will also tend to be regarded as 'simpler'. This social-teleological sense of simplicity is distinct, let us say, from the psychological sense in which Euclidean geometry is regarded as simpler than the non-Euclidean geometries; the Euclidean geometry is easier to understand, and more familiar. This latter type of simplicity has a common psychological basis. The social-teleological principle, on the other hand, introduces variations founded on social and political factors. Its sense of 'simplicity' merges with the logical and psychological ones.

The various cosmological theories that have been proposed in recent times are thus remarkably associated with corresponding political standpoints. The 'steady-state' theory, for instance, according to which the universe, viewed from any frame of reference, remains unchanging both in space and time, was developed by Fred Hoyle, Hermann Bondi, and Thomas Gold during the period after World War II when there was a veering away from evolutionary, progressive ideas. Arnold Toynbee's views on history were then highly influential and popular, with his vision of the eternal sameness of things throughout the rise and fall of civilizations, and of our consequent essential contemporaneity with the Greeks. According to Toynbee, however far back one went in time, the social universe perceived by the macro-historian of civilizations would not differ essentially from ours. Karl Popper's critique of evolutionary, historicist models had also added much at this time to rendering out of fashion the Marxist model of progressive evolution. An evolutionary sociology, according to Popper's analysis, would presuppose that society had an 'origin', a notion he repudiated; the steady-state astronomers indeed made the same criticism of the Big Bang theory of the stellar evolutionists.

Rather surprisingly, Fred Hoyle regarded as important in the genesis of his theory a curious personal incident, 'a strange and still unresolved incident that took place one afternoon'. He and Bondi were having a discussion, 'with a good fire burning . . . One of us dropped a small article. I have forgotten exactly what. I think it was a screw or a nail, but it may have been a pencil sharpener or some such item'. Then 'somewhat idly' they looked about for it, but were unable to find it. Whereupon they looked more intensively, 'for over an hour, finally more or less dismantling the whole room. But the article was simply not to be found. A simple time reversal of this situation leads to the creation of matter'. Hoyle's account naturally would pose a problem for the psychoanalytic mode of inquiry into scientific ideas.[28] But for our purposes, the sociological values and hopes that Hoyle has indicated in his novels and books provide a more verifiable ground for determining the social teleological principle that propels his scientific outlook.

The steady-state theory was not propounded as an answer to some novel observational facts. Rather, as Hoyle tells, when he and his friends, Bondi and Gold, were discussing those matters at the end of the war, they were concerned with one dilemma which arose out of the accepted science. They reasoned that if the theory of the 'expanding universe' were valid, then over long periods of time the density of matter would thin out, and that consequently all the basic constants of physics, such as the charge of the electron, would be altered. This 'difficulty', Gold pointed out, could be obviated if the Universe were not taken as changing with time.[29] Now this 'difficulty', of a change in some physical constants because of historical, 'environmental' effects had not been perceived as a 'difficulty' before World War II. Paul Dirac had proposed in 1937 his famous theory of the changes in the gravitational constant with the aging of the universe. But in the intellectual climate of the thirties, a more positive welcoming attitude toward change, social change especially, prevailed. Hoyle and his friends, however, having experienced the war, and having participated in military research, shared the longing for a more stable restful, unchanging universe.

This conation toward a timelessness in things expressed itself in

Hoyle's fiction. His spokesman tells a scientist of our time in *October the First is Too Late:*

> It will be hard for you to understand our point of view. In your time, everything of importance lay in the future. You worked for the future, you were dominated by a sense of progress ... Our philosophy is quite different. We have strong ideas of how life should be lived ... You see, we do not believe in time as an ever-rolling stream. We believe all times are equally important, the past is not lost.

Hoyle's protagonist affirms as against the progressive evolutionist a doctrine of egalitarian temporalism: 'If there's one thing we can be sure about in physics, it is that all times exist with equal reality.'[30] Apart from the accumulation of knowledge, Hoyle argues, the societies of the last twenty thousand years are 'essentially identical' with respect to their 'social factors'.[31]

The 'perfect cosmological principle' held by Hoyle and his friends prescribed that the universe must not only look the same to every observer from every position in space, but also, it must look the same from whatever time-instant it were regarded, in past, present, or future. To compensate for the difficulty posed by the observational evidence for the recession of the distant galaxies, and the consequent thinning out of their matter, Hoyle and his friends therefore postulated a continuous creation of matter throughout space. But Hoyle felt that their own teleological longing for an unchanging universe reflected the underlying structure of the universe. He characterized the epistemological basis for his perfect principle as something akin to a 'religious sentiment'. '[T]he universe', he wrote, 'wasn't following our logic, it's *we* that are constructed in accordance with the logic of the Universe.' Indeed intelligent life might be defined as 'something that reflects the basic structure of the Universe'.[32] Thus, the teleological designer in us, moved to formulate basic laws in accordance with our emotions and feelings, corresponded isomorphically with the cosmological structure; personal microcosm and cosmic macrocosm were structurally similar.

Meditating on his version of scientific plenitude, the 'millions of different kinds of planets with different kinds of creatures', what he called 'the fantastic zoo', Hoyle affirmed: 'This is the way I would design it if the choice were mine ... '[33] Every scientific theorist indeed is a designer of the world who asks always the teleological question: how would I have designed it? And as was the case with the mechanist Jacques Loeb, there was the conviction in Fred Hoyle that if his cosmological model were accepted, it would make possible the realization of higher human value, 'exposing the futility of nationalistic strife. It is in just such a way that the New Cosmology may come to affect the whole organization of society'.[34]

V

An altogether different teleological principle guided the formation of the theory of the 'expanding universe'. It was conceived during the relatively optimistic nineteen-twenties, when the world was putting war behind it, when treaties of peace and the outlawry of war became the fashion, and when, in America especially, an unprecedented prosperity and pleasure in the goods of life burgeoned. The great best-sellers were books such as H. G. Wells's *The Outline of History* and Will Durant's *The Story of Philosophy* in which the cosmos was interpreted as the setting for the cheerful adventure of the human race. The observational basis for the theory of the 'expanding universe' was propounded in the mid-twenties by Edwin P. Hubble; according to his telescopic data the light that was emitted from distant galaxies exhibited a shift toward the red end of the spectrum; interpreted as a 'Doppler effect', this phenomenon signified that those galaxies were receding, with a velocity, indeed, that was proportional to the galaxies' distance from the given frame of references (whether the earth or a presumed 'center') so that a simple equation $v = ar$, where a was 'Hubble's constant', v the velocity of recession, and r the radial distance to the galaxy, became the most basic cosmological formula.

Edwin P. Hubble had been a lawyer practising in Kentucky. As an undergraduate, he had studied astronomy, but then went to Oxford as a Rhodes Scholar to read Roman law. The vocation to return to the stars proved, however, irresistible, and Hubble renounced the law in order to resume his astronomical studies. Hubble's philosophy of research was later described by his co-worker, H. U. Mayall, as 'somewhat mystical'.[35] It was a frontiersman's teleological principle that guided his observations, a sense of what William James called the 'open universe', growing with increments of reality, and venturesome with unrealized possibilities. Thus Hubble wrote that the emerging cosmology was that of 'an expanding universe that is finite, small, and young'.[36] That the word 'young' would appear in a cosmological statement suggested the background of a characteristically American teleological principle. The young open universe of William James was projecting itself into a cosmogony.

There had, however, been a theoretical harbinger for Hubble's observations. A Russian mathematician, of alien descent, Alexander Friedmann, had, indeed, in 1922 shown that there was a whole class of non-static models, expanding, contracting and pulsating, that satisfied Einstein's basic equations of general relativity, and that Einstein's own proposed solution, that of a 'spherical universe', would be characterized by an unstable equilibrium. Einstein, at first unsympathetic to Friedmann's models, changed his view under the impact of Hubble's work on the extra-galactic 'red shift'. As Hoyle writes, his own steady-state model of the universe, proposed in 1948, 'was an attempt to return to Einstein's pre-Friedmann point of view'.[37] There seems indeed to be an oscillating fluctuation in the careers of teleological principles themselves. What manner of teleological principle had led Friedmann, a distinguished mathematical meteorologist, to immerse himself in Einstein's general relativity, and to seek out the flaw in Einstein's cosmological derivation, we do not know. That he was attached to Marxism in any profound way seems doubtful in light of the fact that in 1914 he was a volunteer for service in an aviation detachment, then for several years a military flyer at the front; in 1917, moreover, Friedmann became a section chief, and later director in Russia's first factory for manufacturing airplane measuring instruments. Furthermore, neither of the alternative models with which Friedmann concluded his paper in 1922, one for a temporally finite world-existence, the other involving a periodic curvature of space, were 'dialectical' in structure.[38] Gamow's autobiography gives us an all-too brief glimpse of the pure mathematician with a love for cosmology, and pioneering in the use of meteorological balloons.[39] At any rate, Hubble found in Friedmann's theoretical models a guide to the development of his theory and for further observational testing. Throughout his work, Hubble recognized explicitly that a 'perfect principle' was underlying to relativistic cosmology, the 'sheer assumption' that 'there must be no favored location in the universe, no center, no boundary; all must see the universe alike'.[40] An acceptable theory of the material universe, it was held, must be consistent with this 'perfect principle' as well as providing for the determination of the geometry of space by its contents. Requiring even more than these principles, the cosmologist could interpret the meaning of the 'red-shifts' as indicating the recession of distant galaxies only by the boldest use of 'the principle of the uniformity [of nature] as governing phenomena in the most remote regions, and making extrapolation possible'.[41]

The theory of the 'expanding universe' led by retrodictive extrapolation to the notion of an initial date at which the matter of the universe existed in a tremendously dense state within a small region, prior to the beginning of the process of expansion. At the hands of the Catholic Abbé Georges Lemaître, who returned from the trenches at the end of the First World War to study for the priesthood and physics, this notion was elaborated in the hypothesis of the 'primeval atom'. In a celebrated passage he described the initial creative explosion, the 'Big Bang' as it became called, with the atom world breaking up into fragments, with successive fragmentations

> to reach the present pulverization of matter into our poor little atoms... The evolution of the world can be compared to a display of fireworks that has just ended; some few red wisps, ashes, and smoke. Standing on a well-chilled cylinder... we try to recall the vanished brilliance of the origin of the worlds.[42]

To such an astronomer as Herbert Dingle all this seemed Aristotelian teleology *redivivus* with T. S. Eliot's *Wasteland* as the guiding cosmological model: 'we must reject the claim, only too often made, that the universe "must inevitably" have begun with a bang and be destined to end with a whimper.'[43] But the thirties with their years of depression and deepening crises seemed to many the diagram of a principle of collapse; some saw the portents of civilization's collapse while others limited the collapse to that of capitalistic systems; scientific theorists, too, were moved by corresponding models of collapse and epochal change.

VI

It was thus in the thirties that J. Robert Oppenheimer became absorbed in the problem of the collapse of capitalist civilization; the questions of unemployment, human misery, and the rising Nazi anti-intellectualism were always with him, the theme, as his then close political friend put it, of 'catastrophic capitalist collapse'. When Oppenheimer, during the summer of 1936, went East, he took 'with him all three volumes of Marx's *Kapital*... reading them through from cover to cover on the train'. Then he 'bought [the] complete works of Lenin and read them'.[44] The impact of personal tragedy too

56   Varieties of Scientific Experience

was felt at this time; Oppenheimer's reflections in physical theory turned to processes that might make for cosmological collapse, to the cumulative effect of the thermonuclear reactions that were being discussed.

> What would happen to a star, he wondered, when its thermonuclear life ran out? . . . The star would redden and contract into a heavy nuclear core smaller than the smallest stone. An observer standing upon it would see it fall in upon itself. In the mathematics of collapsing suns he passed his most nearly cheerful hours of 1937.[45]

Evidently Oppenheimer was the first to examine the theory of 'catastrophic gravitational collapse'.[46] An original theorist's choice of problem and the direction of his desired solution are largely determined by emotions which he is seeking to objectify; his character shapes to the maximum possible his choice of theory, aiming thus to alleviate the discords within himself—an external drama of fulfillment isomorphic with his own longings, this is precisely the outcome sought by the scientist's emotions, expressing themselves in a teleological principle, and operating as an emotional *a priori*.

Also in 1937 Paul Dirac proposed his theory that the value of the gravitational constant changed with the 'age' of the universe; Marxist thinkers welcomed it as the kind of historical physics that confirmed their 'dialectical' world outlook.[47] There were, however, two teleological principles involved in Dirac's analysis. The first was what has been called the 'numerological principle', according to which arbitrary, very large, or very small numerical coefficients 'ought not to occur a priori in the basic laws of physics'.[48] And where such arbitrary numbers do enter through the constants of physical laws, then, according to Dirac, a new principle, that we might call the 'historical principle', should guide our thinking; in Dirac's words, the principle 'that all the very large dimensionless numbers occurring in Nature are simple powers of the epoch, with coefficients of the order unity'.[49]

Dirac directed attention to some remarkable numerological coincidences which adumbrated a historical theory of the gravitational constant. The ratio between the two forces, of elementary electric charges, electron or proton, and their gravitational interaction was 10^{40};[50] meanwhile, the age of the universe according to the Expanding Universe theory was 10^{17} seconds; but the diameters of all elementary particles are about 3.10^{-13} cms, which could be traversed by light in 10^{-23}

seconds, the elementary time unit. But when we divide the number for the present age of the universe, 10^{17} seconds, by the elementary time interval, 10^{-23} seconds, we get the number 10^{40}. This suggests that the arbitrarily large ratio of electric to gravitational forces, also 10^{40}, is founded on the present age of the universe; in Dirac's words: 'Such a coincidence we may presume is due to some deep connection in Nature between cosmology and atomic theory.'[51] And since the elementary electric charge was regarded as probably having been unchanged, it was inferred that the gravitational charge had decreased in proportion to the expansion of the aging universe; thus when the Universe was half its present 'age', the gravitational constant would have been diminished in the requisite proportion.

A new class of cosmological models was defined by the use of Dirac's new principles.[52] They were of a teleological character insofar as they were functions of certain emotional-ethical-esthetic longings of the individual. To men of another temperament another kind of teleological principle would have been more expressive and congenial. William James, relishing every variety of stubborn, irreducible fact, reveling in the contingent and sheer plurality, would have caviled at a principle which aimed at explaining away the arbitrariness of a basic constant. Dirac had been impressed by Arthur Eddington's mystical Quaker conviction that the basic laws of nature derived from the epistemological requirements of the mind, sufficiently so that he thought that some of Eddington's deductions of universal constants were 'probably substantially correct'.[53] Then to Eddington's expanding universe he conjoined as well the 'historical principle' which carried the imprint of the Marxist notions that were agitating the universities in the thirties.

The 'numerological principle', the historical principle, and indeed a third, the principle of coincidence of macrocosm and microcosm, that is between cosmology and atomic reality were all varieties of 'perfect principles'. Indeed, twentieth-century cosmology has been based above all on what is called 'Mach's Principle', likewise indeed, a 'principle of perfection'. It may seem surprising to associate Ernst Mach, the great empiricist with a teleological principle, yet fortunately, Mach, himself a keen psychologist and introspective master, provides the evidence clearly. Two forms of 'Mach's Principle' have been distinguished, the metrical (or operational) sense, and the cosmological (or perfect) sense. In its loose form, the Principle is an admonition 'that the rest of the universe may be relevant to an observation or experiment'.[54] More specifically, in the first sense, it was the methodological notation that experimental statements concerning the earth's rotation are made by reference to the fixed stars. This operational fact in no way implied that the stars are causally, or functionally efficacious in the earth's rotation. Mach was at first making a purely metrical proposition, not one concerning the influence of the fixed stars on centrifugal forces on the Earth. For propositions concerning the latter seemed experimentally unverifiable, as no experiment could alter the fixed stars to observe the consequences for terrestrial phenomena. Nonetheless, Mach affirmed that 'it is not to be antecedently assumed that the universe is without influence on the phenomenon here in question'.[55] Thus the perfect sense of Mach's Principle began to emerge. Over and above the metrical, operational relationship between any individual phenomenon and the rest of the universe, there was also presupposed some basic causal relatedness. Einstein used Mach's Principle as asserting that 'the inertia of a body is due to the presence of all the other matter in the universe'. Mach, as J. Bradley has observed, was 'tacitly introducing ideas from his general philosophical monism'.[56] Indeed, Mach himself had written movingly of his longing for 'making man disappear in the All, in annihilating him, so to speak'; sympathetic to Buddhism for its 'wonderful story', he felt it approximated his own notion that the Ego should be 'nothing at all'.[57] Such passages led Erwin Schrödinger, the discoverer of the Wave Equation, to characterize 'the idea of Mach' as approaching 'as near to the orthodox doctrine of the Upanishads as it could possibly do without stating it *expressis verbis*'.[58]

The metrical, operational form of Mach's Principle was at the basis of the special theory of relativity; the perfect form of the Principle, toward which it evolved, was underlying to the general theory of relativity; it set Einstein on the line of inquiry which began with the removal of the distinction between gravitational and inertial forces.

VII

But even the operational principle in the course of its history has functioned as a teleological one. Most significantly does this emerge in the writings of the great spokesman and theorist of operationism in the United States, P. W. Bridgman. Ostensibly, and on the manifest level, Bridgman wrote his influential book *The Logic of Modern Physics* to accommodate himself to Einstein's ideas and to liberate himself from the doctrine of determinism, to alleviate 'my own disquietude', as he put it.[59] Deeper, emotional factors, however, were involved in the operationalist trend. This was the time of the twenties, when 'relativity' was indeed spreading through American social ideas, as the master-word that signified the emancipation of the American intellectuals from all absolutes. Relativity was the leading movement in ethics, anthropology, economics, history, and philosophy, the theme for discussions from Greenwich Village Bohemia to the Ivy League Halls.

Bridgman has been described as 'something of an individualist': 'He liked to work by himself, spending long hours in the laboratory, and eschewing attendance at academic committees.'[60] Of his more than 260 papers, only two named a co-author. He was indeed the last great individualist in American science; like a frontiersman resenting the encroachments of the settlers, Bridgman watched with sorrow how his sort was a dwindling variety among later generations organized by large-scale, collective, monopolistic science.[61] Operationism for Bridgman constituted more than the precept for eliminating unverifiable entities from physics: 'any concept is nothing more than a set of operations; *the concept is synonymous with the corresponding set of operations . . .*' He had none of Einstein's mysticism, that inclination to ask how God would have solved the problem, none of Dirac's conviction that the beauty of a theory was a guide to truth that might outweigh the given experimental evidence.[62] '[T]he Jew', wrote Bridgman critically, 'is particularly prone to mystical modes of thought, as suggested by the line of Jewish philosophers with their apparent need for an underlying "reality" '[63] He was a laboratory man, rebelling against the ideas upon which he had been raised; he spoke of how hard it was to exorcise childhood beliefs, 'I should hate to confess how long it was before I could get rid of the sneaking conviction that there really is a life after death';[64] he wrote feelingly of the conflict between parent and child which made the individual realize 'with redoubled poignancy his essential isolation and loneliness'.[65] He even reacted against the intrusion on privacy that took place in parlor social games.[66] And he sought in the operational theory the basis for a reinforced individualism. He dissociated himself from his socialistic colleagues who regarded as primary the alleged social, intersubjective character of science: no, what was controlling, Bridgman reiterated, was 'my' own acceptance of the laws of physics; 'no one can force his authority upon me'. and 'if I were on a desert island and had no fellows at all, I think that I still could eventually build up a valid body of truly scientific physics . . .'.[67] He deflated the impersonal style of philosophers, and felt 'we should rather go to the other extreme, and use the first person exclusively, forgetting the conventions of modesty'.[68] And he was convinced that the operational standpoint was the master-key to solving the world's political and social ills. For 'the parallelism in situation between physics and society', he argued, was 'more than a mere analogy, it reveals a logical identity'.[69] The experiences of physics 'in the last few years' were in his view, 'nothing short of revolutionary in their implications for the terribly complex social situation'.[70] One wonders what these revolutionary implications of operationism would have been; Bridgman felt there had to be 'radical political and economic changes'; also he found nothing wrong in Hitler's wanting a homogeneously German society; what he objected to was rather Hitler's 'taste in selecting his incompatibles that must arouse the abhorrence of outsiders'.[71] He wrote unforgettably on behalf of suicide as an act of human freedom, almost foreshadowing his own later action: 'This is the one situation in which society is most impotent to impose its wish on the individual',[72] and he 'contemplated with equanimity' the choice of policies 'that the race may ultimately perish'.[73] Thus Bridgman's operational principle was emotionally tied to a teleological principle which aimed to safeguard the individual against the encroachments of organized society and its 'social ethic'. It was Bridgman's agency for affirming the centrality of his values and attitudes as the only rational ones for science and society. It was a principle of science guided in large part by emotional-teleological drives.

Meanwhile, a novel crisis has arisen for the operational standpoint. For cosmological statements are freely made which violate the operational principle as Bridgman formulated it. To resolve this situation may require that we enunciate a 'perfect operational principle'. Take for instance such a daring speculation as that of George Gamow on the 'Augustinian era' that may have preceded the expansion of the universe:

> [T]he Big Squeeze which took place in the early history of our universe was the result of a collapse which took place at a still earlier era, and . . . the present expansion is simply an elastic rebound which started as soon as the maximum permissible density was reached . . .

Most likely the masses of the universe were squeezed together to such an extent that any structural features which may have been existing during the 'pre-collapse' era were completely obliterated, and even the atoms and their nuclei were broken up into the elementary particles ... Thus nothing can be said about the pre-squeeze era of the universe, the era which may properly be called 'St. Augustine's era', since it was St. Augustine of Hippo who first raised the question as to 'what God was doing before he made heaven and earth'.[74]

Now from the standpoint of Bridgman's operationism, a speculation concerning the existence of a 'pre-collapse era' would have to be adjudged as meaningless precisely because all structural features would have been obliterated. Thus, likewise, a hypothesis of the universe evolving through cycles of alternating expansions and contractions, would have to be excluded as meaningless, since there would be no criterion for differentiating the successive, recurrent phases. Yet our imaginations do find it meaningful to envisage such an oscillating universe. We can then salvage our imagination by combining it with the operational standpoint to enunciate a 'perfect operational principle': that is, if a state of affairs is imaginable or existent, then some operational criterion must likewise exist that would differentiate its existence from all other states. Thus, for a cyclical universe, there would be no recurrence so absolute as to preclude the knowledge that it was serially a later instance than a predecessor; an operation for determining place in a serial order would always exist. In a 'pre-collapse' era, no obliteration of structural features would be so complete as to make it impossible to describe its properties. By affirming such a perfect operational principle, we assert a parallelism of the order of existence and possible operations. It is a teleological principle, to be sure, one which asserts that cosmological processes are so constituted as to make possible knowledge of them. Einstein might have joined with Berkeley to say this is how God would have articulated the operational principle.

VIII

Are teleological principles, however, at most psychological 'aids to the imagination' that are logically irrelevant to both the structure and functioning of science? We have seen in passing that such a noted astronomer as Herbert Dingle warns against all 'perfect principles' such as the 'perfect cosmological principle'. They constitute in his view, a return to the outmoded mode of thought exemplified in

Aristotle's preference for perfectly circular orbits and immutable heavens. The 'perfectarians' (to mint a word, since 'perfectionist' is already used in quite a different sense), he argues, offer no justification for their respective perfect principles 'except that they seem right'; they are indeed, says Professor Dingle, 'atavisms'.[75] Certainly scientific writers have long held that Aristotle's error was in allowing ethical ideas to shape his scientific notions; 'ethical neutrality', as Bertrand Russell once wrote, is the standpoint of the scientist.

To which we might respond: all theorizing, all systematic 'world-building' does, however, as a matter of fact rest on teleological, perfect principles. At any given time, a given 'perfect principle' such as Aristotle's might be discarded, but it is not overthrown to introduce an absence of 'perfect principles'.[76] Rather some new 'principle of perfection', every whit as teleological, guides the interpretation of the contrary evidence. Underlying teleological principles, moreover, are never superseded; they are intellectual invariants that recur in different guises; Plato's principle of plenitude takes novel forms in the theories of Mendeleev and Gell-Mann, while Leibniz's law of economy hovers through the Hamiltonian least action form of the wave equation.

A conflict of teleological principles always exists at an irreducible minimum because it is the corollary of the conflict of scientific generations. When A. N. Whitehead, for instance, in 1916 argued before the Aristotelian Society that 'no one need be bludgeoned into accepting the rather bizarre doctrine of relativity', he was indicating quite clearly that in his view, the values of the Victorian era were at stake, values with which the old theory of the ether seemed congruent:

> The good old homely ether, which we all know, can in this case serve the purpose. Just as any author of genius, if he lives long enough, survives the inevitable accusation of immorality, so the ether by dint of persistence has outlived all reputation of extravagance . . . Scientifically, it has a perfectly adequate use by veiling the extremely abstract character of scientific generalizations under a myth, which enables our imaginations to work more freely.[77]

The eminent physicist Sir Joseph Larmor added, too, his descriptive imagery of the 'quivery permanent object' pervading their universe, and 'its tremblings that travelled to them with the message of sight'.[78]

The vivid language of good old 'homely', 'veiling', 'permanent', 'tremblings', all evoked the maternal emotions that the ether had come to exert, threatened as it was by rebellious 'bludgeoning' forces.

At the least, teleological principles are central to the functioning of science; they are the social psychological source of novel hypotheses, novel questionings, and novel experimental searches. Without them, creative scientific research would wither. But the structure of science reflects this functioning as well. Laws are formulated, axioms are stipulated, observations are adduced, that are congruent with certain teleological principles; then they are limited as special cases in accordance with others. The structure of science is guided not by one architectonic, but by several. We might say that the perfect principle in the history of science is that every teleological principle finds its fulfillment. In the Theory of Theories, there is a principle of theoretical plenitude: for every mode of theory, there is some segment of reality for which it supplies the best explanation. If there were a Periodic Table of Theories, each model would have its niche, and, indeed, some novel types of theory might be predicted from the occurrence of empty niches.[79]

Teleological principles in conflict do not altogether cancel each other; for all agree that values, purposes, are somehow fulfilled by the truths of science they discover. In a world which did not objectify value, an Einstein could not have asked heuristically how God would have arranged this matter, nor would a Dirac have been able to pursue the beauty of a theory as the heuristic guide to its truth. A world that did not objectify value in its constitution would have shown no such improbable basic pattern. And likewise, the adherent to mechanistic explanation finds it somehow the agency for the progress of the human spirit, as if mechanism and purpose were ultimately in harmony, something hardly to be expected in a world of fortuitous motions. Leibniz, wondering on the basis of the principle of scientific plenitude, went so far as to venture the notion that 'every possible is characterized by a striving toward existence', that otherwise a mere possible is a being frustrate, uncompleted.[80] Yet this would seem to be multiplying teleological principles unnecessarily.

In Einstein's work, the centrality of a teleological principle appears as the criterion of the ' "inner perfection" ' of the theory. Einstein's language is indeed much like Leibniz's 'principle of perfection'; he notes the primacy of 'external confirmation' in the validation of a physical theory, but then considers as equally significant the theory's 'inner perfection'.[81] The latter is more than 'logical simplicity', that is, the relative fewness of a theory's logically independent premises. What it is Einstein finds it difficult to say, except that that theory is superior among equally 'simple' ones which 'most sharply delimits the qualities of systems in the abstract'. It is not, however, 'an arbitrary choice'. Clearly, one has here a preference for a certain kind of world, a feeling so deep-rooted that this is the way things should be that it cannot be regarded as arbitrary. Einstein indeed is introspectively describing the teleological principle that has guided his theorizing. Einstein's co-worker, Leopold Infeld, tells that when Einstein 'had a new idea, he asked himself: "Could God have created the world in this way?" or "Is this mathematical structure worthy of God?" '[82] Infeld thought that Einstein's use of God as a heuristic principle reduced to one of logical simplicity. Yet Einstein felt that the principle of 'inner perfection' asked for something more; it might be called, following a usage of John Laird, the 'principle of deiformity'.[83]

We have not ventured thus far to try to characterize the status of teleological principles of science, to assert whether they are *a priori* intuitions, socially or genetically conditioned, or perhaps psychoanalytically determined; or indeed divinations in which we share in God's mind. A powerful argument affirms that such principles are imprinted in us through an 'evolutionary matrix', the outcome of the cumulation of selected variations in the natural history of the human mind.

Naturalistic accounts of teleological principles seem, however, beset by inherent inadequacies. Let us consider, for instance, the psychoanalytical and evolutionary hypotheses concerning their origin.

A 'principle of perfection', from the psychoanalytical standpoint, conjoins the pleasure-principle with the reality-principle; for such a union would indeed be the meaning of 'perfection'. In infancy, the child's wish for the mother's breast is usually followed by its attainment. Presumably then, the unconscious quest to restore the infantile world manifests itself in teleological principles; for the latter reinstate the infantile maternal perfection in the total external world itself. A primal infantile motivation is thus held to underlie teleological principles.

But let us look more closely at the infantile condition. The experience of birth, we can infer, is traumatic because the child finds itself in an external environment that crushes and chokes it, that leaves it cold, foodless, deprived of the all-embracing equilibrium. Experiencing this utter disruption of equilibrium, the child makes its Primal Hypothesis; these experiences so anxiety-engendering with their sundering of pleasure from reality, must arise from some source outside itself. The

Primal Hypothesis is guided by no analogy; the child has no fund of cumulated experiences on which to draw. Every hypothesis in subsequent experience is guided partially by the use of some analogy. But the Primal Hypothesis is a pre-analogical one; the child knows that it did not wish these unpleasant experiences; guided by an inherent disposition, an *a priori* tendency, it projects the hypothesis that they were caused by a reality external to itself.

The child's Primal Hypothesis is then confirmed by what we might call its first experiments, by experiences to which it attaches an experimental significance in a primal manner. It finds that its weeping is followed by food, warmth, and softness. Having inferred that the Perfect Maternal Equilibrium does not have a necessary existence, that an external disturbance has upset it, it also finds that perfection can be recovered partially. And it finds, too, a new pleasure in a freedom of movement not known in the mother's womb.

The child's proto-intellectual life seems then initiated by a hypothesis concerning the existence of an external world which lies beyond prior model or psychoanalytical explanation. To this pre-analogical hypothesis, the child, in its experience of its mother's breast, attaches its first encounters with beauty and simplicity, its first experience of form that is satisfying. But it has reached and grasped for these forms instinctively. In this sense, the later instinctive yearnings, which as Ernst Mach pointed out ramify through our scientific reasonings, recapitulate these early instinctive gropings, and are therefore explicable on a psychoanalytical model.[84] But the initial yearnings are elementary and pre-modeled.

Nor does the survival of these 'primal propensities', the gropings for an external world of beauty and elegance, seem explicable as the outcome of evolutionary history. The dominant impact of the struggle for existence would have been to place a negative survival value upon all 'principles of perfection'. For the zoological world is overwhelmingly one of hostilities, of omnipresent enemies, perpetual fears, unending threats. And the believer in 'perfection' in the earliest stages of evolving intelligence, would, as a believer in the pleasure-principle, have tended to act naively, and have found himself annihilated. One recalls the believers in God's goodness and omnificence who found themselves destroyed by the Nazis. Childhood experience too, with its rude rejections, is an ontological reminder that the perfect does not exist, that all existence is in the imperfect mode. Yet as knowledge-seekers we find ourselves holding to 'principles of perfection' almost dead against the evolutionary veto. The mentality of the scientific theorist should long ago have been penalized to extinction in the struggle for existence. The naive boy, Albert Einstein, puzzled by the most everyday realities of childhood, a stumbling, fumbling dreaming child regarded as stupid, might not have passed the test of biological struggle.

Nature in its animate struggles is so harsh and cruel that an organism tending to think in gentler terms would have been emotionally and intellectually disarmed. In Nature, one always has to fear the unexpected, the ambush, the chance; believers in simple, beautiful laws might have been simple-minded, and they would not have survived. The character traits that contributed to the development of astronomy and physical theory were not always those conducive to primitive survival. Possibly a belief in the mathematical beauty of Nature's laws was somehow a by-product of sexual selection in Nature. The males, however, that triumphed, according to Darwin in conflict over females, and thereby reproduced themselves more frequently, seem to have done so by virtue of physical superiority rather than for their theoretical tendencies.

The belief in the beauty and simplicity of Nature's laws may then be supervenient, or epigenetic, in biological evolution. If so the scientist's experience of sharing in divine thoughts is then no poetic illusion, born in the grandeur and exaltation of discovery. A Kepler, rhapsodizing on the beauty of Nature's laws, would in the zoological world have been no more than a self-deluding and self-destroying creature. That his kind of mind thus teleologically moved, survived is not to be set down altogether as an achievement of biological evolution. The gravamen of scientists' experience, as participants in God's thoughts, has not been merged into the premises of psychogenetic or biogenetic explanations.

Such is the place of teleological principles in the physical sciences. Far different is their situation in the social sciences. For what sociologist trying to formulate a sociological theory would feel that this is how God would have arranged social reality? But this takes us far beyond the aim of this essay.

Bertrand Russell used to ridicule Eddington and Jeans for finding that cosmology respectively and incompatibly fulfilled their conceptions of God's teleology,[85] yet in his most intensely introspective moments, as in 1916, he would write:

The centre of me is always and eternally ... a searching for something beyond what the world contains, something transfigured and infinite—the beautific vision—God—I do not find it, I do not think it is to be found—but the love of it is my life ... it is the actual spring of life within me.[86]

The scientists' emotional longings are not merely personal details extraneous to an understanding of science. Without their animating spirit, the scientists' creative drive would retrograde. Though a textbook might for its purpose exclude them, they are intrinsic to the pre-textualized scientific experience. The common ingredient in the scientific emotion, moreover, delineates an ultimate postulate—that somehow the world of verified fact is congruent, or isomorphic with, the emotionally sought.

Teleological cravings, these conjoint demands of our intellect, feeling, and will, may somehow take us directly to the core of reality.

Notes

1. Olexa-Myron Bilaniuk and E. C. George Sudarshan, 'Particles Beyond the Light-Barrier', *Physics Today*, Vol. 22, May 1969, p. 44.
2. As Lovejoy noted, Leibniz on occasions used the expression 'principle of perfection'. Arthur O. Lovejoy, *The Great Chain of Being: A Study of the History of an Idea* (1936), reprinted, Harper & Row, New York 1960, p. 338.
3. Y. Ne'eman, 'The Symmetry Approach to Particle Physics' in Murray Gell-Mann and Yuval Ne'eman, *The Eightfold Way*, W. A. Benjamin, New York 1964, p. 314. Also cf. Richard Feynman, *The Character of Physical Law,* M. I. T. Press, Cambridge, Mass. 1965, reprinted 1967, p. 155.
4. Wolfgang Pauli, 'Remarks on the History of the Exclusion Principle', in R. Kronig and V. F. Weisskopf (Eds.), *Collected Scientific Papers,* Interscience Publishers, New York 1964, Vol. 2, p. 1074.
5. Jagdish Mehra, ' "The Golden Age of Theoretical Physics": P. A. M. Dirac's Scientific Work from 1924 to 1933', in Abdus Salam and E. P. Wigner (Eds.), *Aspects of Quantum Theory,* Cambridge University Press, Cambridge 1972, pp. 47–48.
6. W. A. Tilden, 'Mendeleeff Memorial Lecture', *Journal of the Chemical Society,* Vol. XCV, Pt. II (1909), p. 2082.
7. *Poems and Political Letters of F. I. Tyutchev,* trans. by Jesse Zeldin, University of Tennessee Press, Knoxville 1973, pp.42–43. Also cf. Fyodor Tyutchev, *Poems of Night and Day,* trans. by Eugene M. Kayden, University of Colorado Press, Boulder 1974, p. 24.
8. *Poems and Political Letters of F. I. Tyutchev,* p 12.
9. *Poems of Night and Day,* p 38.
10. According to one scholar, Tyutchev's poem, 'The Dream at Sea', 'rehearses that part of Schelling's *Naturphilosophie* which posits a chain of being from inert matter to the conscious soul', with the artist's genius, in romantic fashion, alone perceiving the underlying harmony. R. Matlaw, 'The Polyphony of Tyutchev's

Son na More', *Slavic and East European Review,* Vol. XXXVI (1957–8), pp. 198–204. Tyutchev's poems of chaos, with their fear of the *Abgrund* (abyss) and search for the *Urgrund* (original ground), are said to have been written under Schelling's influence. In 'The Last Cataclysm', God's face finally emerges visible out of the chaos. Cf. Richard A. Gregg, *Fedor Tiutchev: The Evolution of a Poet,* Columbia University Press, New York 1965, pp. 94–95, 101.

11. D. Mendeleev, *The Principles of Chemistry,* trans. by George Kamensky, from the sixth ed., Longman's, Green & Co., New York 1897, Vol. 1, p. vii.
12. Cited in William A. Tilden, *Famous Chemists: The Men and Their Work,* J. Routledge, London 1935, p. 243. 'Mendeléeff Memorial Lecture', *Journal of the Chemical Society,* London, Vol. 95 (1909), Pt. II, p. 2079. The Soviet biography of Mendeleev, when quoting this passage, suppresses the sentence as to how often dialectical methods deceive. Cf. O. N. Pisarzhevsky, *Dimitry Ivan Mendeleyev: His Life and Work,* Foreign Languages Publishing House, Moscow 1954, p. 11.
13. W. A. Tilden, 'Mendeléeff Memorial Lecture', p. 2083.
14. Mendeléev, *The Principles of Chemistry,* p vii.
15. Sinclair Lewis, *Arrowsmith,* Harcourt, Brace, New York 1925, p. 39. Cf. Charles E. Rosenberg, *No Other Gods: On Science and American Social Thought,* Johns Hopkins University Press, Baltimore 1976, pp. 123–31. Paul de Kruif, *The Sweeping Wind,* Harcourt, Brace & World, New York 1962, pp. 97–98. Robert J. Griffin (Ed.), *Twentieth Century Interpretations of Arrowsmith,* Prentice-Hall, Englewood Cliffs 1968. The Soviet physicist, Lev Landau, thought that *Arrowsmith* was the 'best' portrayal that had ever been written of the 'psychology of scientific workers'. Lev Landau, 'A Scientist on Films', *Soviet Literature,* No. 3 (1963), p. 151.
16. Report has it that ardent feminists 'who believed human fathers to be a nuisance' hailed Loeb's discovery of the chemical fertilization of sea urchins' eggs as liberating. Cf. Paul de Kruif, 'Jacques Loeb', *The American Mercury,* Vol. V (1925), p. 277. Other women, unmarried, renounced bathing in the salty, ocean waters lest they incur parthenogenesis. The notoriety that surrounded Loeb's discovery is said to have been the reason why Loeb was never awarded a Nobel Prize. Cf. Joseph Bernstein, 'Jacques Loeb and the Reductionist Fallacy', *Reconstructionist,* Vol. XXXII (1966), p. 10.
17. W. J. V. Osterhout, 'Jacques Loeb', *The Journal of General Physiology: Jacques Loeb Memorial Volume,* 8 (1928), pp. xi, xii. Paul de Kruif, op. cit., p. 278.
18. W. J. V. Osterhout, op. cit., pp. lvi, li.
19. Jacques Loeb, 'Mechanistic Science and Metaphysical Romance', *Yale Review,* Vol. IV (1914–15), p. 785.
20. Benjamin Harrow, 'Some Notes on Jacques Loeb', *The Menorah Journal,* Vol. XI (1927), p. 621.
21. Jacques Loeb, *The Mechanistic Conception of Life,* University of Chicago Press, Chicago 1912, p. 31.
22. Jacques Loeb, 'Freedom of Will and War', *The New Review,* Vol. II, Nov. 1914, p. 631.
23. Joseph Dorfman, *Thorstein Veblen and His America,* Viking Press, New York 1934, pp. 196, 303. 'Veblen often chided him [Loeb] with the accusation that his materialism was a metaphysical principle' (p. 200). *The International Socialist Review,* the principal American Marxist journal in the early 1900s, regarded Loeb's work as contributing to socialism (p. 154).
24. Paul de Kruif, op. cit., p. 277.
25. Osterhout, op. cit., p. liii.

26. L. K. Hirshberg, 'Jacques Loeb, The Mechanistic Conception of Life', *The Monist*, Vol. XXIII (1913), p. 154.
27. P. W. Bridgman, *Reflections of a Physicist*, 2nd ed., Philosophical Library, New York 1955, p. 306.
28. Fred Hoyle, *Encounter with the Future*, Simon & Schuster, New York 1965, p. 93. From the psychoanalytical standpoint, the lost pencil-sharpener, for instance, suggests anxieties of lost potency; an ensuing feminine disarrangement (of the room) ensues. The theory of continuous creation might then have emerged as an answer to anxieties of diminishing procreative power; hence the importance given to the recollection of this incident.
29. Hoyle, *Encounter with the Future*, p 104.
30. Fred Hoyle, *October the First is Too Late* (1966), reprinted, Penguin Books, Harmondsworth, Middlesex 1972, p. 70.
31. Fred Hoyle, *Man and Materialism*, George Allen & Unwin, London 1957, p. 71.
32. Fred Hoyle, *The Black Cloud* (1957), reprinted, Penguin Books, Harmondsworth, Middlesex 1971, p. 181.
33. Fred Hoyle, *Of Men and Galaxies*, University of Washington Press, Seattle 1964, p. 42.
34. Fred Hoyle, *The Nature of the Universe*, B. Blackwell, Oxford 1952, p. 8.
35. H. U. Mayall, 'Edwin Hubble, Observational Cosmologist', *Sky and Telescope*, Vol. XIII (1954), p. 79. I. S. Bowen, 'Edwin P. Hubble: 1889–1953', *Science*, Vol. 119 (Feb. 12, 1954), p. 204. Harlow Shapley did not like Hubble, who he feels disparaged and denied the priority of his work. Yet he acknowledges Hubble was the better observer and depicts his total inner absorption: 'Hubble just didn't like people. He didn't associate with them, didn't care to work with them.' Harlow Shapley, *Through Rugged Ways to the Stars*, Charles Scribner's Sons, New York 1969, p. 57.
36. Edwin Hubble, *The Observational Approach to Cosmology*, Clarendon Press, Oxford, p. 21.
37. Fred Hoyle, *From Stonehenge to Modern Cosmology*, W H. Freeman, San Francisco 1972, p. 56.
38. A. T. Grigorian, 'Aleksandr Alexandrovich Friedmann', *Dictionary of Scientific Biography*, Charles Scribner's Sons, New York 1972, Vol. V, pp. 187–8. A. Friedmann, 'Über die Krümmung des Raumes', *Zeitschrift für Physik*, Bd. 10, 1922, pp. 385–6 Loren R. Graham notes 'some effort on Friedmann's part to connect his views of the universe with materialism'. *Science and Philosophy in the Soviet Union*, A. Knopf, New York 1972, p. 497. The book in question, however, written with V. K. Frederiks, and designed as a text for Soviet schools, probably had, as was frequently the case, its editing political commissar.
39. George Gamow, *My World Line: An Informal Autobiography*, Viking Press, New York 1970, pp. 41–45.
40. Hubble, op. cit., p. 53.
41. Hubble, op. cit., p. 54. Mayall, op. cit., p. 79. Hubble's lifework was at odds with those critics who have periodically argued that the physical laws obtaining in distant places and times might differ radically from those that governed our part of the Universe. Presumably such critics would also maintain that there were no common, universal laws from which the various regionally or temporally restricted laws could be derived as special cases. Cf. Norwood Russell Hanson, 'Some Philosophical Aspects of Contemporary Cosmologies', in *What I Do Not Believe, and Other Essays*, Reidel, Dordrecht 1971, p. 62. Morris R. Cohen,

Reason and Nature, Harcourt, Brace & Co., New York 1931, p. 225. O. Chwolson, 'Duerfen Wir die Physikalischen Gesetze auf das Universum Anwenden', *Scientia*, Vol. VIII (1910), pp. 41–53.
42. Canon Georges Lemaître, *The Primeval Atom: An Essay in Cosmogony*, trans. by B. H. and S. A. Korff, Van Nostrand, New York 1950, p. 78.
43. Herbert Dingle, *The Scientific Adventure*, Pitman, London 1952, p. 275.
44. Haakon Chevalier, *Oppenheimer: The Story of a Friendship*, Pocket Books, Inc., New York 1965, reprint, 1966, p. 12.
45. Nuel Pharr Davis, *Lawrence & Oppenheimer* (1968), reprinted, Fawcett Publications, New York 1969, p. 78.
46. Kip S. Thorne, 'Gravitational Collapse', *Scientific American*, Nov. 1967, p. 90. Cf. J. R. Oppenheimer and H. Snyder, 'On Continued Gravitational Contraction', *Physical Review*, Vol. 56 (Nov. 1, 1939), pp. 455–9.
47. J. B. S. Haldane, *The Marxist Philosophy and the Sciences*, Random House, New York 1939, p. 76. The work of Milne and Dirac was said to have 'introduced the historical process into exact physics'.
48. F. J. Dyson, 'The Fundamental Constants and Their Time Variation', in Salam and Wigner, op. cit., p. 214.
49. P. A. M. Dirac, 'A New Basis for Cosmology', *Proceedings of the Royal Society*, Vol. 165 (1938), p. 201.
50. I am following George Gamow's more simplified version of Dirac's argument with its more approximate values of the various ratios. Cf. *Gravity*, Doubleday & Co., New York 1962, pp. 138–40,
51. Dirac, op. cit., p. 201.
52. Fred Hoyle, *From Stonehenge to Modern Cosmology*, p 78.
53. P. A. M. Dirac, 'The Cosmological Constants', Nature, Vol. 139 (Feb. 20, 1937), p. 323.
54. J. Bradley, *Mach's Philosophy of Science*, Athlone Press, London 1971, p. 10.
55. Ernst Mach, *The Science of Mechanics: A Critical and Historical Account of its Development*, trans. by Thomas J. McCormack, 6th American ed. Open Court Publishing Co., La Salle 1960, p. 341. Cited in J. Bradley, op. cit., p. 155.
56. J. Bradley, op. cit., p. 146.
57. Ernst Mach, *Popular Scientific Lectures*, 4th ed., trans. by Thomas J. McCormack, Open Court Publishing Co., Chicago 1910, p. 88. Ernst Mach, *The Analysis of Sensations and the Relation to the Physical to the Psychical*, trans. by Sydney Waterlow, 5th German ed., Open Court Publishing Co., Chicago 1914, p. 356.
58. Erwin Schrödinger. *My View of the World*, trans. by Cecily Hastings, Cambridge University Press, Cambridge 1964. p. 37. Also cf. Lewis S. Feuer, 'Ernst Mach: The Unconscious Motives of an Empiricist', *American Imago*, Vol. 27 (1970), pp. 12–40. The noted Viennese philosopher, Heinrich Gomperz, also called Mach 'the Buddha of science'. 'Autobiographical Remarks', *The Personalist*, Vol. XXIV (1943), p. 258.
59. P. W. Bridgman, 'P. W. Bridgman's "The Logic of Modern Physics" After Thirty Years', *Daedalus*, Vol. 88, No. 3 (1959), *Proceedings of the American Academy of Arts and Sciences*, p 519.
60. Rupert Stevenson Bradley, 'Percy Williams Bridgman', *The New Encyclopaedia Britannica, Macropaedia*, Vol. 3, University of Chicago Press, Chicago 1974, pp. 191–2.
61. P. W. Bridgman, *Reflections of a Physicist*, 2nd ed, Philosophical Library, New York 1955, pp. 432–5.

62. Dirac concluded his account of Schrödinger's discovery of the wave equation by saying: 'I think there is a moral to this story, namely, that it is more important to have beauty in one's equations than to have them fit experiment.' P. A. M. Dirac, 'The Evolution of the Physicist's Picture of Nature', *Scientific American*, Vol. 208 (May 1963), p. 47.
63. P. W. Bridgman, *The Intelligent Individual and Society*, Macmillan, New York 1938, p. 247.
64. Ibid., p. 209. Also pp. 210, 295.
65. Ibid., p. 151.
66. Ibid., p. 149.
67. Ibid., pp. 156–7.
68. Ibid., pp. 154–5.
69. Ibid., p. 7.
70. Ibid., p. 11.
71. Ibid., p. 293.
72. Ibid., p. 226.
73. Ibid., p. 301.
74. Gamow, *The Creation of the Universe*, rev. ed.,Bantam Books, New York 1965, p. 28.
75. Herbert Dingle, 'Cosmology and Science', *The Scientific American*, Vol. 195 (Sept. 1956), pp. 234–6.
76. While Galileo as against Aristotle found the stars scattered 'without any rule, symmetry and elegance', he built upon a new teleological principle, 'that matter is unalterable ... and that because of its eternal and necessary character' it is possible to describe its nature in mathematical demonstrations. Matter was thus raised from its lowly estate to that which had hitherto been reserved for Pure Ideas. No longer was earthly matter corruptible and mutable; it was unchanging and mathematical. This was the teleological principle of an Italian hedonist and man of the Renaissance. Cf. Leonardo Olschki, 'Galileo's Philosophy of Science', *Philosophical Review*, Vol. LII (1943), pp. 354, 359, reprinted in Franklin L. Baumer (Ed.), *Intellectual Movements in Modern European History*, Macmillan, London 1965, pp. 46, 61.
77. A. N. Whitehead, 'Space, Time, and Relativity', *Proceedings of the Aristotelian Society,* Vol. XVI (1915–16), p. 127.
78. Sir Joseph Larmor, 'Relativity: A New Year Tale', ibid., p. 131. It might be noted that Mendeleev in his old age held on to the conceptions of the ether and the atom; he felt 'the notion of a splitting up of atoms into "electrons" only complicates' matters, and that no contribution to this problem would result from 'a study of electrical and optical phenomena'. So he wrote in the edition of his textbook in 1903. Cf. Tilden, op. cit., p. 2104.
79. Thus the history of mathematics in relation to science exhibits what might be called a principle of perfect applicability. Whenever a branch of mathematics is discovered, the longing arises in time to discover for it a physical application. William Kingdon Clifford describes how the longing arose in him for a non-Euclidean world: 'I do not mind confessing that I personally have found relief from the dreary infinities of homaloidal [Euclidean] space in the consoling hope that, after all, this other [Riemannian space] may be the true state of things.' (*Lectures and Essays,* ed. by Leslie Stephen and Frederick Pollock, Macmillan, London 1886, p. 230). Ernst Mach as early as 1872 thought that a satisfactory theory of electricity would require the use of a geometry not restricted to three

dimensions. *History and Root of the Principle of the Conservation of Energy,* trans. by Philip E. B. Jourdain, Open Court Publishing Co., Chicago 1911, p. 54. Karl Pearson in 1885 also suggested that the notion of a curved space might be more useful to physicists than that of a subtle medium in homaloidal space. William Kingdon Clifford, *The Common Sense of the Exact Sciences* (1885), D. Appleton, New York 1894, p. 226. Also J. J. Sylvester, *The Laws of Verse,* Longmans, Green, London 1870, pp. 113–14. Boolean Algebra for long was regarded as the impregnable stronghold for pure forms, that would never be sullied by practical application. Boolean Algebra, however, became the logic of computer science. As Leibniz once wrote, the pure idea longs to actualize itself.
80. Lovejoy, op. cit., p. 177.
81. Albert Einstein, 'Autobiographical Notes', in *Albert Einstein: Philosopher-Scientist,* ed. by Paul Arthur Schilpp (1949), Harper & Row, New York 1959, Vol. 1, p. 23.
82. Leopold Infeld, *Quest: The Evolution of a Scientist,* Doubleday, Doran, New York 1941, p. 267.
83. John Laird, *Theism and Cosmology,* Philosophical Library, New York 1942, p. 290.
84. 'The greatest advances of science have always consisted in some successful formulation ... of what was instinctively known long before ... ' (Ernst Mach, *Popular Scientific Lectures,* trans. by T. J. McCormack, 4th ed., Open Court Publishing Co., Chicago 1910, p. 191.)
85. 'Eddington deduces religion from the fact that atoms do not obey the laws of mathematics. Jeans deduces it from the fact that they do.' Bertrand Russell, *The Scientific Outlook,* W. W. Norton, New York 1931, p. 108. Yet the notion of free will could make sense only in a world in which laws prevailed though not with a complete determinism. Eddington and Jeans were not holding to contradictory views. Thus Max Born, developing the standpoint of quantum indeterminacy, found a teleological principle in the seeking of a place for free will against a causal background. Cf. *The Born-Einstein Letters,* trans. by Irene Born, Macmillan, London 1971, pp. 154–5.
86. *The Autobiography of Bertrand Russell, 1914–1944,* Little, Brown, Boston 1968, p. 100.

3

God, Guilt, and Logic: The Psychological Basis of the Ontological Argument

Every mode of philosophical argument indicates something of the underlying consciousness of its user, and where a mode of argument is dominant in a given time, culture, or social segment, its attractive power is founded on corresponding emotional components in the social setting. When during the eighteenth century the argument for God's existence was increasingly founded on evidence of a benevolent designer, the mode of argument acquired its plausibility from the growing optimism in the social consciousness. An argument indeed differs from a demonstration or proof insofar as an argument always appeals to some inarticulate emotional premise congenial to the users; a demonstration, on the other hand, is cross-cultural because it makes no appeal to covert emotional premises.

I

That the ontological argument for God's existence has seemed convincing only to philosophers of a certain emotional temperament has long been known. John Locke, for instance, found it utterly unconvincing to a man of his sort; '... in the different make of men's tempers and application of their thoughts', he wrote, 'some arguments prevail more on one, and some on another...' He himself preferred not even to examine the notion that the existence of God can be proved from the idea of a most perfect being because it seemed to him such 'an ill way of establishing this truth and silencing atheists...'[1] Indeed, to John Locke, the medical practitioner, such 'divinity disputations' were an 'ill way' of arguing because they literally arose from a kind of illness. He described the eager monks in such argument as engaged in what later would be called a sublimated sexual assault: 'Poor *materia prima* was canvassed cruelly, stripped of all the gay dress of her forms, and shown naked to us, though, I must confess, I had not eyes good enough to see her. The young monks (which one would not guess by their looks) are subtle people, and dispute as eagerly for *materia prima* as if they were to make their dinner on it . . .'[2] The ontological argument seemed to Locke to derive from an utter confusion of psychical with external reality; 'any idea', he wrote in a private paper, 'barely by being in our minds, is no evidence of the real existence of anything out of our minds, answering that idea. Real existence can be proved only by real existence . . .'[3] Evidently then a certain impairment of the sense of reality lies at the basis of the ontological argument; men's emotional temperaments, under certain historical circumstances, predispose them to become (what we may call) 'ontologians'. Certain social circumstances are required as the supporting ground to the ontologians' argument; where they are lacking, the argument either lacks appeal or fails to arise. 'No trace of it exists in Greek thought', observes Étienne Gilson, the eminent historian of philosophy, 'but it does not seem to have occurred to anyone to ask either why the Greeks never dreamt of it, or why, on the contrary, it was perfectly natural that Christians should be the first to conceive it.'[4] Clearly there was a psychological ingredient in the Christian experience which provided the ontological argument with an inarticulate major premise. Especially are we led to infer the presence of such a Christian ingredient by the report of Harry Austryn Wolfson on his inventory of the proofs for the existence of God among Jewish philosophers: 'As for the ontological argument . . . it is entirely absent, though some of the ingredients of which it is made up were not unknown . . .'[5] What then was the particular emotional ingredient, recurring in Christian experience, which made the ontological argument seem so convincing to its users?

When we proceed with this inquiry, a rather astonishing fact comes into view. A common emotional ingredient does indeed characterize the ontologians of the most diverse places and times. Anselm, Abbott of Bec in Normandy in the eleventh century, Josiah Royce, a Californian who grew up in a mining town and became a Harvard professor of philosophy toward the end of the nineteenth century, Norman

Malcolm, an American disciple of Wittgenstein and linguistic philosophy in the twentieth century, and Karl Barth, the Swiss theologian who in 1931 revived the ontological argument, all shared a common concern with the experience of individual guilt. And as we shall see, this component of guilt-preoccupation is an essential one in the consciousness which perceives the ontological argument as convincing. It is the source of a mode of thinking which might be called 'logical masochism'. To assuage guilt, the ontologian is prepared in all humility to bow his logical powers submissively before an entity which is transcendentally exceptional to them. Let us see how this pattern reveals itself in the utterances and writings of the several thinkers.

II

An emotional base underlying a mode of philosophical argument is, of course, best grasped in the complexities of the philosopher's personal life. We know that for Anselm, the inventor of the ontological argument, mortification was the chief joy almost all his life.[6] In his maturity, he still longed to be a little boy again, quailing under his master's rod. He conceived of God as very much like the punishing schoolmaster: '... it is certain that God "scourgeth every son whom he receiveth", the same love in a wonderful manner... For we ought to consider, beloved', he wrote to Hernostus, 'what consolation these sufferings bring with them...' 'We willingly yield to our chastiser', he said, 'by acts of thanksgiving to the chastising hand.'[7] Left in his early years with an imprint of hatred for his father, whose harshness he resented, Anselm, on the other hand, loved his mother devotedly, and when he became a priest, it was in accord with his mother's wish but in defiance of his father's. When he dreamt, it was evidently of physical union with his mother sanctified by God; he dreamed (allegedly as a child) of climbing a mountain to the Court of God, then crossing a river, grieving at the careless King's maidens, then reaching the Invisible King who had him eat 'bread of an exceeding whiteness', so that with 'corporal mouth' he had eaten the Bread of God. The dream is eloquent with its theme of the rejection of the earthly father for the Heavenly One.[8] The careless King, disporting with maidens, is the earthly harsh father, so unkind to the mother; as Anselm climbs the mountain, achieving the height of manhood, he crosses the river, the watery womanly canal, and God has him eat 'bread of an exceeding whiteness', the roundness of his mother's breast or buttock, and he is blessed. To mark his devotion to his mother, Anselm wished to be a monk. Christ in his eyes became a mother, and it was in such terms that Anselm addressed his prayer: 'And Thou, Jesus, dear Lord, art Thou not a mother too? . . . Indeed, Thou art, and the mother of all mothers, who didst taste death in Thy longing to bring forth children unto life.'[9] A strong maternal fixation thus persisted in the Prior of the monastery at Bec; it suggests indeed a passive homosexuality, but we shall not speculate further on the sources of Anselm's guilt and masochist experience. Instead, we shall emphasize the impressive fact that no philosophical system has made guilt so central in its notion of the universe as did Anselm's.

For man's guilt is the cardinal metaphysical fact, according to Anselm, from which the details of the Universal Drama necessarily follow. Anselm indeed professed to prove with deductive logic the necessity of the Incarnation and Atonement of Jesus. The logical steps, to his mind, were simple and rigorous. Man, in his disobedience, had committed a sin which was infinite; to atone for an infinite guilt, no finite sacrifice could be adequate; therefore God Himself had to become Man, so that an Infinite Atonement of Infinite Guilt could be achieved; therefore Jesus had become Man and was crucified.[10] We may omit some of the intervening corollaries in the deduction; such in its essentials was Anselm's theology of guilt which became known in the history of theology as the 'satisfaction doctrine'.[11] In its time, it represented a new departure in the theory of Man's Redemption. For us, it is remarkable for its projection on a cosmic scale of its central metaphysical notion of man's guilt. This is the mythology, above all, of guilt-consciousness.

It was precisely during an intense experience of guilt that Anselm's logical resistances gave way, and he yielded to the validity of the ontological argument. He 'embraced' thoughts which he had previously been 'repelling'. He had struggled to find a convincing proof of God's existence to overcome his own doubts. Let us remember that Christianity was relatively new to Normandy; paganism and scepticism quietly survived in the interstices of feudal society.[12] In his own doubt, Anselm cried: 'Lord, if thou art not here, where shall I seek thee, being absent?' His own personal guilt tormented him: 'My iniquities have gone over my head, they overwhelm me; and, like a heavy load, they weigh me down. Free me from them; unburden me, that the pit of iniquities may not close over me.' The elusive proof sometimes, he wrote, 'seemed to be just within my reach, while again it wholly evaded my mental vision'. Then came what we might characterize as a capitulation of logical masochism: '. . . at last in despair I was about to cease, as if from the search for a thing which could not be found . . . So one day, when I was exceedingly wearied with resisting its importunity, in the very conflict of my thoughts, the proof of which I had despaired offered itself, so that I eagerly embraced the thoughts which I was strenuously repelling.'[13] This emotional current was expressed in Anselm's maxim: 'I do not seek to understand that I may believe, but I believe in order to understand.'[14] Under the dominance of such feelings, Anselm formulated his ontological argument:

> Something exists, in the understanding, at least, than which nothing greater can be conceived . . . And assuredly that, than which nothing greater can be conceived, cannot exist in the understanding alone. For, suppose it exists in the understanding alone: then it can be conceived to exist in reality; which is greater . . . Hence, there is no doubt that there exists a being, than which nothing greater can be conceived, and it exists both in the understanding and in reality.[15]

Now the ontological argument at the end of the eleventh century remained very much the individual standpoint of Anselm. Despite the high position which he reached as Archbishop of Canterbury, despite his staunch leadership in opposing the claim of the King of England to the right to invest bishops, Anselm's writings, as M. J. Charlesworth observes, had little influence, and during the next hundred years were 'rarely cited or discussed'.[16] There was no doubt a sociological basis for the failure of Anselm's argument to win a widespread acceptance in his time. And such a sociological explanation would have to show how Anselm's logical masochism failed to find its counterparts in the social climate, why, in other words, Anselm's standpoint remained largely an individualist one. What this sociological explanation is seems to me suggested by certain facts to which various scholars, including Karl Barth, have alluded. Anselm wrote his *Cur Deus Homo* in 1094–8; the *Proslogion* was written much earlier in 1077–8, but Anselm did not achieve the Archbishopric of Canterbury until 1093. In 1095, however, Pope Urban II at the Council of Clermont sounded the call to Europe for what became the First Crusade. Anselm's argument fell on unwilling ears and feelings precisely because Europe was being urged to direct its aggressions outwards rather than inwards; the campaign against the Saracens, the military slogans against the infidel, undermined the emotional base for logical masochism. 'This land which you inhabit', said the Pope, 'shut in on all sides by the seas and surrounded by the mountain peaks, is too narrow for your population . . . Hence it is that you murder and devour one another . . .' Those who joined the Crusade were promised 'the remission of sins'.[17] Cultural masochism declined as Europeans were stirred to embattle the Saracens. Dissenters such as the Jews were now to be assaulted physically rather than met intellectually.[18] The ontological argument with its self-abasement of the logical faculties did not correspond to the adventurous pride of the expanding towns and the recruiting ardor of the land-seeking, sin-shedding warriors and their guiding Church.

Before this time, however, Anselm evidently in 1078 was obliged to confront the powerful criticisms which Gaunilo, a monk of Marmoutier, directed against the ontological argument. The exchange was interesting especially for the light it cast on the emotional base of the argument. Gaunilo, a man of noble birth, then in his eighty-fourth year, had known the world and a wife; he wrote his *Pro Insipiente* (on Behalf of the Fool), as Barth notes, with none of Anselm's passion and compulsion.[19] Gaunilo appeals above all to the sense of reality. One can think, he says, 'of all sorts of things whose existence is uncertain, or which do not exist at all . . .' One can think of a lost island, for instance, 'more excellent than all other countries'; and yet if one argued that this hypothetical island necessarily existed on the ground that if it did not exist, some actual island would be more excellent than this Perfect Island, which would be contrary to the latter's perfection, then clearly, says Gaunilo, we would join with the Fool in rejecting this specious logic. 'For he [the ontologian] ought to show first that the hypothetical excellence of the island exists as a real and indubitable fact . . .'[20]

We cannot help but be struck by Gaunilo's remarkable imagery of the 'lost island'. This uninhabited 'lost island' is a classical maternal symbol, with its suggestion of the womb with all its comfort, surrounded by water, and indeed lost, like our own primal uterine happiness. Freud once observed how the Portuguese discoverers of a lost densely forested island in the Atlantic Ocean called it Madeira, the Portuguese word for wood, derived at once from *mater* (mother) and *materia*.[21] Gaunilo, we may venture to suggest, with his serene sense of reality, was a man who had mastered his longings for the Lost Island; having mastered his Oedipal longings, his sense of the distinctness between psychical and physical reality was not impaired; he was free from the neurotic's pattern of what is called 'the omnipotence of thought'. In his recognition of the non-existence of the Lost Island, he insists above all on the 'reality-principle'. Thus, in his vivid choice of example, Gaunilo has probably penetrated into a basic psychological source of Anselm's argument.

Gaunilo's criticism impelled Anselm, however, to introduce a drastic modification in the ontological argument, and one which partially explains the emergence of two interpretations of the ontological argument, the first purely logical and universal, the second social and psychological. Quite apart from any argument from the definition of God, the most Perfect Being, to His existence, Gaunilo says that, in the first place, he does not even have an idea of a most Perfect Being to begin with:

> I ... am as little able to conceive of this being, when I hear of it, or to have it in my understanding, as I am to conceive of or understand God himself; whom, indeed, for this very reason I can conceive not to exist. For I do not know that reality itself which God is, nor can I form a conjecture of that reality from some other like reality ... but of God, or a being greater than all others, I could not conceive at all, except merely according to the word. And an object can hardly or never be conceived according to the word alone.[22]

Gaunilo, a precursor of Lockean empiricism, simply finds no imaginable content associated with the word 'God'. One wonders what sort of man he was, what he was doing in a monastery. His criticism raised the spectre of genetic analysis for the idea of God itself. And at this point, threatened with a psychological criticism that the word 'God' is meaningless, Anselm made his appeal to the tradition of Catholic usage. He appeals to Gaunilo, not as spokesman for the fool, but as a 'Catholic': 'I call on your faith and conscience to attest that this is most false.'[23] Anselm was like a desperate professor of metaphysics replying to an obdurate logical positivist: since both are members of the philosophical fraternity, the professor appeals to the tradition of their profession, its many metaphysical arguments; surely they were not all nonsense, or about words; otherwise why would the positivist himself still be a member of the philosophical association? Anselm's argument was cultural, *ad hominem,* more precisely, *ad conscientiam,* or *ad traditionem,* but it signified a retreat from the original logical universality of the argument. Henceforth, fools, unbelievers, believers in a finite God, could never be expected to be persuaded by it. Logical masochism found itself looking for support to a cultural masochism, an unquestioning acceptance of the basic terms of one's religious tradition.

Interpreters of the ontological argument in recent times have stressed the socio-cultural premise, the cultural masochism in Anselm's argument, by emphasizing the text of the reply to Gaunilo. The more universalist interpreters, however, have emphasized the plain meaning of Anselm's first statement before he was set upon by the fool. Among ontologians today there is similarly an oscillation between logical and cultural masochism.

III

The American culture of the nineteenth century was scarcely one in which the ontologians' mode of thought could flourish. William James spoke for the dominant mood when he said that to any person who had doubted the existence of a good God, such scholastic arguments sounded 'simply silly'.[24] James utterly rejected the temperamental standpoint of what we have called logical masochism; the world for him was a 'republican banquet' in which each person was sacred, and no 'monster' of 'intellectual despotism' could compel all existence to 'bend the knee to its requirements'.[25] The American mood was one of a basic optimism and confident self-reliance; despite a Calvinist background, the guilt-experience was always becoming recessive. Only in the philosophy of Josiah Royce, especially toward the end of his life, did the ontological argument find an original expression. It provided, he said, the 'escape from egocentricity'; if a rebuttal of solipsism was to be made, Royce said, 'its kernel will be found in what the ontological argument essays to state'.[26]

Curiously, Josiah Royce was also the only American thinker to write both extensively and probingly on the philosophical significance of treason and guilt.[27] The whole bearing of guilt-consciousness for the nature of reality troubled him deeply. When he wrote his memorable portrayal of the traitor, his words recalled those of Nathaniel Hawthorne's 'unpardonable sin':

> He had a cause ... And he was false to his cause ... That is, in so far as in him lay, he destroyed by his deed the community in whose brotherhood, in whose life, in whose spirit, he had found his guide and his ideal.
> This unpardonable sin would be betrayal.
> In brief, by his own deed of treason, the traitor has consigned himself,—not indeed his whole self, but his self as the doer of the deed,—to what one may call the hell of the irrevocable.
> The hell of the irrevocable: all of us know what it is to come to the border of it when we contemplate our own past mistakes or mischances.
> ... by my own traitorous deed, I have banished myself to the hell of the irrevocable.[28]

We do not know why guilt and the unpardonable sin so preoccupied Royce. We do know that he disliked his contemporaries, Ibsen, Shaw, and Nietzsche, who were in revolt against guilt-anxiety.[29] As with Anselm, there was a strong maternal fixation in Royce, beside a traumatic concern with his lack of masculinity. Royce literally personified the Community as a feminine person: 'The mother which is a republic is a community which is also a person, and not merely an aggregate, and not merely by metaphor a person.'[30] When, during his boyhood, his classmates rejected him for his awkwardness and clumsiness in playing games, for being a 'sissy', Royce found it a traumatic experience of philosophical moment.[31]

When he forsook California for the East, Royce did not regard his re-location as a rational step for his advancement and philosophical work. Rather, it became an occasion for all sorts of brooding on guilt with respect to the community. He wrote a novel, *The Feud of Oakfield Creek*, to justify himself against the Californian community. Amidst its dissipations and corruptions, Royce wrote, one would 'feel quite out of place'; California was not a place where the intellect could be at home, but rather a sordid environment for a 'man of character', entrapped among 'puzzling and doubtful' people. Such a man felt himself 'banished as if by the curse of Cain' from his old Californian friends. He railed against the state as hollow with the tombs of cultural ventures, and feared that its enmity to the ideal would drive one to suicide. He noted how its students and newspapers were moved to cruel 'outbursts' against an erring professor, how its stockraisers stole horses from Indians, how its pioneer families quickly degenerated physically, how its millionaires and clergymen conspired to wreck a college. California, a 'world of the bitterest contradictions' corrupted the strongest of men. It bred intellectual hypocrisy; the Jewish communist 'ceased to talk socialism', and forgot about the social revolution, while Tom Eldon, Royce's protagonist, chose prudently to suppress his socialist and single-tax views, reassuring himself that someday 'he would confess to the public that he had been a socialist all along . . .'[32] Royce used to tell his friends that he was utterly ignorant in matters political. Nevertheless, if ever a novel was composed in an emotional medley of hatred, guilt, and rejection of one's own political community, that book was *The Feud of Oakfield Creek.* Royce tried in later years to become the philosopher of the Beloved Community; evidently a guilt for his hatred of his own community posed for him a basic emotional problem which had to be resolved. As a philosopher, seeking for his guilt-consciousness a higher pardon in the Community, he was impelled to believe that the ontological argument brought an escape from the loneliness of egocentricity.

The ontological argument thus came to the fore among American philosophers only when unusual circumstances intensified guilt consciousness. Charles Peirce, contemporary with James and Royce, and formulating the principle of pragmatism in his mid-thirties in 1878, found the ontological argument alien to the scientific spirit. He held with Vacherot that it was more 'agreeable to his reason' to assert that the perfect was necessarily non-existent rather than, as Anselm and Descartes would have it, necessarily existent. Thirty years later, however, in 1910, having sustained guilt and social ostracism, Peirce found it hard to doubt God's existence. His emotions and reasoning became those of a logical masochist and ontologian. His instinctive belief was reinforced, he wrote, when he observed 'the superhuman courage which such contemplation has conferred upon priests who go to pass their lives with lepers and refuse all offers of rescue . . . ' This was not 'silly fanaticism' but akin to 'the power of the passion of love which more or less overmasters every agnostic scientist . . .'[33] In such a mood, Peirce averred that the idea of God could be no delusion, saying: 'I cannot think a thing is black if there is no such thing to be seen as black.' To Peirce the ontologian the unknowable became a 'nominalistic heresy'. Thus the founder of pragmatism, moved by guilt-consciousness, was drawn in his 'musements' (that is, his free associations) to an ontological argument for God's existence. Peirce called it the 'Humble Argument';[34] 'every heart will be ravished by the beauty and adorability of the Idea, when it is so pursued'. The language of logical masochism replaced the pragmatic.

IV

The year 1931 was the decisive one in the revival of the ontological argument for the contemporary world. That year Karl Barth published his *Anselm: Fides Quaerens Intellectum* that set forth a new interpretation of the ontological argument which through devious routes found its way into the writings of analytical, linguistic philosophers of the Wittgensteinian school. The neo-ontologian mode of thought was part of the 'theology of guilt' which emerged after World War I. The new 'crisis theology' or 'dialectical theology', as it was called, was permeated with a sense of man's guilt and lowliness; it conceived of God as a 'wholly other', from whom man was separated by a chasm which only God's grace could bridge.[35] Man's reason by itself could not hope either to prove God's existence or to delineate his attributes. But the ontological argument, as Barth saw it, was not a device of pure logical demonstration; rather, as Henri Brouillard said, it expressed the logical structure of the act of adoration.[36] Anselm, according to Barth, is giving an account of the experience of prophetic insight; he is not providing a universally valid logical demonstration.[37] Anselm's despair, he holds, is an intrinsically necessary prelude to the ontological argument; the knowledge that the latter imparts 'has to be sought in prayer; the fool, the 'insipiens', is clever, and not logically deficient, but he 'does not know the fear of the Lord'.[38] Post-Adamic man, afflicted with guilt, must seek God in prayer side by side with the application of his intellectual powers. 'Christian humility' is a necessary ingredient for the ontologian's thought; Kierkegaard, says Barth, rightly perceived that Anselm's way of proving God's existence requires God's help. The believer and the fool are thus symbols of two radically different modes of human existence ;[39] the fool, we might say, is one not dominated by the consciousness of guilt and sin.

A new phase in the history of the ontological argument began, a curious stage in the history of 'logical masochism', for now the ontological argument was asserted to require certain hitherto inarticulate emotional premises—humility, guilt, prayer. It was no longer regarded as an argument valid for all rational persons, unbelievers and believers alike; rather it was said that only believers would be able to understand it, that only those with faith would grasp its validity. In vain, such scholars as Étienne Gilson pointed out that Anselm's text says as clearly as possible that he intends to prove God's existence in a way which would be convincing to all rational persons, including unbelievers, that he was not simply describing the experience of faith in the word of God.[40] What was taking place in the cultural climate of the 1930's was a reformulation of the ontological argument as the logical structure of a faith which involves a prior capitulation of the human reason. Karl Barth's book had a tremendous influence on subsequent philosophical theology.

The new version of the ontological argument expressed the ambivalences of its generation. On the one hand, it took into account the influence and high status of logical analysis and mathematical logic. The new ontologians were aware that logicians generally were a reinforcing phalanx to Kant's and Hume's rejection of the ontological argument.[41] And yet they experienced poignantly a despair of knowing God akin to Anselm's. They saw an unparalleled spectacle of mass human evil in the Nazi movement; they saw a loneliness and estrangement of men from each other and God more intense than that which the latter half of the nineteenth century had known. Sensitive to man's unprecedented guilt, they wanted to bow all the more before the still all-powerful, Living God. But their logical argument could never, after centuries of logical criticism, have the primitive, pristine confidence of Anselm's, addressed to all rational human beings. Hence, the new version of the ontological argument was a halfway house, a compromise between competing forces. It would be immune to Kantian and analytic criticisms because it would not profess to be a proof in the logical sense. But it would still be a 'proof' in some strange new sense, a 'proof' to those who either already had faith or had experienced God; it expressed the need for logical argument to fortify one's faith. It retained an ill-repressed anxiety lest logical argument prove antagonistic to faith. It reiterated Anselm's warning against a theology of 'bats and owls', of logic-choppers, when what was needed was child-like obedience; the bats and owls had squabbled with the eagles about the reality of the beams of the mid-day sun.[42] Thus, under the impact of guilt-consciousness, new inexplicable senses of 'argument' and 'proof' were contrived to make possible a revival of the ontological argument.

V

The revival of the ontological argument among contemporary analytic philosophers had its precursor in Ludwig Wittgenstein; he set forth a conception of God which was filled with the emotions and temper of logical masochism. The guilt-motif was Anselmian in Wittgenstein. He had the conviction 'that he was doomed', says his friend, von Wright. In gloom, he regarded modern times as a dark age, and human beings helpless as in a sort of predestination: 'The thought of God', he said, was above all for him 'the thought of the fearful judge.'[43] In Anselmian fashion, he sought a monastic existence, and worked for monks as an assistant gardener; trying to assuage an inner guilt, he fled to obscure vocations, working as village schoolteacher, living among primitive fishermen, and wishing to merge himself into the anonymity of a Soviet collective farm. Guilt-consciousness haunted him in manifold ways, in his fear of insanity, and in the anxiety of suicide, which had destroyed three of his brothers.[44] He told his pupil Malcolm that he thought that he could understand the conception of God, in so far as it is involved in one's awareness of one's own sin and guilt. He added that he could *not* understand the conception of a Creator. 'I think', writes Malcolm, 'that the ideas of Divine judgment, forgiveness, and redemption had some intelligibility for him, as being related to feelings of disgust with himself, an intense desire for purity, and a sense of the helplessness of human beings to make themselves better. But the notion of a being *making the world* had no intelligibility for him.'[45]

Thus in Wittgenstein one can see the formation of the emotional decision-base of 'logical masochism'. 'Wittgenstein', Malcolm believes, 'was prepared by his own character and experience to comprehend the idea of a judging and redeeming God.' On purely logical and linguistic grounds, a Divine Judge should, from Wittgenstein's standpoint, have been every whit as unintelligible as a Divine Creator. But while the gloomy temper of Wittgenstein found 'repugnant', as Malcolm says, 'any cosmological conception of a Deity', rejecting the optimistic belief in such a God, he needed, on the other hand, a God fearful, cruel, despotic, as harsh with himself as he was with his disciples. In such a mood, he drew close to an interpretation of faith similar to Barth's. On hearing Kierkegaard's remark, 'How can it be that Christ does not exist, since I know that He has saved me?', Wittgenstein exclaimed: 'You see ! It isn't a question of *proving* anything!'[46]

What is guilt-consciousness? And how does it determine a philosophy? The experience of guilt is one of self-aggression, self-punishment; the conscience issues decrees against the ego; the tormented self, the 'contrite consciousness' arises. The individual tries to atone by humbling himself in his work and thought, by declaring both his reason and will to be weak and impotent. 'Logical masochism' is the outcome of such humility; the individual's reason, abdicating and humiliating itself before the condemnation of his super-ego, atones with a subservience, with an unquestioning faith in the Super-Ego's God. Thus guilt-consciousness becomes the emotional groundwork for an ontological argument for God's existence.

Aggressive emotions, as their respective directions change, issue in different varieties of philosophies. Guilt-consciousness involves a directing of aggression against one's self. But aggressive drives can be channeled in several ways. When the aggressive energies are deflected from one's self, and turned against other people, the unconscious fashions for itself a destructive philosophy, dissolving things and persons, so that as with Wittgenstein in his first phase, the 'causal nexus' itself can be called a 'superstition'. Not only the nexus but all the involvements with persons; for the other-aggressor's world finally leaves him lonely as all ties are ruptured. The aggressive person often resorts to 'mutism', sitting silent, not talking with people, the 'silent treatment', making them feel uncomfortable. And Wittgenstein's personal relations were filled with this mutism, which could take a cosmic direction of keeping silent toward the world.[47] In this other-aggressive mood, he took offence when a student showed signs of not being a submissive disciple.[48] When the direction of aggressive energy, however, is turned from outward to inward, when it is directed against one's self, then logical masochism prevails rather than the analytic sadism. If the *Tractatus Logico-Philosophicus* was a logical statement of externally-directed aggression, then the later reflections on a fearful God were those of an internally-directed aggression in guilt-consciousness.

The writing of Wittgenstein's disciple, Norman Malcolm, explicitly indicates the source of the ontological argument in guilt-emotions. In the United States of 1960, the Anselmian mood of the eleventh century recurred in Malcolm's essay; guilt-consciousness, whether in an industrial or feudal society, a modern or medieval one, tends to produce the same mode of thought. Norman Malcolm wrote:

> Why is it that human beings have even *formed* the concept of an infinite being, a being a greater than which cannot be conceived? ... There is the phenomenon of feeling guilt for something that one has done or thought or felt or for a disposition

that one has. One wants to be free of this guilt. But sometimes the guilt is felt to be so great that one is sure that nothing one could do oneself, nor any forgiveness by another human being, would remove it. One feels a guilt that is beyond all measure, a guilt 'a greater than which cannot be conceived'. Paradoxically, it would seem, one nevertheless has an intense desire to have this incomparable guilt removed. One requires a forgiveness that is beyond all measure, a forgiveness 'a greater than which cannot be conceived'. Out of such a storm in the soul, I am suggesting, there arises the conception of a forgiving mercy that is limitless, beyond all measure. This is one important feature of the Jewish and Christian conception of God.[49]

Here we are once more in the guilt-dominated world of Anselm and Royce. Like Anselm's infinite guilt calling for infinite atonement, like Royce's unpardonable sin, Norman Malcolm transmutes an experience of incomparable guilt into the source of his concept of God. We may well doubt the version of the history of religion which the guilt-consciousness projects. The idea of a great God indeed arose among the Hebrew tribes because they wanted One strong enough to defeat the gods of the Canaanites and Philistines, and One, too, who would forego totemistic human sacrifice. Whatever the historical origin, however, of the concept of God, the notion of proving His existence by an ontological argument is connected with a self-immolation in an experience of guilt. An emotional drive, issuing from the latter, exerts its pressure on deductive processes, twisting them so that the conclusion can be said in some sense, deviant from the logical, to 'follow' from the premises and definitions even when it is superimposed with a nonlogical need to abase one's self before a God.

VI

Ontologians, aware that an emotional drive is the inarticulate premise underlying their argument, have therefore understandably moved toward a Barthian-like standpoint. They claim that a religious experience is an essential condition for giving validity to the ontological argument; given that religious experience, the ontological argument, they say, will be perceived as valid. One well-known exponent of this view, for instance, writes:

> Thus God's existence does after a fashion, 'follow' from the definition of the term, not in any deductive sense, but in the sense that that body of experience indispensable to an adequate understanding of this (experience-presupposing) word is necessarily also adequate to validate reasoned assent to the argument.[50]

A person devoid of religous experience, says Professor Rescher, simply cannot understand the word 'God'; and if he has had such experience, he will assent to the ontological argument for God's existence. Such statements exhibit something of the emotional strain whence the ontological argument derives. For if religious experience is in some fashion perceptual in character, then the experiencer would find the ontological argument repugnant to him. If a person, for example, perceives a table, he would regard any ontological argument for its existence from its defined essence as ridiculous. Similarly if a person were to experience God, that is, perceive Him in some manner, he would regard an ontological argument, which is purely definitional and devoid of experiential content, as utterly alien to his experience. That is why religious personalities from Pascal to William James, who have felt themselves to experience God with some directness, have felt no force in the ontological argument.[51] Where the ontologians have erred is not only in their logic but in their introspective reporting on the character of their 'religious experience'. For the latter confers an apparent validity on the ontological argument not by bringing to bear a perceptual awareness of God; rather the 'religious experience' is one of guilt for the person's doubt or denials of the existence of God, guilt for a rebellion against religous traditions and morality, guilt which may even have Oedipal sources as in guilt for death-wishes against one's father. When Professor Rescher appeals for a sense of the word 'follow' different from the formal, deductive sense, he is appealing for a privileged emotional 'logic' in which the conclusion will not follow from the conscious premises but will satisfy the needs of an unconscious emotional longing. But once we have brought all this to consciousness, we cannot credit the view that God's existence 'follows' in some fashion from His definition.

The revival of the ontological argument on the part of twentieth-century linguistic philosophers brought with it, too, a replication of by-product corollary arguments which had been in disuse for three centuries. One writer, C. K. Grant, found it remarkable that Anselm had failed to see that the ontological argument also entailed 'the nonexistence of the Devil'. For the Devil was, by definition, 'an absolutely imperfect being,' and 'a completely imperfect being which exists is a contradiction in terms'; hence, 'non-existence must also be a property of a completely imperfect being.'[52] If Anselm, however, failed to record this ontological disproof of the devil, Spinoza, on the other hand, advanced it three centuries ago in precisely the same form. The Devil, said Spinoza, cannot exist; 'he is precisely identical with nothing". 'As the Devil has not the least perfection in him', wrote Spinoza, 'how should he then, I think to myself, be able to exist? . . . For whatever duration a thing has results entirely from the perfection of the thing, and the more essence and godliness things possess, the more lasting are they . . .'[53] The rigorous ontologian was thus driven to an impasse typical of all reasoning controlled by an unconscious projective drive; by submitting in self-abnegation to violence against his own sense of reality and logical processes, he impaired his sense of reality and logic to such an extent that he could not then account for his own evil choices and the evil influences on him; his wholly evil act, his Fall, was indeed nothing, and an Infinite Guilt was guilt over nothing. The ontological 'logic' finally made it impossible to deal logically with the facts of guilt-consciousness itself.

VII

We have traced the recurrence of the ontological argument in different cultural place-times, and in each case have found its emotional necessary condition in guilt-consciousness. The few available statistical facts concerning the argument's revival are in keeping with our hypothesis. The contemporary discussion of the ontological argument reached its highest peak of publication during April 1960 to March 1962, when according to the enumeration (incomplete to be sure) in the *International Index* twelve articles were devoted to the subject; only one article had appeared in each of the intervals from April 1955 to March 1958, and from April 1958 to March 1960.[54] After the maximal interval, the number of articles on the ontological argument dropped by half to six between April 1962 and March 1964, and equally three each in the periods from April 1964 to March 1965, and April 1965 to March 1966. If we allow for approximately a year's lag between writing and publication, then the highest interest in the ontological argument was during the time between the spring of 1959 and 1961. It coincided roughly with the height of the agitation against the testing of nuclear bombs then abruptly defeated in September 1961, when the Soviet Union resumed such testing in the atmosphere; also it was in May 1960 that Adolf Eichmann, the Nazi organizer of the destruction of European Jewry, was captured, and after a subsequent trial of four months, hanged. The problems of guilt and responsibility were very much in the consciousness of the intellectuals, and probably nurtured unconsciously the revival of the argument.

The history of the ontological argument, moreover, in its broad outlines confirms the correlation between guilt-consciousness and the ontological argument. Descartes and Spinoza, thinkers in whom guilt-emotions were strong, were great exponents of the ontological argument.[55] Leibniz, the metaphysical optimist, is a transitional figure toward the eighteenth century which finally renounces the ontological argument.[56] Leibniz's ingenious additions tended to de-ontologize the ontological argument, for he made it dependent on recognizing the existence of ordinary, contingent things; it was no longer a matter of arguing from a pure idea of a perfect being. The ontological argument, says Leibniz, rests on the modal proposition 'that if necessary being is possible, it exists'. But to prove that necessary being, or 'being of itself' is possible, Leibniz argues that 'if necessary being is not, there is no being possible'; but since obviously, contingent facts do exist, then 'necessary being is possible'. The latter proof thus is equivalent to a cosmological argument for God's existence; God the necessary being is required to sustain contingent things in their existence. Leibniz has thus introduced a premise concerning the empirical world into the heart of the argument for God's existence.

Bishop Berkeley, a gentle, serene idealist, whose basic maxim was 'that he is the true possessor of a thing who enjoys it', found however that the ontological argument both absurd and uncongenial.[57] His private *Philosophical Commentaries* said plainly: 'Absurd to Argue the Existence of God from his Idea. We have no Idea of God. It is impossible!' More publicly, through his dialogue-character Alciphron, Berkeley declared: 'I am not to be persuaded by metaphysical arguments; such, for instance, as are drawn from the idea of an all-perfect being ... This sort of arguments I have always found dry and jejune; and as they are not suited to my way of thinking, they may perhaps puzzle, but never will convince me.' 'God's existence', he said, was 'a matter of fact', and as such, 'not to be proved by notions, but by facts.'[58] Berkeley's spirit was drawn to an optimistic conception, a cosmological argument for a benevolent God. Kant, a man of the Enlightenment, ridiculing metaphysics in the manner of Voltaire and Shaftesbury in his *Dreams of a Spirit-Seer,* became the great critic of

God, Guilt, and Logic   91

the ontological argument; he himself, however, had once been attracted to a Leibnizian version of the argument.[59] With Kant's complex emotional basis we shall hope to deal on another occasion. Let us recall, however, that it was Kant who, counseling a woman oppressed by guilt over her adultery, and contemplating suicide, wrote: 'But to brood over one's remorse and then, when one has caught on to a different set of attitudes, to make one's life useless by continuous self-reproach on account of something that happened once upon a time and cannot be anymore—that would be a fantastic notion of deserved self-torture (assuming that one is sure of having reformed).'[60]

With Hegel, there was a return to the ontological argument, and a rejection of the Kantian critique.[61] Hegel recognized a connection between the ontological argument and the consciousness of guilt. Kant's common logic, he said, was founded on a 'very false idea of Christian humility' and 'abjectness'; 'for the honor of true humility', wrote Hegel, 'we must not remain in our misery, but raise ourselves above it by laying hold of the Divine'. And the guilt of the 'contrite consciousness' was surmounted as Thought realized itself necessarily in Existence. 'Thought, the Notion, of necessity implies that the Notion does not remain subjective . . .' Where Kant insisted on the distinction between psychical and external reality, between one hundred imaginary and one hundred real dollars, Hegel drew on examples of intense needs fulfilling themselves—the hungry man, for instance, imagining food, sets about finding it to satisfy his hunger; 'the imaginary hundred dollars becomes real, and the real ones imaginary: this is a frequent experience, this is their fate';[62] subjectivity is abrogated. Thus Hegel feels somehow that the Notion of God, like a human need, must fulfill itself in existence. His examples are specious, and would indeed allow, by analogy, the Real God to become imaginary, to die; a human need, however, is impelling the subjective notion of God to become objectively real; a human psychological drive is thus the motive power of this ontological 'deduction' rather than any valid deductive argument. What these ingredients of 'contrite consciousness' were in Hegel, we cannot say. The Hegel of the *Phenomenology,* however, writes Professor Walter Kaufmann, is 'far closer to the world of Dostoevsky's novels . . . the full measure of his torment has not yet been suggested'. The nervous breakdown of Hegel's sister, Christiane, her suicide a few weeks after Hegel's death as if in obedience to his statement: 'the loss of the brother is therefore irreplaceable for the sister', his view that the brother-sister relationship is the most ethical in the world because 'they do not desire each other', his discussion of guilt revolving around Sophocles' *Antigone,* suggest a possible experiential basis for Hegel's pages on the 'contrite consciousness'.[63] The latter, indeed, as Hegel writes, takes in its thinking 'the form of Adoration', what we have called 'logical masochism': 'Only by such a genuine self-surrender could consciousness prove its own resignation ... Such offering does indeed strip from the individual all independent might, and ascribes all the glory to the heavenly Giver.'[64]

The association of the ontological argument with guilt-consciousness continues into modern times. When Robin Collingwood more than a generation ago renewed the argument at Oxford, the central role of guilt in his thinking was evident as he rebuked those who regarded as a 'legal quibble' the notion of the Atonement of Christ for the responsibility of Adam's sin: 'This idea is an integral part of the ordinary moral consciousness, at least in Christendom; it is perplexing only to a man who is too weak in the head to follow the logic of a case where an obligation is distributed over three agents.' 'The believer', wrote Collingwood, 'thinks of himself as saddled with responsibility for Adam's sin, and as freed from it through assumption of it by God Himself in the person of Christ.'[65] One had come back, in the mid-twentieth century, to Anselm's satisfactionist theology. The neo-Anselmites were men of diverse modes of guilt-consciousness; among them even former logical positivists, repenting their youthful metaphysical parricide, found sustenance. It was an argument expressive of one mood of our purported neo-Dark Ages in which the nature we see, in Harry Ruja's vivid words, is one 'of fang and claw, of the unthinking libido, of entropy, genocide, and the destructiveness of atomic energy'.[66]

Yet the temper of our era is such that the religious-minded will mostly find the ontological argument at variance with their feelings and insight.[67] If man bears his burden of guilt, he also today like Job indicts God for His guilt. And if God is absolved, it is because He is finite in his powers, even when His goodness is unalloyed; some say He is dead. An age which cries out in anger that God is dead is not one which will be generally logical masochist in its thinking. Interestingly enough, the eighteenth century, which led the way in rejecting the ontological argument, was also the one which could join with Voltaire in blaming God for the needless evil and torture of men in the disastrous Lisbon earthquake of 1755.[68] The concentration camps, the extermination of the Jews, have revived the Jobian standpoint; the revival of the ontological argument remains confined to the logical masochist segment of the philosophical spectrum.

If the ontological argument thus emerges as a recurrent episode in the patho-philosophy of humanity, we do not wish to deny that the ethical experience of guilt encloses a divine component seeking to disenthrall and realize itself. Whether this affirmation is founded on intuition or hypothesis, our religious sanity gains by its not being associated with a self-inflicted violence to our logical faculties. No just God ever called on man to believe in Him through an act of logical self-immolation.

Bertrand Russell has written that modern logic has quite decisively proved that the ontological argument is invalid, and that 'this is not a matter of temperament or of the social system' but 'purely a technical matter'. What our psycho-sociological survey has shown is that whenever social or personal conditions make for the emergence of a logical masochist temper, one may expect the recurrence of the ontological argument among logicians and non-logicians alike. 'Technical' considerations will never prevail against compulsive emotions; an ontologian will listen to some technical argument that logical analysis comes closest to the common sense of reality when 'exists' is taken not as a predicate but as a quantifier, so that one should say 'there is an X such that X has all the predicates'; the ontologian replies that his sense of reality is best served by taking all abstract entities as subjects, and 'exists' as a predicate superimposed on them, with the Perfect Being as the exceptional one which 'exists' necessarily. Even Russell in his youth not only believed that Homeric gods and chimeras all had being; he accepted the validity of the ontological argument. A child or a man of common sense will continue to laugh at the ontological argument; when they hear it said: 'a Perfect being exists necessarily', they will say: 'Yes, but only if such a being exists in the first place.' The normal sense of reality finds the ontological argument absurd, verbal legerdemain.[69] The logical masochist, on the other hand, finds its denial inconceivable, that is, psychologically inconceivable. His guilt-emotions have enlisted logic and the sense of reality to do his penance of self-castigation and self-mutilation.

Notes

1. John Locke, *An Essay Concerning Human Understanding*, A. Routledge, reprint, London (undated), pp. 529–30.
2. Thomas Fowler, *Locke,* Harper, New York 1880, p. 14.
3. John Locke, 'Descartes's Proof of a God, from the Idea of necessary Existence, examined 1696,' in Lord King, *The Life and Letters of John Locke,* A. Bell, London 1894, p. 316.
4. Étienne Gilson, *The Spirit of Mediaeval Philosophy,* trans. by A. H. C. Downes, Scribner, New York 1936, p. 59.
5. Harry Austryn Wolfson, 'Notes on Proofs of the Existence of God in Jewish Philosophy', *Hebrew Union College Annual,* Vol. I (1924), p. 575. Also cf. Shubert Spero, 'Judaism and the Ontological Argument', *Judaism,* Vol. 14 (1965), No. 1, p. 41.
6. Martin Rule, *The Life and Times of St. Anselm,* Vol. I, Paul, London 1883, pp. 57–58.
7. St. Anselm, *Cur Deus Homo,* J. Arant, Edinburgh 1909, pp. 116–17.
8. Martin Rule, op.cit., pp. 57, 63, 12–13. Anselm, *Cur Deus Homo,* p. xi.
9. M. J. Charlesworth, *St. Anselm's Proslogion,* Clarendon Press, Oxford 1965, p. 16.
10. Cf., especially, Saint Anselm, *Basic Writings,* Trans. by S. N. Deane, 2nd Ed., Open Court, La Salle 1962, pp. 231–9, 247–50.
11. F. S. Schmitt, 'St. Anselm of Canterbury', *New Catholic Encyclopedia,* Vol. I, McGraw-Hill, New York 1967, pp. 582–3.
12. Cf. Anselm, *Cur Deus Homo,* p. xii.
13. Saint Anselm, *Basic Writings,* pp. 2, 6.
14. Ibid., p. 7.
15. Ibid., p. 8.
16. M. J. Charlesworth, op. cit., p. 20. Thomas Aquinas later magisterially rejected the ontological argument. He was, as Josef Pieper tells us, a thinker whose optimism reflected the growing universities and cities of the thirteenth century. He was also part of the 'youth movement' which flourished in the urban setting.Thomas thus held to the goodness of all created things, and 'had no thorn in the flesh', that is, was relatively free from masochism, Josef Pieper, *Guide to Thomas Aquinas,* trans. by Richard and Clara Winston, New American Library, New York 1964, pp. 27–28, and 110.
17. James Harvey Robinson, *Readings in European History,* Vol. I, Ginn, Boston 1904, pp. 312–16.
18. Villehardouin and De Joinville, *Memoirs of the Crusaders,* trans. by F. Marzials, Dent, London 1926, pp. 147–8.
19. M. J. Charlesworth, op. cit., pp. 82–83. Karl Barth, *Anselm: Fides Quaerens Intellectum,* trans. by Ian W. Robertson, SCM Press, London 1960, p. 96.
20. St. Anselm, *Basic Writings,* pp. 150–1.
21. Sigmund Freud, *A General Introduction to Psychoanalysis,* trans. by Joan Riviere, Garden City Publishing Co., New York 1938, p. 143.
22. Saint Anselm, *Basic Writings,* op.cit., pp. 148–9.
23. Ibid., pp. 153–4. Cf., Harry Austryn Wolfson, *The Philosophy of Spinoza,* Vol. I, Harvard University Press, Cambridge, Mass. 1934, p. 171. Wolfson tries to combine both a 'social' interpretation of the ontological argument and one which sees it as expounding the 'immediacy' of our knowledge of God.

24. William James, *The Varieties of Religious Experience,* Longmans, Green, New York 1904, p. 448.
25. William James, *The Will to Believe,* Longmans, Green, New York, 1897, pp. 270, 272, and 291. James's article 'On Some Hegelisms' was published in *Mind,* April, 1882. Walter Kaufmann, therefore, can hardly be correct in his view that when James polemicized against Hegel, he 'really meant Royce', for Royce at this time was still in California, not at Harvard. Cf., Walter Kaufmann, *Hegel: A Reinterpretation,* Doubleday, New York 1966, p. 287.
26. William Ernest Hocking, 'The Ontological Argument in Royce and Others', in *Contemporary Idealism in America,* ed. by Clifford Barrett, Macmillan, New York 1932, p. 58.
27. Royce's 'great contribution' to the philosophy of religion was said to be 'in the way he apprehended the magnitude and tragedy of evil'. Henry Nelson Wieman and Bernard Eugene Meland, *American Philosophies of Religion,* Willett, Clark, New York 1936, p. 102. For his early American predecessor, cf. Alfred Owen Aldridge, *Jonathan Edwards,* Washington Square Press, New York 1964, p. 8.
28. Josiah Royce, *The Problem of Christianity,* Vol. I, MacMillan, New York 1913 (new ed. University of Chicago Press, Chicago 1968), pp. 293–4, 254, 263, 264, 265, and 266.
29. Josiah Royce, *The Philosophy of Loyalty,* MacMillan, New York 1908, pp. 94–99.
30. *Papers in Honor of Josiah Royce,* reprint of *The Philosophical Review,* Vol. XXV (May, 1916), p. 67.
31. Ibid., pp. 281–3.
32. Josiah Royce, *The Feud of Oakfield Creek,* Houghton, Mifflin, Boston 1887, pp. 168, 164, 185, 482, 200, 171, 168, 211, 69, 180–1, 241, 54, 357, 289–90, and 358. 'Individualistic communities', said Royce, 'are extremely cruel to individuals.' The Californians, in his opinion, never would have given Phillips Brooks a hearing; they were mobsters easily led by demagogues. Royce even disliked his Californian students. Cf. Josiah Royce, *Race Questions, Provincialism and Other American Problems,* MacMillan, New York 1908, pp. 217–23. William Belmont Parker, *Edward Rowland Sill,* Houghton, Mifflin, Boston 1915, p. 137. *Papers in Honor of Josiah Royce,* p. 282.
33. *Collected Papers of Charles Sanders Peirce,* ed. by C. Hartshorne and P. Weiss, Vol. VI, *Scientific Metaphysics,* Harvard University Press, Cambridge, Mass. 1935, pp. 284, 300, and 347.
34. Ibid., pp. 327, 333, 334, 338, and 339. Peirce, it should be noted, throughout his life had a propensity for ontological modes of argument. Even when he was doing his notable work on the frequency theory of probability, he tended to derive invalid ontological conclusions from formal premises. Take, for instance, Peirce's eloquent probabilistic pessimism: 'All human affairs rest sure of seeing the day when everything in which he had trusted should betray his trust, and, in short, of coming eventually to hopeless misery. He would break down, at last, as every good fortune, as every dynasty, as every civilization, does. In place of this we have death.' We might call this 'probabilistic masochism'. Actually nothing in the frequency theory of probability leads to the conclusion that all in which we trust must break down. The most improbable occurrence, with a probability of zero, may still be realized, for its zero probability refers only to a series of increasingly rare occurrences, and conversely, the most probable event may still not be realized; its probability of one is only a limit of a series of increasingly frequent occurrences. An ontological certainty of the defeat of every human hope does not

follow from any of the definitions or axioms of probability. Cf. Charles S. Peirce, *Chance, Love, and Logic,* ed. by Morris R. Cohen, Harcourt, Brace, New York 1923, pp. 72 and xxxiii.
35. Will Herberg, 'The Social Philosophy of Karl Barth', in Karl Barth, *Community, State, and Church,* Doubleday, New York 1960, pp. 14–16, and 19–21.
36. Vincent G. Potter, 'Karl Barth and the Ontological Argument', *The Journal of Religion,* Vol. LXV (Oct., 1965), p. 322. James D. Smart, *The Divided Mind of Modern Theology: Karl Barth and Rudolf Bultmann, 1908–1933,* Westminster Press, Philadelphia 1967, pp. 111, 113, 114, and 194.
37. Karl Barth, *Anselm: Fides Quaerens Intellectum,* p. 76.
38. Ibid., pp. 16, 102, and 104.
39. Ibid., pp. 40, 26, 39, and 105. Also, cf. Hugo A. Meynell, *Grace Versus Nature: Studies in Karl Barth's Church Dogmatics,* Sheed and Ward, London 1965, p. 65.
40. Cf. Étienne Gilson, 'Sens et nature de l'argument de saint Anselme', *Archives d'histoire doctrinale et littéraire du Moyen Age,* Vol. IX (1934), p. 26. Cited in Vincent G. Potter, op. cit., p. 311. A. Nemetz, on the other hand, holds that 'St. Anselm did not intend to make a formal proof for the existence of God'. Cf. 'Ontological Argument,' *New Catholic Encyclopaedia,* Vol. X, p. 701.
41. On Hume, cf. Charles William Hendel, Jr., *Studies in the Philosophy of David Hume,* University Press, Princeton, 1925, pp. 29, 366, 389. It is noteworthy that when Hume as a young man was attracted to the ontological argument, it was precisely at a time when he was emotionally drawn to the pessimism of an 'original sin' doctrine.
42. Karl Barth, op. cit., pp. 27 and 34.
43. Norman Malcolm, *Ludwig Wittgenstein: A Memoir,* Oxford University Press, London 1962, p. 20.
44. Ibid., pp. 10, 3, 94, and 52.
45. Ibid., p. 71. Wittgenstein also said 'the notion of immortality can acquire a meaning . . . through one's feeling that one has duties from which one cannot be released, even by death'.
46. Loc. cit.
47. Ibid., pp. 53 and 40. Also cf. Karl Britton, 'Recollections of L. Wittgenstein', *The Cambridge Journal,* Vol. VII (Sept. 1954), p. 712.
48. Ibid., p. 32.
49. Norman Malcolm, 'Anselm's Ontological Arguments,' *The Philosophical Review,* Vol. LXIX (1960), p. 60. For an interesting counterpart, and *a posteriori* argument for God's existence, cf. Frederick B. Fitch, 'On God and Immortality', *Philosophy and Phenomenological Research* Vol. VIII (1948), p. 691.
50. Nicholas Rescher, 'The Ontological Proof Revisited', *Australasian Journal of Philosophy,* Vol. 37 (1959), pp. 138–48.
51. 'The metaphysical proofs of God are so remote from the reasonings of men, and so complicated, that they make little impression . . . an hour afterwards they fear they have been mistaken.' 'I cannot forgive Descartes. In all his philosophy he would have been quite willing to dispense with God.' 'Descartes useless and uncertain.' 'The heart has its reasons which reason does not know.' Blaise Pascal, *Pensées and The Provincial Letters,* trans. by W. F. Trotter, Random House, New York 1941, pp. 172, 29–30, and 95.
52. J. K. Grant, 'The Ontological Disproof of the Devil,' Analysis, Vol. 17 (1957), pp. 71–72.
53. *Spinoza's Short Treatise on God, Man, and His Well-Being,* trans. by A. Wolf, A. & C. Black, London 1910, p. 143.

54. Cf. *International Index: A Guide to Periodical Literature in the Social Science and Humanities,* H. W. Wilson Co., New York, Vols. XIV, XV, XVI, XVII, XVIII, and XIX.
55. I have omitted considering Descartes and Spinoza in this essay only because I have elsewhere dealt at length with the psychological basis of their philosophies. Cf. 'Anxiety and Philosophy: The Case of Descartes,' *The American Imago,* Vol. 20 (1963), pp. 411–49. *Spinoza and the Rise of Liberalism,* Beacon Press, Boston, 1958, pp. 84–85, 216 ff., and 241–2.
56. Henry Vyverberg, *Historical Pessimism in the French Enlightenment,* Harvard University Press, Cambridge, Mass. 1958, pp. 230–1.
57. Cf. *The Works of George Berkeley,* ed. by Alexander Campbell Fraser, Clarendon Press, Oxford 1871, Vol. III, pp. 158–61.
58. *The Works of George Berkeley,* ed. Alexander Campbell Fraser, Clarendon Press, Oxford, 1901, Vol. II, *Alciphron or the Minute Philosopher,* pp. 155 and 158. *The Works of George Berkeley,* ed. by A. A. Luce and T. E. Jessop, Vol. I, *Philosophical Commentaries,* T. Nelson, London 1948, p. 94. Cf. Edward A. Sillem, *George Berkeley and the Proofs for the Existence of God,* Longmans, Green, London 1957, pp. 31 and 44.
59. Immanuel Kant, *Religion Within the Limits of Reason Alone,* tr. T. M. Greene and H. H. Hudson, Harper, New York 1960, p. xiii.
60. Kant, *Philosophical Correspondence, 1759–99,* ed. by Arnulf Zweig, University of Chicago Press, Chicago, p. 190.
61. It should be noted parenthetically that Schelling, too, accepted the ontological argument, while Fichte, after first denying it, accepted it in 1806. Cf. René Wellek, *Immanuel Kant in England, 1793–1838,* University Press, Princeton, 1931, pp. 191 and 294.
62. *Hegel's Lectures on the History of Philosophy,* trans. by E. S. Haldane and Frances H. Simson, Vol. III, Routledge & Kegan Paul, London 1955, pp. 454–5. Also cf. *The Logic of Hegel,* trans. by William Wallace, 2nd Ed., Clarendon Press, Oxford 1892.
63. Walter Kaufmann, *Hegel,* pp. 91, 125, 127, and 176.
64. It should be observed that 'Contrite Consciousness' was the section of the *Phenomenology of Mind,* which Josiah Royce translated. Cf. Hegel, *Selections,* ed. by Jacob Loewenberg, Charles Scribner's, New York 1929, pp 79–79 and 95–96.
65. R. G. Collingwood, *An Essay on Philosophical Method,* Clarendon Press, Oxford 1933, pp. 124–33. R. G. Collingwood, *The New Leviathan: or Man, Society, Civilization and Barbarism,* Clarendon Press, Oxford 1942, reprinted 1958, p. 121. Also Alan Donagan, *The Later Philosophy of R. G. Collingwood,* Clarendon Press, Oxford 1962, p. 307. Also cf. Gilbert Ryle, 'Mr. Collingwood and the Ontological Argument,' *Mind,* Vol. XLIV (1935), pp. 137–51.
66. Harry Ruja, 'The Ontological Argument and a "Living Faith" ', *The Personalist,* Vol. 44 (1963), p. 300. David Rynin, 'On Deriving Essence from Existence,' *Inquiry,* Vol. VI (1963), pp. 144–5, and 154.
67. Cf. Thomas J. J. Altizer and William Hamilton, *Radical Theology and the Death of God,* Bobbs-Merrill, Indianapolis 1966, p. 5, where the restudying of the ontological argument is taken as a continuation of the older, 'liberal theology'.
68. T. D. Kendrick *The Lisbon Earthquake,* Methuen, London 1956, pp. 124–5. To the God who does not answer, Voltaire said in 'Poème sur le désastre de Lisbonne': 'On a besoin d'un Dieu qui parle au genre humain.' The young Immanuel Kant, still imbued with Leibnizian arguments for God's existence, was shaken by the Lisbon earthquake into writing three papers in 1756 on the causes of earthquakes

in which, still struggling to hold on to the Leibnizian apriorism, he bade men to acquiesce without questioning to God's inscrutable wisdom. Cf. T. D. Kendrick, op. cit., pp. 132–3.
69. Max Eastman, a student of laughter who found the typical American reaction to metaphysics to be laughter, was intrigued as an undergraduate with the ontological argument. Cf. Max Eastman, *Enjoyment of Living,* Harper, New York 1948, p. 226.

4

Sociological Aspects of the Relation Between Language and Philosophy

Language is the primary fact which concerns contemporary philosophy. Men have been speaking and writing for a long time, but it is only recently that the task of philosophy has been said to be the analysis of language. Ethical perplexities, social anxieties, the nature of scientific knowledge, religious speculations, are held not to be directly the problems of the philosopher. They enter his study by way of a domain of languages and sub-languages. This preoccupation with language is itself an unusual phenomenon in our intellectual history. It challenges the sociologist of philosophic ideas for an explanation, and it leads one to wonder upon what evidence philosophers have accepted the doctrine of linguistic primacy.

Bertrand Russell has for many years been the leading proponent of the importance of the analysis of language. More than a half-century ago, he wrote: "That all sound philosophy should begin with an analysis of propositions, is a truth too evident, perhaps, to demand a proof." (19, p. 8). None the less, Russell sought to defend this "truth" with a sociological argument concerning the causes of philosophic beliefs. A large part of Western philosophy, Russell affirmed, was determined by the notion that every proposition must consist of a subject and a predicate. "Any philosophy," he said, "which uses either substance or the Absolute will be found, on inspection, to depend on this belief. Kant's belief in an unknowable thing-in-itself was largely due to the same theory." (19, p. 15). In his later writings, Russell often recurred to the sociological argument for the linguistic determination of philosophic ideas. "We have to guard," he wrote, "against assuming that grammar is the key to metaphysics," an assumption which he believed had been made by traditional philosophy. He referred with approval to the views of the philologist Sayce. "Sayce maintained that all European philosophy since Aristotle has been dominated by the fact that philosophers spoke Indo-European languages, and therefore supposed the world, like the sentences they were used to, necessarily divisible into subjects and predicates." (18, p. 212). The linguistic interpretation of philosophic history was central in Russell's thinking, but it was never supported with the requisite historical data. It took on the character of a postulate whose proof was commended to others. He thus declared that "a great book might be written showing the influence of syntax on philosophy; in such a work, the author could trace in detail the influence of the subject-predicate structure of sentences upon European thought, more particularly in the matter of 'substance.' " (21, p. 243).

In this essay, I shall try to show why this great book could not be written.

Our problem is a sociological one. The linguistic school adheres to the sociological doctrine that the structures of languages have been the primary determinant of philosophies. I shall try to show that the available evidence is against this sociological theory. When writers like Russell speak of the influence of syntax on philosophy, they are presumably speaking of a verifiable sociological causal relation. And in what follows, I shall approach the problem as one in the sociology of philosophic ideas. There is overwhelming evidence that the structure of languages has had no primary, determining effect on men's philosophies.

The philologist Sayce, upon whose authority Russell relies, declared: "Had Aristotle been a Mexican, his system of logic would have assumed a wholly different form." (23, p. 329). We cannot perform this experiment, but we can show how the same metaphysics has arisen among peoples with radically different languages, and how the most diverse types of philosophies have arisen among men who used the same language. And, to begin with, we can show how syntactical variations have not been accompanied by corresponding differences in metaphysics.

Syntactic Variation not Correlated with Changes in Philosophic Ideas

The linguistic interpretation of the history of philosophy affirms that the forms of language, the mode of syntax, tend to be projected by thinkers as the substance of reality. Linguistic determinants, of which the thinker is often unconscious, are thus held to shape his metaphysical outlook. Part of the contribution of philosophical linguistics is then believed to be its bringing into clear consciousness the hitherto unconscious influence of syntactical forms. What philological facts are at variance with the linguistic interpretation of philosophy? There are many languages, in the first place, in which gender is a syntactical category, for instance, Latin, French, German, Spanish. (13, p. 55). In English, we don't assign gender to "the table," but the French say "la table." We cannot, however, affirm that this difference in syntax reflects itself in a corresponding distinction between English and French philosophy. Perhaps the distinction in gender early reflected motivations toward metaphysics which Freud has illumined. It may be that primitive philosophers once projected the masculine and feminine principles throughout nature. Perhaps the progenitors of the feminine *l'étendue* and the masculine *le temps* conceived of space as a female receptacle, and time as a cosmic male organ. A philosophical psychoanalyst might perceive in Alexander's theory, that time is the "mind" of space, a weakened version of the more forthright primitive metaphysics. He might even regard the emergent qualities as the offspring of the sexual union of space and time. The origins of the classifications of gender, however, remain obscure. If one finds that the "earth" and the names of trees are conceived as feminine after the fruit-bearing analogy, there are still such strange facts as that words for "table, thought, fruit" are masculine in one language and feminine in another. As Jespersen says, "It is certainly impossible to find any single governing principle in this chaos." (13, p. 228). Whatever the origin of gender, its grammatical structure has within historic times been no determinant of metaphysics. If grammar itself was once founded on an unconscious metaphysic, this linkage is now so vestigial as to have no appreciable bearing on the structure of philosophic ideas.

In many languages, to take another group of facts, there is a tendency to get rid of the subjunctive mood. "In Danish and in Russian there are only a few isolated survivals; in English the subjunctive has since Old English times been on retreat, though from the middle of the nineteenth century there has been a literary revival of some of its uses." (13, p. 318). The trend is toward stating the content of traditional contrary-to-fact conditional sentences in the indicative mood.

To what is this decline of the subjunctive mood to be attributed? It is difficult to say, for grammarians are not clear as to what characterizes the use of the subjunctive in Aryan languages. It seems to denote a certain hesitation, or doubt, or anxiety concerning the reality of the proposed situation. Perhaps it was once culturally more important to indicate the doubts, anxieties, uncertainties of the speaker. With the decline of social anxieties, the cultural need for a differentiated subjunctive mood may have lessened. But whatever the causes of syntactical change, one cannot say that the fortunes of the subjunctive mood have been reflected in corresponding philosophic doctrines. The character of Danish and Russian philosophic speculations is not correlated with the low estate of the subjunctive. The anguish and anxiety of Kierkegaard could express itself in an idiom which was discarding the subjunctive mood.

Ring Lardner's baseball player, like his fellow Americans, was unaware of the existence of the subjunctive mood. His syntax conformed to that law of laziness which, according to Sayce, is the basic cause of change in language. Perhaps the promptings of laziness are involved in the decline of the subjunctive. There is no evidence, however, that the American public has been influenced by an unconscious metaphysical revolt against contrary-to-fact conditional significances. (14, pp. 425, 437).

French metaphysicians and French positivists use the same idiom with the same structure; the grammar they use has not predetermined them to a common mould. Descartes, Comte, and Bergson used the same language with the same grammatical structure. Their modes of thought are diverse, but each could state his philosophy with lucidity in the same syntax.

Again, we might ask what has been the philosophical bearing of the phenomenon of double or cumulative negation. Languages as varied as those of Russian, Spanish, Magyar, and Bantu exhibit this syntactical form. (13, pp. 332–333 & 12, pp. 118–119). Two negatives in these languages do not cancel each other; on the purely linguistic level, it is not true in their syntaxes that not-not p = p. Do these languages then promote some special alternative logic or metaphysics?

Not at all. As Jespersen says, the function of double negation is not as a logical, but as a psychological device. A layer of negative coloring is spread throughout a whole sentence instead of being localized in one part. (13, p. 337). Where repeated negatives are not used for emphasis, they may serve to convey attitudes of hesitancy. "This is not unknown to me" conveys a reticence not present in "I know this." Double negation can thus express either a strengthened negative or a weakened positive.

In the common American language, as H. L. Mencken said, "the double negative is so freely used that the simple negative appears to be almost abandoned." (14, p. 468). The title of a once-popular song, "I ain't never done nothing to nobody no time," abundantly illustrates the vulgar syntactical form. There is no evidence, however, that the American proletariat has developed a metaphysic of dialectic negation to higher powers.

The syntax of cumulative negation has never in any historic language been the cause for the formulation of some non-Aristotelian logic. In all languages, moreover, when it is made clear that negative expressions are to refer to the same word, such expressions do cancel each other. The manifest, syntactical rules do not reach down into an unconscious metaphysic of "negation" other than the two-valued logic of truth and falsehood. Grammar does not legislate its corresponding metaphysics.

Especially do we see this to be the case with respect to the syntactical differences among languages in the expression of time-relations. The different tense-systems do not involve corresponding differences in the metaphysics of time. The usage of the "dramatic present" illustrates well how a syntactical form is without metaphysical consequence. It has been observed that the language of uneducated people narrates past events by using the "dramatic present." "One need only listen to the way in which people of the humblest ranks relate incidents that they have witnessed themselves to see how natural, nay inevitable, this form is." (13, p. 258). The style of Damon Runyon is indeed contrived around such a use of the "dramatic present." There is no evidence, however, that either the humblest persons or the sophisticated denizens of Broadway have been led by their elementary syntax to deny the distinction between past and present, or to insist that the present alone is real.

Many languages, furthermore, have no future tense, and make use of such devices instead as the use of the present tense to convey futurity. (13, p. 160). There is no evidence that the people who use these languages have therefore confounded the present with the future in a metaphysical sense.

Similarly, there are some American Indian languages in which the terms for day-before-yesterday and day-after-tomorrow are the same. (8, p. 652). This usage has not led to a metaphysics in which the immediate past and the immediate future are identified. The time-experience of certain Indian tribes differs in some respects from our own. Their days are not punctuated with clocks and calendars. They have no exact chronologies, and they place events in approximate order by reference to memorable events and periods in their own life-histories. They have no weekly recurrence of a Sabbath. Withal, there are basic aspects to the human experience of time which are universal,—the distinctions between past, present and future, the irreversibility of time; and no linguistic usage or cultural outlook can negate these common properties of human experience.

The evidence of primitive languages is especially crucial in the consideration of the bearing of syntax on philosophy. The syntax and vocabulary of primitive languages are an outgrowth of dealings with concrete objects. One would then expect that there would be a syntactical resistance on the part of primitive peoples to Western philosophies and religious ideas, a resistance which, in view of the tremendous linguistic differences, might well ensure the non-comprehension of those ideas. Such, however, is not the case. Franz Boas' conclusions on this matter are directly to the point:

"In primitive culture people speak only about actual experiences. They do not discuss what is virtue, good, evil, beauty; the demands of their daily life, like those of our uneducated classes, do not extend beyond the virtues shown on definite occasions by definite people, good or evil deeds of their fellow tribesmen, and the beauty of a man, a woman, or of an object. They do not talk about abstract ideas. The question is rather whether their language makes possible the expression of abstract ideas. It is instructive to see that missionaries, who in their eagerness to convert natives have been compelled to learn their languages, have had to do violence to the idioms in order to convey to the natives their more or less abstract ideas, and that they have always found it possible to do so and be understood. Devices to develop generalized ideas are probably always present and they are used as

Sociological Aspects 105

soon as the cultural needs compel the natives to form them." (2, pp. 141–142). "It is true," Boas writes, "that in many languages it would be difficult to express the generalized statements of philosophic science, because the categories imposed by the structure of the grammar are too specific." However, he adds, "It is not true that primitive languages are unable to form generalized concepts."

Languages are flexible instruments. When the cultural need for the expression of abstract ideas is felt, the appropriate devices are used to achieve this expression. But it is the cultural need which is decisive. The character of the language-instrument does not itself shape the idea to be expressed. A given language, for instance, may seem to be well suited by its syntactical forms to lead its users to a belief in metaphysical universals. But in the absence of certain cultural demands, such a belief will not be found. Empiricists, for instance, often hold that the expression of adjectival situations through the use of nouns gives rise to the belief that there are abstract entities, universals, denoted by these nouns. When one says "this flag has whiteness" instead of "this flag is white," one is supposed to be on the path to Platonic realism. For "whiteness" as a noun is thought to mislead one into the belief that it symbolizes an abstract entity. Primitive languages, however, often have this grammatical usage without giving evidence of a tendency toward Platonic metaphysics. Franz Boas describes this mode of primitive syntax:

"It is not unimportant to recognize that in primitive languages, here and there, our adjectival ideas are expressed by nouns. A poor person may be conceived rather as a person who has poverty; a sick person as a person who has sickness; and it is not necessary that these qualities should be conceived as concrete objects. Sickness is often so conceived, but poverty or size is not. We also find cases in which the structure of the sentence demands the frequent use of abstract nouns, as when the Kwakiutl Indian of Vancouver Island says to a girl, 'Take care of your womanhood'; when the Eskimo speaks of the smallness or largeness of an object, or when the Dakota Indian speaks of strength and goodness." (2, p. 142).

The grammar of abstract nouns does not seem to have predisposed the Kwakiutl, Eskimo, or Dakota Indians toward a Platonic metaphysic. The causal sources of Platonism are still obscure, but it is noteworthy that the distinguished historical scholar, F. M. Cornford, finds its basis in socio-cultural rather than in linguistic phenomena: "you could build the whole structure of Platonism round that central scheme for the reform of society. What we call the theory of Forms or Ideas, and the whole conception of the universe that goes with it, could be represented, not unfaithfully, as deducible from the moral thesis." (3, p. 38). The motivation toward a "participationist" metaphysics seems also to have derived from the early influences of the totemistic social order. Totemism, as Jane Harrison said, was also an epistemology (7, p. 122). Value attitudes, emotional projections, social perspectives,— these are the founts which nurture the diverse modes of metaphysical ideas. The language one uses is shaped and misshaped to express the metaphysical ideas which its users are trying to project. Some languages may be more malleable and expressive in certain respects than others. But every language lends itself to the appropriate tinkering required to convey the new meanings. At most, a language may be a minor agency of syntactical resistance or syntactical propensity toward certain philosophic views. It has none of the proportions of importance, however, with which it is endowed by linguistic philosophers.

The Diffusion of a Common Philosophy through Different Syntaxes

The Aristotelian metaphysics has not been the exclusive property of the Indo-European languages. It is often overlooked that it was propounded and highly developed by Arabic and Hebrew thinkers even before it was espoused by the medieval Christian philosophers. The syntax of the Semitic languages differs markedly from the European tongues, but Semitic syntactical rules proved no insuperable obstacle to the formulation of the Aristotelian ideas. Words were invented to express the novel metaphysical ideas. The Talmudists had been singularly averse to speculative notions, but the medieval Jewish philosophers contrived the necessary forms to convey the foreign concepts. The Jewish philosophers derived their knowledge of Aristotle through the intermediary of Averroes. Averroes himself did not know Greek; his commentaries were based on Arabic translations from either the original Greek or from Syriac translations. When, during the twelfth century, the fanatical sect of Almohades in Spain forbade the study of philosophy, Averroes' own works in their original Arabic were almost completely lost. (10, p. 2). His writings, however, had meanwhile

Sociological Aspects 107

been translated into Hebrew, and it was in this form that they first spread through Europe. Aristotle was imported into Western Europe with Semitic bills of lading.

There were some misrepresentations of Aristotle's ideas which arose as they were refracted from one linguistic medium to another, Syriac, Arabic, Hebrew, Latin. The outstanding fact, however, is that problems of syntax and terminology were surmounted with relative ease to achieve an expression of Aristotelian metaphysics in the Semitic languages. Slight differences in metaphor had no bearing on the effective statement of common metaphysical ideas. The medieval Jewish philosopher would translate the "self-existent" essence of substance with the phrase "sheyiyeh omed biphnai atsmoh," which affirms literally, "that which will be standing by itself." (10, p. 40). "Self-existent" was conveyed by a phrase "standing by itself." The metaphor of "standing" did not, however, obstruct the formation of the idea of self-existent substance. Words took on new significances; to ask "whether the universals are only in the mind," the medieval Jewish philosophy would say, "im haklallim haym b'saychel l'vad," and this use of the word "haklallim," though derived from a familiar root, was not to be found in Biblical or Talmudical sources. Some translations would have behavioristic overtones,—"rational" was expressed by "m'daber," which literally is the participle "speaking," "quantity" was translated by "hakamah," "the how much." When Crescas discusses the existence of God, his terminology has an empirical flavor. "God exists" is translated "haeloh nimtzah," that is, "God is found." (27, p. 130) There is no simple verb "to exist" in classical Hebrew; in that language, things are found or not found, and the verb "matzoh," to find, is one of homely human action. The absence of the abstract verb "to exist" was no barrier to the discussions of the proofs for God's existence. The diffusion of metaphysical culture thus takes place across boundaries of syntax.

The notions of "being," indeed, were explicated in Hebrew despite the fact that ordinary noun-clauses in that language are without a copula of any kind. The syntactical relation between subject and predicate in such sentences is expressed, as in mathematical logic, by mere juxtaposition. The use of the verb "hayoh" in a sense near to that of a copula becomes frequent only in the later Books of the Bible. (6, pp. 476–477). The Hebraic sensitivity to problems of existence does not, however, seem to have been affected by their syntactical omission of "being" as a copulative form.

The Social Sources of Linguistic Metaphors

When we leave the Mediterranean basin, and move our study to the Far East, we are in lands which were relatively unaffected by any diffusion of philosophic ideas. The Chinese language and writing are profoundly different from the European. We would expect, if the linguistic theory were true, that the Chinese culture would therefore be incapable of evolving philosophic ideas similar to those of Western civilization. Bertrand Russell, after his travels and teaching in China, wrote a book on that country in which he speculated briefly on the probable influence of the Chinese language on its philosophy. He inclined to agree with the view of a Chinese thinker that the alphabetical civilizations are fickle and lack solidity. He quoted the latter's words: "Certainly this phenomenon can be partially explained by the extra-fluidity of the alphabetical language which cannot be depended upon as a suitable organ to conceive any solid ideas". . . (As for the Chinese language,) "it is invulnerable to storm and stress. It has already protected the Chinese civilization for more than forty centuries. It is solid, square, and beautiful, exactly as the spirit it represents." (20, p. 37).

If the Chinese language possesses such sociological powers of durability, we would expect that it would have nourished such a doctrine as the substance-attribute philosophy. Free from the atomicities and unstable transiences of the alphabetical languages, it would be predisposed to see unchanging solidities in things, whereas Western philosophy would, by contrast, be characterized by an underlying aversion to enduring substances. Actually, no such correlations hold.

There have been Chinese philosophers who developed doctrines akin to those of classical Western philosophy. Kung-sun Lung, early in the third century B. C., elaborated a doctrine, for instance, which was very much like the Platonic theory of universals. His word "chih", which meant "pointers", conveyed a notion similar to "universals". Kung said: "Things in every instance involve universals, but universals do not point to the material world. If there were no universals, things could not be described as 'things' . . . The fact that the material world has no universals springs from each single thing having a name. A name is itself not a universal." (9, pp. 125–126). And there have been Chinese philosophers such as Mo Ti, who formulated an empirical outlook: "The words of our Master Mo: the universally true way of learning by investigation whether a thing exists or not, is, without question, by means of the actual knowledge (on the evidence) of everybody's ears and eyes. If it has been heard and seen, then it is undoubtedly to be taken as existing. If no one has heard of it or seen it, then it is undoubtedly to be taken as non-existing." (9, p. 51).

Chinese philosophy, on the whole, has not, however, been preoccupied with the problems which have been the especial concern of Western philosophers. Theory of knowledge, for instance, has not developed in China; anxieties concerning the existence of the external world, the haunting presence of things-in-themselves and unknowables, are not characteristic of Chinese thought. Evidently this indifference to epistemology does not derive from the structure of its language. The Chinese language has no inflections, no declensions or conjugations, in our sense of the terms. Nevertheless, as Max Müller once said, "there is no shade of thought which cannot be rendered in Chinese." (16, p. 118).

The distinctive characteristics of Chinese philosophy are an expression, according to Feng Yu-lan, its recent leading historian, of the social, economic, and technological traits of Chinese civilization. Feng agrees that epistemologic questions are not "problems" to the Chinese thinker. "Whether the table that I see before me is real or illusory, and whether it is only an idea in my mind or is occupying objective space, was never seriously considered by Chinese philosophy." (5, p. 25). Not that the Chinese had somehow learned the positivist critique of non-experimental questions. Rather, the farmer's outlook had uniquely conditioned the outlook of the Chinese philosopher. "What farmers have to deal with, such as the farm and crops, are all things which they immediately apprehend. And in their primitivity and innocence, they value what they thus immediately apprehend. It is no wonder then, that their philosophers likewise take the immediate apprehension of things as the starting point of their philosophy." (5, p. 25). The agricultural setting of Chinese thought similarly impressed itself upon John Dewey during his sojourn in the Far East: "China is agrarian, agricultural; everyone knows that fact. But while we know it, we forget how long and how stable is their agriculture. The title of a book by an American agriculturalist, *Farmers of Four Thousand Years,* is infinitely significant when we reflect upon it." (4, p. 205).

We may venture the hypothesis, furthermore, that the intensity of Chinese family life, the "we-feeling", kept the sense of reality and involvedness with things strong. The Chinese individual was not afflicted with loneliness or emotional isolation; it is striking that epistemologic subjectivism begins to torment philosophers with the onset of an age of economic individualism.

Linguistic influences in the modes of formulation within Chinese philosophy, provide the lines of transmission for more underlying socioeconomic determinants of modes of thought. Expressions of the Chinese language reflect the agricultural life of a great land-domain. Feng states their close relationship: "China is a continental country. To the ancient Chinese their land was the world. There are two expressions in the Chinese language which can both be translated as the world. One is 'all beneath the sky' and the other is 'all within the four seas'. To the people of a maritime country such as the Greeks, it would be inconceivable that expressions such as these could be synonymous. But that is what happens in the Chinese language, and it is not without reason." (5, p. 16). The metaphors, however, which are embedded in the linguistic expressions are without any significant impact on the varieties of Chinese philosophy. The metaphors in time lose even any recessive efficacy; people become unaware of them. They are then curiosities for the social philologist, but they are not the source of an unconscious metaphysic which permeates the minds of their users.

Occasionally a metaphor will help give a philosophy a vogue which it might not otherwise have enjoyed. The syntax of the English language was common to both William James and Josiah Royce, to the pluralistic empiricist and the monistic idealist. Syntax did not impose its inevitable impress upon their philosophies. Both of them were aware that their doctrines were in different ways at variance with ordinary syntactical forms, the plurality of self-existent qualities as much so as the Reality which was the presumable immanent subject of all propositions. But James made use of a metaphor of the American language which brought friends to his philosophy among the citizenry at large. He spoke of the pragmatic method as a way of realizing the "cash value" of words. Sensitive to the American idiom, he had also described his God as one who does a retail, not a wholesale business. And James' call for confidence in God, for a faith akin to that which wins a man "promotions, boons, appointments" smacks of a salesman turned theologian, who writes his theology with the help of phrases from a manual on "How to Win Friends and Influence People". These were linguistic metaphors, but James found to his consternation that his philosophy was henceforth linked to them. In vain, he protested against the description of pragmatism "as a characteristically American movement, a sort of bobtailed scheme of thought, excellently fitted for the man on the street, who naturally hates theory and wants cash returns immediately." (11, p. 185).

What is remarkable, furthermore, is that James' financial metaphors were so infectious that they infiltrated the language of his avowed opponents. Josiah Royce thus found himself criticizing James in James' own vocabulary of business enterprise. Pragmatism, Royce argued, would be unable to meet the calls for cash payments; it would have to confess bankruptcy. (17, p. 347). An Absolute Pragmatism, a central cosmic reserve bank, with hard metaphysical bullion, would be required, he argued, to guarantee the otherwise inflationary, wild-cat finance of the plural, independent banks. The American business culture thus embraced its philosophers in its idiom. Pragmatist and absolute idealist both found themselves speaking its language. But such influences do not pertain to the syntactical forms of the English language. The pecuniary metaphors were rather part of the content of the American idiom, more immediately responsive to cultural changes than syntactical forms. And as such, these expressions are the direct vehicles for socio-economic influences on philosophy. By contrast, the grammatical forms remain in the background, invariant, unchanging while the diverse philosophies arise to compete with each other under diverse cultural circumstances.

Inadequacy of the Doctrine of Linguistic Relativity

The linguistic interpretation of the history of thought has given rise to a special epistemological doctrine which may be called "linguistic relativity". According to this doctrine, each language defines for its users a unique cultural universe, one which includes within itself a unique physical universe as well. There is, according to this view, no world of common physical uniformities and facts which is the same for all cultures. As B. L. Whorf, a leading proponent of this theory, says: "We are thus introduced to a new principle of relativity, which holds that all observers are not led by the same physical evidence to the same picture of the universe, unless their linguistic backgrounds are similar, . . ." (26, p. 215). Every language is thus held to impose a kind of linguistic *a priori* on the forms for the analysis and description of nature. The grammar of each language, Whorf states, "is not merely a reproducing instrument for voicing ideas but rather is itself the shaper of ideas, the program and guide for the individual's mental activity, for his analysis of impressions, for his synthesis of his mental stock in trade.... We dissect nature along lines laid down for us by our native languages." (26, p. 214).

On what kind of evidence does the so-called principle of linguistic relativity rest? Various facts are adduced such as the difference in Eskimo linguistic usage from ours. The Eskimo would regard our use of "snow" as too large and inclusive. We use the same word to denote falling snow, snow on the ground, snow packed hard like ice, slushy snow, flying snow,—in short, all the varieties of snow. But an Eskimo has different words for falling snow and slushy snow, "he would say that falling snow, slushy snow, and so on are sensuously and operationally different, different things to contend with; he uses different words for them and other kinds of snow." (26, p. 217).

Do such facts, however, establish a principle of linguistic relativity? Far from it. Rather, they illustrate the principle that where the specific differences among similar events are more important in the life of the people than the membership in the general class, the people's language may not include the abstract, general term for the class in question. Specific functions are then more emphasized than the abstract class membership. (2, p. 130). To the Eskimo, in his struggle with his material environment, the differences in the varieties of snow are what are important. "Snow" in the abstract would be co-terminous with his known physical universe, an unchanging background to his struggle for existence. What does concern him in vital ways are the variations of his world,—is the ice, for example, such as can be carved for a house? His purposes differ from ours, and perceived differences, which are unimportant for us, are those to which he must most attend. The Eskimo's language does not, however, impose a specific metaphysics. He works with his day-by-day objects in an empirical spirit, he acknowledges the uniformities and causal processes of nature. As for his religious beliefs in spirits and the powers of the charms of his shamans, they bear no relationship to the structure of his language. (24, p. 391).

Categories that are unessential to a given culture, says Boas, will on the whole not be found in its language. Categories, on the other hand, that are culturally important will be found in detail. (2, p. 141). A society in which the distinction between paternal and maternal lines is socially important will have a suitable distinctive terminology. Our society lacks such a vocabulary. The obligatory categories similarly vary from one people to another. We say: "the man killed the bull", but in one Indian language, epistemological niceties are insisted upon. One has to specify the source of information, its location, and whether the data are seen, heard, or inferred. One would say: "this man kill as seen by me that bull." But, as Boas further points out, "the relational functions of grammar have certain principles in common all over the world." (2, p. 133). Every language will set forth determinate relations between a subject and predicate, verb and adverb, noun and attribute. Whatever the differences in environment, a common world of spatio-temporal objects confronts all men, and there are common properties to the matter-of-fact dealings of all peoples with nature.

Different languages emphasize different relations of men to nature, but the segments which are emphasized do not define incommensurable universes. Each culture can be informed in its own language concerning the limitations of its experience; if it wishes to, if it needs to, it can add to the resources of its language. The Hebraists in Israel have thus taken an ancient language, and expanded its vocabulary so that textbooks of electro-dynamics can be written in the idiom of the prophets. The "principle of linguistic relativity" argues that there are incommensurable cultural universes. An incommensurable cultural universe would be an unknown one. The fact of linguistic communication, the fact of translation, belies the doctrine of relativity.

The "will to be untranslatable" grows during an era of cultural regression and ethrocentrism. An old Italian proverb has it, *Traduttori, traditori* ("Translators are traitors"). And the Talmud likewise said that the oral law was not written down, because God feared that otherwise it would be translated into Greek, and he wished to keep it the special mystery of his people. The later rabbis mourned the Septaugint translation of the Bible with a day of fasting (la, p. 28). Linguistic relativity is the doctrine of untranslatability in modern guise.

We might indeed regard the "principle of linguistic relativity" as a pseudo-sociological hypothesis. A doctrine which is stated in the language of sociology but which no sociological observation could possibly confirm is a pseudo-sociological hypothesis. For instance, since we use English, we are held at once to be precluded from understanding, let us say, the Chinese civilization and world-standpoint. When we show that a Chinese can understand us, and we the Chinese, it is argued that such an Asian has been contaminated by Western European culture. And since the very fact of communication is "evidence" of such contamination, it follows that we couldn't possibly introduce a negative instance to this hypothesis. In short, it is not an empirical, scientific theory. How it itself could be discovered, how the relativist has escaped his own linguistic *a priori* to know, for example, the details of the Hopi *a priori* is incomprehensible. For he does manage to state the Hopi perspective in English, and this, if his doctrine were true, he should be unable to do.

The "principle of linguistic relativity" is an instance of a phenomenon among thinkers which we might call "illegitimate diffusion". When the theory of physical relativity acquired its world renown, there was a tendency for theorists in other departments to run riot with the words of "relativity". The psychological and social sciences, for instance, began to discover "frames of reference" everywhere. Economic classes, social observers, different philosophies,—all of these were variously denoted as "frames of reference". The immense prestige of the physical theory of relativity was the covert, emotive argument for the adoption of these "relativities". As a matter of fact, none of these usages have any significant analogy to the physical theory.

In physical science, a body which is described as a "frame of reference" conforms to certain conditions. Thus it is stipulated that laws of nature will be invariant for all frames of reference. Although time measurements and distance measurements vary with respect to different observers, there is a space-time interval between events which is invariant for all frames of reference, moving uniformly relative to one another. But linguistic relativity, with its incommensurable universes, has forgotten the invariant world which is common to all observers. And whereas the physical theory was an attempt to explain certain unusual experimental findings, the linguistic theory arose from no counterpart of a Michelson-Morley experiment in communication. It is rather a projective superimposition upon the facts, and indeed, essentially contrary to their sense.

Despite his theory, the linguistic relativist always has had a marked ability for transcending his own linguistic a priori. "Our language," says Whorf, "gives us a bipolar division of nature. But nature herself is not thus polarized." (26, p. 216). We are thus informed that: "Nature herself, apart from any definition or determination by an *a priori* linguistic scheme, is not polarized." We evidently have no difficulty in stating propositions which escape the net of the linguistic *a priori,* and the relativist himself joins in this easy transcendence. A language is not a "frame of reference" in any sense which can give rise to a doctrine of linguistic relativity.

Common Human Experience and Universal Categories

A common, universal, scientific mode of thinking manages to express itself in all languages. It is the linguistic aspect of the common struggle of men everywhere for survival in the midst of their environment. There is the imprint on all languages of those categories and distinctions which facilitate the matter-of-fact causal reasoning without which men could not cope with their problems of biologic existence. As Malinowski said, "The categories derived from the primitive use will also be identical for all human languages, in spite of the many superficial diversities. For man's essential nature is identical and the primitive uses of language are the same.... The fundamental grammatical categories, universal to all human languages, can be understood only with reference to the pragmatic Weltanschanung of primitive man,..." (15, p. 266). Language can undertake vagaries only after it has fulfilled the minimum tasks of assistance in living. Jespersen has vividly described the common, human logic which is found in different linguistic guises: "When we examine languages, we do indeed find many things which imply the existence of a fundamental common nature in human beings all the world over. Some features of language are due to a common humanity: they show themselves just because the individuals who speak the languages are human beings...." "this is certain that we see everywhere in the history of languages a uniform striving to be quit of the same superfluous distinctions and to reduce the grammatical apparatus to the simplest possible, to a system in which the great innersyntactical, logical, or rational categories are denoted sharply and unmistakeably. Such distinctions as those between one and more, he and she, between animate and inanimate, between past, present, and future, between the three persons—distinctions which in the infancy of language were chaotically coupled with one another and with obscure ideas of a quite different kind—come by this means to stand out sharper and sharper, while a logic common to all mankind breaks radiantly through the barriers of linguistic expressions" (12, p. 206, 218).

The syntax of any given language may be regarded as composed of two kinds of forms and distinctions, those which correspond to natural realities and those which are projected by the languages upon realities. We might call the former "realistic syntactical distinctions" and the latter "projective syntactical distinctions". The distinction of gender, for instance, as applied to inanimate objects is an example of the latter. Now Russell believes that logical analysis of "the properties of language may help us to understand the structure of the world." He holds that study of syntax enables us to "arrive at considerable knowledge concerning the structure of the world", and as an example of such knowledge, he mentions our acknowledgment of relations as part of the non-linguistic constitution of the world. (22, p. 429). Where Russell errs is in his belief that it is the study of syntax which has led to such knowledge. The occurrence of syntactical forms is no guarantee that there are corresponding forms in nature. The distinction of gender, as we have said, is one such non-corresponding form. Our acknowledgment of relations is one founded on our actual experience of relations, and if a language were to lack relational forms, we would say so much the worse for that language. We must turn to extra-linguistic facts in order to decide which syntactical forms correspond to natural realities, and which don't. The natural world is the extra-linguistic criterion with which we judge the extent to which linguistic forms accurately conform to the structure of things. Language is not an autonomous metaphysical organon, and Russell's view that language by itself is a guide to the world's structure is an example of that faith in the "omnipotence of thought", to use Freud's expression, which is the hallmark of the unrealistic, unscientific, magical mode of thought.

Linguistic philosophers have expressed the view that incorrect grammar leads to erroneous metaphysical belief. Correct grammar, it is added, is neutral as to rival metaphysical beliefs. (1, pp. 114, 125, 126). What, however, do we mean by an "incorrect grammar"? It is a classification of linguistic forms which is founded not upon the forms in function but upon an antecedent metaphysics. The Latin grammarians, influenced by Aristotle's metaphysics and not by Latin as it was used, held that every sentence must consist of a subject and a predicate. Academic prejudice applied the dominant metaphysics to the facts of speech, and grammar, which, indeed, is an empirical science of speech habits, was transformed into an a priori set of metaphysical rules. The prestige of Roman civilization promoted a species of linguistic imperialism. As Jespersen states: "Latin has been considered for centuries the supreme language, and extolled for its logic. It was easy then for people to make the mistake of thinking that everything in Latin grammar was pure logic, and that in the other languages only what agreed with Latin could be logically defended." One grammarian in 1861 wrote that to say "It is me" rather than "It is I" was "nothing else than the grossest sin against the first and simplest and most incontrovertible laws of thought and grammar". (12, p. 114). Latin provincialism overlooked the fact that nothing in logic makes it mandatory for the predicative to stand in the same case as the subject; the Russian and Finnish languages, for instance, sometimes employ different cases.

The formulations of the Latin grammarians were an intrusion, a superimposition of metaphysical dogma upon the empirical forms of speech. Their procedure was analogous to the intrusion of Aristotelian metaphysic, or Hegel's metaphysic, upon the forms and laws of physical science. We can say that incorrect grammar leads to erroneous metaphysics only in the same sense as we can say that incorrect physics leads to erroneous metaphysics. Moreover, there is an important distinction between linguistic forms in use and the grammarian's version of these linguistic forms. As an empirical matter, we have seen that empirical linguistic forms do not have a primary role in determining people's philosophies. And we have seen that the grammarian's description of a language errs when it projects some favorite metaphysics upon the empirical forms of speech. But the source of the grammarian's metaphysics is not his language; it is an extrapolative use of metaphysics which had extra-linguistic origins.

The great linguistic revolutions in the history of philosophy have come when men turned from the scholarly, honorific language to the language of common people. The archaic classical language was the depository of an accumulated vested interest in meaningless terms. The vernacular language, by contrast, was closer to the homespun verifiable realities. Meaningless terms stand out more conspicuously when they are introduced into the vernacular because the latter is closely connected with the everyday necessities of biologic existence. The literary classical language, on the other hand, has led a life immured in monasteries, dissociated from the practical concerns of men. No wonder that this dissociation of language from the controls of action finally produces a situation, as Veblen says, where "classical learning acts to derange the learner's workmanlike aptitudes." (25, p. 395). So that Descartes wrote in his momentous linguistic manifesto: "If I write in French which is the language of my country, rather than in Latin which is that of my teachers, this is because I hope that those who avail themselves only of their natural reason in its purity may be the better judges of my opinions than those who believe only in the writings of the ancients." (*Discourse on Method, Part IV*). The change from Latin to the vernacular, as Morris Cohen once said, revealed the emptiness of received systems.

The significance of these historic linguistic decisions has not been by way of a new choice of syntactical determinants. It is also true that the vernacular idiom could in time become encrusted with its own growth of meaningless verbiage. At the crucial juncture of historic change, however, the vernacular was the medium of empirically minded people, those whose minds had not been stamped with the schoolmen's metaphysics. The use of the vernacular was an appeal to fresh social and scientific determinants for philosophy, to open-mindedness rather than tradition. The significance of the vernacular was sociological, not syntactical; its significance was that it opened the windows of science to the fresh sight and evidence of all men, that it cleared the air of clerical vested interest.

The contemporary notion that the special task of philosophy is the analysis of syntax is, we suggest, an ideological formula, a projective definition. Such a definition promotes the separation of philosophy from the actualities of life-decisions; it defines a philosophical realm which has a scholastic autonomy, and proposes indeed that philosophers are lexicographers, the formulators of a next edition of dictionaries. If philosophies are linked to social institutions, if every philosophy is the perspective of a socio-cultural universe, it follows that the critique of philosophies will tend to lead the philosopher toward a criticism of their socio-cultural sources. The analysis of philosophies then becomes their socio-analysis. By contrast, the linguistic interpretation of philosophic thought provides a haven for those who would avoid commitment to matters of fact or social standpoint. Philosophic thought then becomes an unconscious filibuster in which the underlying aim is not to decide problems. Men's thinking, their agency as problem-solving animals, is thus inverted from its biologic role; it becomes a professional device for indefinitely postponed decision. The linguistic interpretation of philosophy is an ideological, projective formula in the sense that it would repress those cultural and psychological conflicts which are the generative source of philosophic history.

It used to be said that man is a speaking animal. Probably it would be more truthful to speak of him as a stammering, stuttering one. Clarity of speech will come to him when the deep conflicts within him are resolved, when his anxieties are reduced. That is why philosophy cannot rest with the critique of language.

References

1. Black, Max, *Language and Philosophy,* Ithaca, 1949.
1a. Bentwich, Norman, *Philo-Judaeus of Alexandria,* Philadelphia, 1910.
2. Boas, Franz, ed., *General Anthropology,* Boston, 1938.
3. Cornford, Francis MacDonald, *The Unwritten Philosophy,* New York, 1950.
4. Dewey, John, "The Chinese Philosophy of Life,". *Characters and Events,* I, New York, 1929.
5. Feng, Yu-lan, *A Short History of Chinese Philosophy,* New York, 1948.
6. Gesenius, William, *Hebrew Grammar,* Edited and Enlarged by E. Kautzsch, tr. by G. W. Collins and A. E. Cowley, Oxford, 1898.
7. Harrison, Jane Ellen, *Themis: A Study of the Social Origins of Greek Religion,* Cambridge, 1912.
8. Hallowell, A. Irving, "Temporal Orientation in Western Civilization and in a Pre-Literate Society," *American Anthropologist,* XXXIX 1937.
9. Hughes, E. R., ed. and tr., *Chinese Philosophy in Classical Times,* London, 1950.
10. Husik, Isaac, *Judah Messer Leon's Commentary on the "Vetus logica,"* Leyden, 1906.
11. James, William, *The Meaning of Truth,* New York, 1909.
12. Jespersen, Otto, *Mankind, Nation and Individual from a Linguistic Point of View,* Oslo, 1935.
13. Jespersen, Otto, *The Philosophy of Grammar,* New York, 1924.
14. Mencken, H. L., *The American Language,* Fourth Edition, New York, 1936.
15. Malinowski, Bronislaw, *Magic, Science, and Religion,* Boston, 1948.
16. Müller, Max, *Lectures on the Science of Language,* New York, 1878.
17. Royce, Josiah, *The Philosophy of Loyalty,* New York, 1908.
18. Russell, Bertrand, *The Analysis of Mind,* London, 1921.
19. Russell, Bertrand, *A Critical Exposition of the Philosophy of Leibniz,* Cambridge, 1900.
20. Russell, Bertrand, *The Problem of China,* London, 1922.
21. Russell, Bertrand, *Philosophy,* New York, 1927.
22. Russell, Bertrand, *An Inquiry into Meaning and Truth,* New York, 1940.
23. Sayce, A. H., *Introduction to the Science of Language,* II London, 1880.
24. Stefansson, Vilhjalmur, *My Life with the Eskimo,* New York, 1913.
25. Veblen, Thorstein, *The Theory of the Leisure Class,* New York, 1926.
26. Whorf, Benjamin Lee, "Science and Linguistics," in *Readings in Social Psychology,* ed. by T. H. Newcomb and E. L. Hartley, New York, 1947.
27. Wolfson, Harry Austryn, *Crescas' Critique of Aristotle,* Harvard, 1929.

5

The Principle of Simplicity

The Scientific Principle of Simplicity

We are all acquainted with persons who seem to have a talent for making things over-complex, persons who invent exceedingly devious explanations for what can be simply explained. Such individuals strike us as hardened violators of Occam's Razor: *Entities are not to be multiplied unnecessarily.* We shall say briefly that such persons *goropise.*

To goropise. The word has never come into use, though it should, since it fulfills a need. It was invented by Leibniz under interesting circumstances. Goropius, a scholar of the sixteenth century, maintained that Adam spoke a dialect of German. He tried to prove that all other languages were derived from their Teutonic prototype. Goropius' hypothesis was not unlike another widespread view according to which all languages originated from Hebrew. Leibniz regarded all such theories as virtual nonsense; they led into "strange and often ridiculous etymologies", with "too many leaps from one nation to another far distant without having good verifications." He added that these extremely complicated etymologies were without "concurrent evidence." Goropius' name became to Leibniz synonymous with a type of scientific malpractice (12, p. 303) To multiply auxiliary hypotheses is to goropise.

The Meta-Scientific Principle of Simplicity

On what grounds do we adhere to the admonition not to goropise? There are thinkers who hold that the principle of simplicity rests on a metaphysical foundation (33, p. 218). The faith that nature is governed by simple laws has been held by the greatest scientists, and is held to underlie the principle of simplicity. Newton declared that the first rule of philosophizing is "that Nature does nothing in vain, and more is vain when less will serve; for Nature is pleased with simplicity, and affects not the pomp of superfluous causes" (20, p. 398). For Einstein, this faith took on an almost religious quality. The idea of mathematical simplicity, he believed, is actualized in nature (5, p. 282). "In every important advance," Einstein wrote, "the physicist finds that the fundamental laws are simplified more and more as experimental research advances" (24, p. 11). When Einstein had a new idea, he is said to have asked himself: " 'Could God have created the world in this way?' or 'Is this mathematical structure worthy of God?' Translated into ordinary language this sentence means: 'Is this theory logically simple enough?' " (6, p. 267)

Now this belief in the simplicity of Nature has a long philosophical history. I shall call it the meta-scientific principle of simplicity in order to distinguish it from the scientific, methodological principle known as Occam's Razor. The meta-scientific belief was held by Leibniz to be the principle of Divine economy: that God achieves a maximum of effects with a minimum of means. This meta-scientific creed has had an immense heuristic influence on the history of science. Under its guidance, the French Newtonian, Maupertuis, was led to the first formulation of the mechanical principle of least action. Since God achieves his effects in the most economical way, reasoned Maupertuis, there must be some function whose value is at a minimum for the actual path taken by a moving object as compared to other possible paths. Maupertuis went on to show in 1744 that both the laws of reflection and refraction of light follow from the principle that light takes the path for which the quantity of action is least. "It is this quantity of action", he affirmed, "which is here the true expense of nature, and which she economizes as much as possible in the motion of light" (9, p. 418). A subsequent memoir in 1748 entitled *Les Lois du Mouvement et du Repos, déduites d'un Principe Métaphysique* linked God's existence to the law of least action.

To the metaphysician, Occam's Razor is simply the methodological expression of an underlying simplicity in the nature of things. This principle, says a spokesman of this standpoint, "takes the form of the law of least action in physics. In philosophy it is known as Occam's

Razor... The truth of Occam's razor depends on one's cosmology and not on logic" (2, p. 207). According to this view, the scientific principle of simplicity is pragmatically successful because the laws of Nature are, as a matter of fact, simple. If God had been more roundabout in his ways, then the principle of simplicity would have been of no use whatsoever. Such is the metaphysical derivation of Occam's Razor.

The meta-scientific creeds of scientists differ, however, enormously. Some have rejected the notion that Nature is governed by simple laws. The naturalist, Buffon, rebutted the Linnaean classification on the ground that its simplicity constituted a slander against God (21, p. 221). Was God, he intimated so simple-minded that his creations could be grasped in so simple a scheme?

The plain fact, however, is that the metascientific creed of simplicity has nothing to do with the scientific, methodological principle of simplicity. Occam's Razor, in other words, doesn't owe its validity to any doctrine that Nature is simple. This can be readily perceived with the aid of an easy imaginary experiment. Let us suppose that the laws of Nature are highly complex, that God, in his Majesty, has chosen to befuddle mankind thereby, reminding them of the pettiness of their intellect. Let us suppose then that the meta-scientific principle of simplicity is false. Would this in any way affect the validity of Occam's Razor? Would the complexity of Nature be a mandate to goropise? Clearly not. In every given problem, we should still look for the simplest possible solution. For every problem in our hypothetical universe, it would indeed be the case, that the simplest possible solution was an extremely complex law, but the latter would none the less be the simplest explanation of the phenomena. Social scientists, for instance, often hold that many variables are required to account for the behavior of social groups. To explain the complex social realities, they enquire into the interrelations between economic, political, religious, sexual, and intellectual factors. Such scientists are as loyal to Occam's Razor as the more simple-minded economic determinist, who insists that there is only one independent variable in social history. The laws which they seek are still the simplest which fit the facts. They are sceptical toward the array of auxiliary hypotheses which follow in the wake of economic determinism. Occam's Razor does not presuppose the simplicity of sociological laws.

The meta-scientific doctrine of simplicity is, indeed, an indeterminate creed. Though it stimulates many scientists to research, the doctrine involves the kind of extrapolation into the unknown which takes one beyond science. Statements such as "Nature is simple" are confronted by a sampling predicament when you try to verify them. If our intelligence is limited, if we are all basically simple-minded, then the laws which we shall be able to discover will be those only which are comparatively simple. "Simplicity", as Jevons remarks, "is naturally agreeable to a mind of limited powers" (7, p. 625). But we cannot infer that our sampling of laws is representative of nature as a whole. We cannot read what may be due to our mode of selection as a universal attribute of nature's laws.

Residual phenomena that are unexplained by the known laws are found in every branch of science, and there is no way of knowing whether the ultimate laws which might explain them are simple or complex in form. Sometimes it is argued that the complexity of proximate laws yields in time to the simplicity of ultimate law. The various elliptic, hyperbolic, parabolic, and circular orbits of different heavenly bodies were thus subsumed under Newton's laws of motion and gravitation. Simple formulae of conic sections were, however, adequate to the "complex" laws in question, and their simplicity was undoubtedly one of the reasons for their early discovery. And, in any case, scientific work often takes us from simple equations to complex functions. The laws of quantum mechanics have supplanted relatively simpler mathematical formulations.

Meta-scientific simplicity is sometimes held to be the limit toward which the known laws of nature converge as scientific systems approach toward completeness. Meta-scientific simplicity is thus regarded as an asymptote to which the sciences draw ever closer. This is indeed the view of Einstein which we have cited. We cannot however affirm that a scientific system is approaching toward completeness. We do not know whether the series of novel phenomena which science discovers is convergent in character. The advent of new and unexplained facts may perhaps be more similar to a divergent series, "so that the more we have explained, the more there is to explain" (7, p. 753). We have no way of asserting the form which sciences will take in their maturity, because there is no test for whether a science is mature. John Stuart Mill once affirmed that the laws of value were so well mastered that nothing more remained to be done in the subject; Lord Kelvin regarded the "properties of matter" as fully comprehended, so that the task of physics was to fill in the details of a chart whose general outlines were known. Both made their assertions of maturity on the eve of a basic overhauling of their respective sciences. Relative to the complexity of the universe, the most mature science may be in its infancy; there is no way of telling. Speculation on futures in the history of science is a game without assignable values. Nor do we have any way of knowing whether a science has explained the major phenomena of its domain, so that minor phenomena only remain unexplained. The explanation of so-called minor "deviations" may require the most drastic reconstruction of the existent theoretical system. Furthermore, new instruments, from the telescope to the mass accelerator, open up new continents of fact, and usually shatter the systematic status quo. We, therefore, have no way of forecasting in principle the direction, whether towards or away from simplicity, which the history of scientific laws will take. Consequently, any assertion concerning the simplicity of nature, as a meta-scientific principle must remain indeterminate. Knowledge is a groping process in which the direction remains indeterminate and the goal unknown.

Einstein's faith in the simplicity of nature is, from our standpoint, a kind of "scientific optimism" (to use Philipp Frank's words). A decision of temperament is involved which is extra-scientific. A "scientific pessimist" will incline toward the faith that "there are more things in heaven and earth, Horatio, than are dreamt of in your philosophy." The notion of a simple universe elicits a rebellion, a claustrophobia, in some people. Whitehead's aphorism, "Seek simplicity, and distrust it", expresses the fruitful interaction of both types of scientific temperament. Simple laws, we often feel, assume a constancy of conditions, which fuller knowledge converts into relevant variables. Boyle's law of gases, for example was superseded by Van der Waals' law, which has an added number of variables and constants. But at no time are we warranted in affirming our metascientific creeds as scientific truths of the universe.

Then on what ground does the methodological principle of simplicity rest?

Occam's Razor as a Special Case of the Principle of Verifiability

Occam's Razor, I shall try to show, is not an independent principle of methodology. It is a special case of the principle of verifiability—

that no theory is meaningful unless it leads to some observable consequence. The goropisor is a person who has embarked upon scientifically meaningless speculations. Let us consider a series of examples drawn from common sense and the sciences; we shall perceive how the principle of simplicity merges into that of verifiability.

Let us suppose that Robinson Crusoe and his man Friday see the footprint of a man. Robinson infers that a man was walking on the beach. But Friday is a goropisor. He declares that the man was undoubtedly carrying someone. Robinson replies that in that case, the footprint should have been deeper. Friday answers that probably the second man was buoyed up by a parasol. Robinson replies there have been no winds these last days. Friday answers that probably there were brief winds just during those moments when they were not looking. Robinson replies that in that case, the open pages of the Bible he left outside should have been blown from their place. Friday answers . . .

Friday, the goropisor, is always prepared to invent an auxiliary hypothesis to explain away evidence. A goropisor is one who is never ready to abide by the weight of a specified crucial experiment. He has a principle of theory construction all his own: to generate an infinite sequence of auxiliary hypotheses rather than surrender his initial hypothesis. And an indefinite sequence of auxiliary hypotheses is pragmatically equivalent to an unverifiable theory.

Occam's Razor explains what is meant by an unverifiable theory, wherein a theory is defined as a system of related statements. An unverifiable statement is one such as "every object is now contracting to half its size." That assertion is unverifiable in and by itself. An unverifiable theory, on the other hand, is one wherein each of the constituent statements may be verifiable, but where the system taken as a whole is unverifiable. For instance, Nazi theologians proposed the hypothesis that Jesus was a German. To sustain this doctrine, they were obliged to invent a series of auxiliary hypotheses. Each member of this series was challengeable; there was no evidence, for example, of a Gothic migration to Galilee in Augustus' time. But refutation was postponed at each step by the invocation of a successor auxiliary hypothesis. Negative evidence can be indefinitely evaded by auxiliary hypotheses. Sometimes the series of auxiliary hypotheses is finite in the sense that it terminates in a final unverifiable auxiliary hypothesis. But, in principle, the series can be non-terminating. The goropisor is afflicted by a tendency to methodological cancer; his principle of theory construction generates an endless series of auxiliary hypotheses, and thereby enables him to repress any confrontation with fact. Occam's Razor excludes a mode of theory construction which eventuates in unverifiability.

The principle of simplicity is, in effect, a pragmatic challenge to the goropisor. It asks him: "Are you prepared at any step to abide by experimental decision? And if so, would you state the conditions for that test? Or are you always going to invent auxiliary hypotheses?" If experimental determination is accepted at some finite stage, the contrary hypotheses can be rejected or confirmed. We may come to believe, however, that the goropisor is intransigent, that he has made a commitment to an article of faith, and will never, in that state of mind, yield to negative evidence. The principle of simplicity defines the species of unverifiability which is exemplified in this ultimate defiance of scientific method. It tells the goropisor: "If this is the kind of procedure you are going to follow, scientific discussion with you is not worthwhile."

In practice, of course, we are never presented in scientific research with a choice between a simple hypothesis and a rival infinite sequence of hypotheses. We are always confronted with conflicting finite systems of hypotheses with varying degrees and modes of complication. What determines our choice, as scientists, among the rival claimants is our practical judgment concerning the character of their underlying respective methods of theory construction. If we believe that the kind of procedure which a theory's proponent is using is the sort which, if adhered to indefinitely, would lead towards hypothesis-systems of uncontrolled complexity, then we reject the proposed theory. The hypothesis-system in question may, at the given time, contain only a small number of auxiliary complications. Nevertheless, we reject it because our practical judgment of its underlying principle of procedure is adverse; we recognize that this kind of thinking would, in the long run, lead us into morasses of unverifiable infinite complexity.

The principle of simplicity in scientific research is thus not primarily a matter of counting the number of variables or evaluating the complexity of a mathematical function. It is basically a pragmatic judgment upon underlying motives in scientific work; it discriminates between motives which welcome experimental test in contrast to those which, enthralled by an *idée fixe,* are compelled in a neurotic fashion to evade a decision by factual evidence. The principle of simplicity is a warning against the psychological repression of fact by recourse to indefinitely extensible systems of hypotheses.

The debate between Malinowski and the Freudian school illustrates vividly the use of Occam's Razor in locating unverifiable theories. According to Freud, hatred against the father should be found in every culture, because the Oedipus Complex, in his opinion, is a primordial, biological phenomenon. But Malinowski found that the Trobriand Islanders were ignorant of the biological facts of paternity, and that in their matrilineal society, the son's hatred was directed not against the father but against the mother's brother. Malinowski concluded that the authority structure of the Melanesian family determined a direction of emotional response different from that of a Western middle class family.

How did the Freudian school reply to the evidence against the hypothesis of the Oedipus Complex? They did not abandon their theory; they invented auxiliary hypotheses. Ernest Jones asserted that the father was hated so much in the Trobriand Islands that his role in procreation was denied. He also said that the maternal uncle was a scapegoat for alleviating the tensions of this great hatred. Malinowski's reply to these auxiliary hypotheses is noteworthy:

". . . a complex can always be empirically reached by the practical methods of psychoanalysis, by the study of mythology, folk-lore and other cultural manifestations of the unconscious. If, however, as Dr. Jones seems fully to admit, the attitudes typical of the Oedipus complex cannot be found in the conscious or unconscious; if, as has been proved, there are no traces of it either in Trobriand folk-lore or in dreams and visions, or in any other symptoms; if in all these manifestations we find instead the other complex—where is then the repressed Oedipus complex to be found? Is there a sub-unconscious below the actual unconscious and what does the concept of a repressed repression mean? Surely all this goes beyond the ordinary psycho-analytic doctrine and leads us into some unknown fields; I suspect moreover they are the fields of metaphysics!" (17, p 144).

The method of goropising in this instance evolved an unverifiable sequence of auxiliary hypotheses. An auxiliary hypothesis can attain a legitimate status within science, only if it is supported by some crucial independent evidence relative to its rival initial hypothesis. But no such evidence was forthcoming from the orthodox Freudian school.

When the great Lavoiser made his celebrated criticism of the phlogiston hypothesis, his essential point was that it dealt with contrary evidence by means of an ever expanding series of auxiliary hypotheses (11, p. xx–xxi; 22, p. 362; 1, p. 90).

"All these reflections," said Lavoiser, "confirm what I have advanced, what I intended to prove, what I am going to repeat again, that chemists have made phlogiston a vague principle which is not strictly defined and which consequently fits all the explanations required of it; sometimes the principle has weight, sometimes it has not; sometimes it is free fire, sometimes it is fire combined with earth; sometimes it passes through the pores of vessels, sometimes these are impenetrable to it; it explains at once causticity and non-causticity, transparency and opacity, colour and the absence of colour. It is a veritable Proteus that changes its form every instant." (18, p. 230).

The principle of simplicity is thus a straightforward basis for rejecting theories as unverifiable. It emphasizes that a theory ceases to be scientifically meaningful if negative evidence cannot possibly at any stage be adduced against it. There are infinite sets of auxiliary hypotheses possible. Einstein, in rejecting a proposal to modify Newton's law of gravitation with an auxiliary hypothesis, remarks pointedly: "Of course we purchase our emancipation from the fundamental difficulties mentioned, at the cost of a modification and complication of Newton's law which has neither empirical nor theoretical foundation. We can imagine innumerable laws which would serve the same purpose, without our being able to state a reason why one of them is to be preferred to the others; . . ." (3, p. 127) Once we have abandoned the criterion of verifiability, no basis for scientific choice exists. The verified theory is the simplest because every unnecessary component is practically an unverified item.

Critique of the Analogy of Mathematical Simplicity

Occam's Razor is sometimes held to be a principle of mathematical elegance, of aesthetic economy. We prefer the simplest hypothesis, it is said, for the same reason that we prefer that postulate-set for a given field which has the smallest number of independent postulates. There are many postulate-sets, for example, for the algebra of logic. Some are preferred to others by the rigorous minded because they derive their theorems from a smaller number of postulates. That set which has the minimum number of postulates, or which when compared to its equals, has the smallest number of constituents in its premises, is commonly regarded as the simplest. And Occam's Razor, the quest for the simplest hypothesis which explains the facts, is said to be grounded on the same aesthetic motive which leads us to prize elegance in a mathematical system.

But mathematical elegance has nothing to do with the scientific principle of simplicity. The discussion of scientific simplicity has been confused by the use of the mathematical analogy.

All the postulate-sets for Boolean algebra, elegant and inelegant, are formally equivalent, that is, every postulate or theorem in one appears as a theorem or postulate in the others. The same propositions are ordered in different ways with respect to alternative postulate-bases. Each, as a formal system, is consistent and complete, and any interpretation of one will be applicable to the others. This is not, however, the case with the rival hypotheses of the scientist and the goropisor. Their hypotheses are not formally equivalent. They share some common consequences, but if they are different hypotheses, they will lead to some different ones as well. Rival hypotheses are non-equivalent systems; otherwise, the operation of verification would make no sense. Two historians who argue about a phenomenon are not setting up equivalent interpretations. They differ with respect to their predictions concerning documentary or archaeological evidence which may be discovered. Indeed, the inelegant historian, the one who knows the complexity of human motives, may tell the "truer" story than his simplicist schematic colleague. In any case, their respective interpretations are not comparable to alternative postulate-sets for equivalent formal systems. The goropisor has a neurosis for evading experimental disproof. But his mechanism of self-defense is not the contrivance of a formal equivalent but rather the generating of an unverifiable theory. The man Friday, in our example, was proposing that many events took place which Robinson did not acknowledge; the proposed sequence of events, however, was constituted into a successive chain of unverifiability.

The acute Jourdain held that Occam's Razor is "the maxim that logical analysis is to be carried as far as possible" (10, p. 451). According to this view, the replacement of Kepler's laws by the Newtonian law of gravitation is akin to the construction of a system with one postulate rather than three. Again, however, the mathematical analogy misses the real point of the Newtonian achievement. The fact was that

Newton's law applied to many more phenomena than Kepler's laws—to the tides, to the moon's motion, to the trajectories of comets, and so on. Kepler's laws were special cases of Newton's law for planetary distances and velocities. This was not a purely formal reduction in the sense in which Nicod's postulate is simpler than the four primitive propositions required by *Principia Mathematica*. Newton's generalization had an *empirical* content far richer than Kepler's laws. Formal elegance was not the crucial factor.

But don't we choose the sun as a frame of reference for the motions of the planets rather than the earth? And is not our choice of frame of reference determined solely by aesthetic and practical motives, the desire to work with simple formulae rather than complex ones? And, in the light of the theory of relativity, is not our decision for a Copernican rather than a Ptolemaic system motivated essentially by aesthetic reasons? These arguments are taken as establishing the independent significance of mathematical simplicity as a principle in scientific work; the use of "simplicity" seems in this example irreducible to the principle of verifiability. As a matter of fact, however, the discovery that either a Copernican or Ptolemaic frame of reference can be used was an empirical one. It is an error to say, as so many text-books do, that the possibility of such a choice is much the same as the choice among alternative equivalent languages. If the principle of relativity were not valid, the earth's direction of motion would have entered into the laws of nature. The equivalence of frames of reference is an experimental, not a linguistic fact. Einstein thus emphasizes: "the most careful observations have never revealed such anisotropic properties in terrestrial physical space, i.e., a physical non-equivalence of different directions. This is a very powerful argument in favor of the principle of relativity" (3, p. 15–17). The theory of relativity, in other words, on experimental grounds, established the physical equivalence of different frames of reference. It set forth the appropriate formulae of transformation from one coordinate system to another moving uniformly with respect to it. We then do prefer to use that frame of reference which gives us the simplest formulae. But we do so only under the warrant of an experimental argument for the physical equivalence of different frames of reference. We do work with the more easily manipulable equations but only after this empirical demonstration of equivalence; we have experimentally shown that one systemic reference is equivalent to another. And this is an entirely different matter from regarding the goropisor as proposing a highly complex but formally equivalent hypothesis to the one we actually choose. There is no experimental demonstration of equivalence between the goropisor and the scientist. We are led back to Mill's insight that the Law of Parsimony is "a case of the broad practical principle, not to believe anything of which there is no evidence . . . The assumption of a superfluous cause is a belief without evidence;" . . . (19, p. 526).

A genuine use of a criterion of aesthetic elegance occurs in science when two formally equivalent systems can both be used to describe the same phenomena. Aesthetic and symbolic preferences, concerning which persons may differ, can assert themselves legitimately in such an instance. But the scientific principle of simplicity, applied in crucial controversies such as that over the phlogiston hypothesis, is not concerned with the respective aesthetic merits of equivalent systems. The latter are indeed alternative languages for the same phenomena, but crucial experiments are decisions of fact, not of arbitrary linguistic choice. The goropisor wishes to create a world of pseudoscientific fantasy.

The Biological Basis of the Principle of Simplicity

As biological organisms, we can cope only with the verifiable. If we propose in any degree to control our environment, to plan our lives, to struggle with obstacles, we must confront the behavior of things insofar as we can learn about them through verifiable sequences of events. As biological organisms, we must leave aside the infinite number of unverifiable hypotheses. We might characterize the principle of verifiabililty as a biological *a priori* requirement, to which we must adhere if we are to sustain ourselves in our natural environment. And the principle of simplicity is a partial expression of this functioning of man as a biological organism.

Only through its rejection of unverifiable hypotheses does Occam's Razor control the work of science. The principle of simplicity is not an anthropomorphic intrusion into science. Ernst Mach, for instance, did interpret the principle of economy so that it verged upon the anthropomorphic. He held that science was a labor-saving device, a short-hand summation of Nature's operations which enables you to deal with situations more easily (15, p. 6, 78, 490). Occam's Razor was traced to the laziness of men, to their propensity to do things with the least effort. The "line of least resistance", formulated as a principle of scientific logic was taken to coincide with the principle of economy. Occam's Razor, from this standpoint, is a projection of laziness into scientific method. F. C. S. Schiller therefore declared that the principle "occurs to us because we have a brief span of life in which to effect our scientific purposes; to a non-human mind that was not pressed for time but disposed of all eternity it would be unmeaning or repugnant" (28, p. 402).

Now there is an undeniable motivation to achieve required goals with a minimum of effort. Experiments with rats have shown that they will seek their goals by spatially and temporally shorter routes, and that they probably prefer paths which involve a minimal expenditure of energy (30, pp. 110–111, 448). A principle of least effort may be a fundamental one in psychology (32, p. 19–20) Nevertheless its workings are so overlaid in various cultures by historical circumstances that it does not legislate a specific mode of action. There have been cultures and sub-cultures which have prized the conspicuous wastefulness of labor. Most of us prefer a style which is simple and direct. Lyly's *Euphues* were a landmark of courtly circumlocution, and Molière satirized the women who chose the most roundabout way of saying anything. Some cultures in other words, reject psychological simplicity as a norm, and in that sense, a psychological foundation for Occam's Razor becomes precarious. Indeed, a cultural historian might argue that the vogue of Occam's Razor is uniquely linked to the rise of capitalist civilization. As Thorburn's researches showed, the maxim of Occam's Razor, in its present form, first began to be used in the seventeenth century (29, p. 351). It appeared exactly in Clauberg's *Logic,* published in Holland in 1654 and was called by its proponents the Principle of Frugality or Parsimony. The ethics of the early capitalist class, if we are to credit Weber and Tawney, was one of asceticism, thrift, and frugality. Was Occam's Razor a logical expression of the Protestant Ethic, a manifestation within science of the motives towards asceticism and frugality? Mach himself said that the principle of economy was identical with the economist's Law of Thrift (16, p. 88).

The principle of simplicity, however, has a trans-cultural significance. Its source is the human struggle with its environment which is common to all societies, no matter what their cultural differences. Even in a society which was predisposed to inefficiency, prolixity, goropising, and wastefulness, it would still remain true that any effort at scientific research would have to conform to the principle of simplicity. A culture can make its choice for mythology and indirection; but if it does choose scientific study, it has no alternative but to adhere to the principle of verifiability and its corollary, Occam's Razor. Even if we had unlimited time at our disposal, we should still have to conform to the principle of simplicity. Otherwise we would spend eternity with infinite sequences of auxiliary hypotheses.

There are symptoms of neurosis in the behavior of the goropisor and the roundabout-man. Persons who think over-complicatedly are usually fixated upon certain institutions and dogmas. When they look upon objective facts and problems, they must at the same time bear in mind institutional dogma. The prolix person is trying, as he writes, to solve some inner anxiety concerning his ability to communicate. The more institutional fixations and anxieties he has, the more complex will be his way of thinking and writing. From the psychoanalytical standpoint, the goropisor and circumlocutionist are also trying to work out a "simplest" possible solution. But their problem is more complex, because they have added unverifiables and unknowables to the equations which they aim to solve.

The Pseudophilosophical Principle of Simplicity

It was Bertrand Russell who, in the twentieth century, emphasized the significance of the principle of simplicity, and argued for its central importance in philosophy. All the more important it is, therefore, to see that the principle of simplicity is turned to illicit use in Russell's philosophy of logical atomism. Russell claimed that "the maxim which inspires all scientific philosophising" is Occam's Razor: Entities are not to be multiplied without necessity (25, p. 107). This meant, in practice, the principle: "Wherever possible, substitute constructions out of known entities for inference to unknown entities" (27, p. 362–363). The central problem of philosophy, as Russell conceived it was "the relation between the crude data of sense and the space, time, and matter of mathematical physics." The distinctive contribution of his method, Russell said, was "the whole conception of the world of physics as a construction rather than an inference" (25, p. v). It was this type of analysis which led Lenin to reject Occam's Razor vehemently as a device of "bourgeois" philosophy: "The principle of economy of thought, if made a basis of the theory of knowledge, cannot lead to anything else than idealism. That it is more 'economical' to 'think' that only I and my sensations exist—is beyond dispute. once we have introduced such an absurd principle into epistemology.' (13, p. 138).

Russell's principle of constructionism, however, has nothing to do with Occam's Razor. Russell's maxim uses the language of the simplicity principle, but its significance is transformed to fulfill motives of psychological projection. A solipsistic intent runs through much of Russell's thought. In memorable autobiographical passages, he has documented its origin in the loneliness and suicidal fantasies of his childhood and adolescence, and his horror at the cruelties of the political world; his free man was a symbol of solitary defiance of the tyranny of time. An emotional narcissism seems to be one underlying component of solipsism. A hostility to other persons and the external world seems to be another. And a philosopher can do in thought what he cannot do in practice: he can annihilate the external world, and make its reality dependent on himself; he can turn all things into collections of his own sense-data. Such non-logical motives, Russell stated with candor, were propelling his philosophy. "It would give me the greatest satisfaction," he wrote in maturity, "to be able to dispense with the sense-data of other people, and thus establish physics upon a solipsistic basis, but those—and I fear they are the majority—in whom the human affections are stronger than the desire for logical economy, will, no doubt, not share my desire to render solipsism scientifically satisfactory" (26, p. 158).

The motivation of Russell's maxim is not to work with the simplest hypothesis. Rather, it's an ideological projection, a standpoint which seeks a system in which narcissistic, insecurity-haunted, and aggressive drives will be satisfied. The sense of reality (to use Janet's term) can be impaired in various ways, and Russell's emotive translation of Occam's Razor is designed not to solve a scientific problem, but rather to express a highly personalized conception of "reality". Russell is therefore impelled to his own species of goropising. His underlying problem can be stated clearly: given a narcissistic dogma, how shall scientific propositions be translated so as to be compatible with these basic anxiety-demands? If we would understand Russell's philosophy, we must render its unarticulated emotional premise explicit. To analyze the meaning of Russell's maxim, we are indeed compelled to psychoanalyze it. And Russell's method expresses the following: assuming yours were the only experience in the world, and your experience the only existents, how would you express the truths of science in a way compatible with this assumption? This is a problem of rationalization, and the formulation of auxiliary hypotheses. In the latter, Russell's imagination abounds; his world of unperceived perspectives, of unsensed sense-data distributed throughout space, of electrons which are composed of sensations, and brain-atoms which are constituted of thoughts and feelings—these are among the motley group of auxiliary hypotheses with which the solipsistic design to build the world out of "hard data" must be bolstered.

In science, Occam's Razor is employed as a guide in the formulation of theories, and in the choice among rival hypotheses. But Russell's maxim of constructionism is *ex post facto*. It is applied *after* the scientific work is over, as a way of "interpreting" its findings. "A moderately advanced science," says Russell, is required for the logician's activity; he is then to interpret their propositions "in such a way as to preserve their truth while minimizing the element of inference to unobserved entities" (27, p. 367). Again, this concern for a "minimal vocabulary" is rather the outcome of non-scientific narcissistic motives rather than scientific logic. As infants, we have all made the inference that there is an external physical world. The child's libidinal interests move from its self to external things and other persons. The influence to an external world is primitive and deep-rooted (23, p. 232–233). A kind of anxiety, however, characterizes many philosophers with respect to this inference. Does it arise from some early difficulty in weaning? Schopenhauer once said: "He to whom all men and all things have not at times appeared as mere phantoms and illusions has no capacity for philosophy." (31, p. 97). Schopenhauer was the despised, unwanted child of a mother with sexual ambition. And Russell has recorded the story of his own orphaned, doctrine-tossed childhood. Russell would have us undo the primitive inference of childhood, and regress, as it were, to the stage of narcissism. This is the emotive imperative which lies behind the maxim: "Wherever possible, substitute constructions out of known entities for inferences to unknown entities."

The heart of science consists of inferences to unknown entities. Whether it is an inquiry into the interior of an atom or the internal constitution of a star, the theories that are proposed are inferences to unknown entities. This is the principle which Einstein has enunciated plainly: "The belief in an external world independent of the percipient subject is the foundation of all science" (4, p 66). Science can make inferences to unknown entities, provided that it verifies the consequences of such hypotheses. To this belief, Russell's maxim of constructionism is opposed. Russell, in his attempt to narcissify science, is led into a misleading mathematical analogy. Russell regards the "data of sense" as related to physical entities in much the same way as rational numbers are related to irrational and complex numbers. Within a logical system of mathematics, the rational numbers are the elements from which other types of numbers may be constructed. But the "data of sense" are not the elements of a self-sufficient system. Scientific, inductive inference is ample witness to the fact that sense-data are taken as the terminal events of causal sequences which have their origin in external, often unknown entities. The sun is never exhaustively defined in terms of our sense-data in the sense in which irrational numbers are defined by series of the system's elements. It may give one a sense of psychological reassurance to feel that one's "data of sense" are the building-blocks of the universe, but the causal relations between physical events and our sense-data are not relations of tautological deducibility.

All the security which science can bring comes through the principle of verifiability, of which Occam's Razor is part. The "security" which Russell seeks through the maxim of constructionism is rather pseudophilosophical, a narcisstic-bound rule of system formation, which is guided by neurosis rather than reality.

In the foregoing, I have tried to show that the scientific principle of simplicity does not rest on the assumption that the laws of nature are simple. We choose the simplest hypothesis because to do otherwise would commit us to a path which leads toward unverifiable theory. The scientific principle is not one of aesthetic preference for mathematical elegance. It is founded in the biological necessities of our dealings with the environment. If, on the one hand, the principle of simplicity is independent of any metaphysics of the universe, on the other, it does not bind us to a solipsistic theory of knowledge. Russell's maxim is an emotional distortion of the scientific significance of Occam's Razor. In conclusion, I might add that there is need for a political logician to do for our time what Jeremy Bentham did for his in the "Book of Fallacies." The violation of the principle of simplicity, flight into goropising, has in late years been the principal occupational fallacy of some well-known politicians (14, p. 267).

References

1. Conant, James Bryan, *On Understanding Science,* New Haven, 1947.
2. Demos, Raphael, "Doubts About Empiricism", *Philosophy of Science,* Vol. 14, 1947.
3. Einstein, Albert, *Relativity: The Special and General Theory,* transl. by Robert W. Lawson, New York, 1920.
4. Einstein, Albert, *James Clerk Maxwell: A Commemoration Volume,* Cambridge, 1940.
5. Frank, Philipp, *Einstein: His Life and Times,* transl. by G. Rosen, New York, 1947.
6. Infeld, Leopold, *Quest,* New York, 1941.
7. Jevons, W. Stanley, *The Principles of Science,* Second Ed., London, 1892.
8. Jourdain, P. E. B., "The Economy of Thought," *The Monist,* Vol. XXIX, 1919.
9. Jourdain, P. E. B., "Maupertuis and the Principle of Least Action", *The Monist,* Vol. XXII, 1912.
10. Jourdain, P. E. B., "The Logical Significance of Occam's Razor", *The Monist,* Vol. XXIV, 1914.
11. Lavoisier, Antoine, *Elements of Chemistry,* transl. by R. Kerr, London, 1802.
12. Leibnitz, Gottfried Wilhelm, *New Essays Concerning Human Understanding,* transl. by A. G. Langley, Second Ed., Chicago, 1916.
13. Lenin, V.I., *Materialism and Empirio-Criticism,* transl. by D. Kvitko, New York, 1927.
14. Levenstein, Aaron, "The Demagogue and the Intellectual", *The Antioch Review,* Vol. XIII, 1953.
15. Mach, Ernst, *The Science of Mechanics,* transl. by T. J. McCormack, Chicago, 1893.
16. Mach, Ernst, *History and Root of the Principle of Conservation of Energy,* transl. by P.E.B. Jourdain, Chicago, 1911.
17. Malinowski, Bronislaw, *Sex and Repression in Savage Society,* New York, 1927.
18. McKie, Douglas, *Antoine Lavoisier,* London 1936.
19. Mill, John Stuart, *An Examination of Sir William Hamilton's Philosophy,* Third Ed London, 1867.
20. Newton, Sir Isaac, *Mathematical Principles of Natural Philosophy,* transl. by Andrew Motte, edited by Florian Cajori, Berkeley, 1934.
21. Nordenskiöld, Erik, *The History of Biology,* transl. by L. B. Eyre, New York, 1932.
22. Partington, J. and McKie, Douglas, Historical Studies on the Phlogiston Theory, *Annals of Science,* Vol. II, 1937.
23. Piaget, Jean, *The Language and Thought of the Child,* transl. by Marjorie Warden. New York, 1926.
24. Planck, Max, *Where is Science Going?* transl. by James Murphy, New York, 1932.
25. Russell, Bertrand, *Our Knowledge of the External World,* second ed., New York, 1929.

26. Russell, Bertrand, *Mysticism and Logic,* London, 1921.
27. Russell, Bertrand, "Logical Atomism", *Contemporary British Philosophy,* First Series, edited by J. H. Muirhead, London, G. Allan & Unwin, 1924.
28. Schiller, F. C. S., "Review of Russell's 'Scientific Method in Philosophy' ", *Mind,* Vol. XXIV, 1915.
29. Thorburn, W. M., "The Myth of Occam's Razor", *Mind,* Vol. XXVII, 1918.
30. Tolman, Edward Chace, *Purposive Behavior in Animals and Men,* Berkeley, 1951.
31. Wallace, William, *Life of Arthur Schopenhauer,* London, 1890.
32. Waters, R. H., "The Principle of Least Effort in Learning", *The Journal of General Psychology,* Vol. XVI, 1937.
33. Whitehead, Alfred North, *The Aims of Education,* New York, 1929.

6

The Genetic Fallacy Re-examined

The Divergence Among Empiricists Concerning Psychoanalysis

Genetic considerations are those bearing on the "extra-logical" circumstances under which belief in a given proposition arises. In the view of many philosophers, such considerations are irrelevant in determining the truth or falsehood of any philosophical or scientific proposition. The "genetic fallacy" is taken to be a class name for all arguments that assume that the extralogical circumstances—social, historical, psychological—under which belief in a given proposition may have originated are relevant to deciding its truth or falsehood.

In this essay, I shall inquire as to whether indeed there is such a fallacy as a "genetic fallacy" and whether its invention was indeed meant to obstruct philosophical inquiry. Ever since Henry Sidgwick asserted that to trace "the Origin of the psychical facts which we call Intuitions" has no bearing "on what I have called their Validity,"[1] the "genetic fallacy" has acquired the status of a philosophical axiom.

Genetic considerations, however, thrust themselves upon every student of contemporary philosophical ideas. Take, for instance, the divided opinion among contemporary positivists and scientific philosophers concerning the scientific validity of Freud's psychoanalytical theories. Virtually all the Austrian and German logical positivists and scientific philosophers have been sympathetic to psychoanalytical work. Hans Reichenbach, for instance, "had a serious interest in psychoanalytic theory" and became an honorary member of a psychoanalytic institute;[2] Richard von Mises, the distinguished theorist of the foundations of probability, declared: "The extremely vigorous objections raised in many circles to the acceptance of the psychoanalytic theory are in large part of a nonlogical nature . . . There can be no doubt, however, about factual agreement with many observations, and hence about a certain practical usefulness of the theory."[3]

American empiricist philosophers, on the other hand, especially those in New York City, though sharing some of the cultural background of their European colleagues, were hostile toward psychoanalytical ideas. How did it happen then that philosophers adhering to a common standpoint differed so sharply when it came to questions of mental phenomena, conscious and unconscious? Had the Europeans been affected by a literary-cultural milieu of sexual emancipation that deflected their scientific reasoning? Or had the New York students been determined in their late adolescence by their teacher, Morris R. Cohen, who all his life disliked vehemently any intrusion of psychological considerations, above all, psychoanalytical ones? I addressed these queries to Sidney Hook after reading how surprised he had been in 1928–1929 upon first discovering Reichenbach's psychoanalytical interests. Professor Hook had narrated:

> I recall also a matter of mutual shock at the time. He [Reichenbach] was startled when I asserted that the theory of psychoanalysis, especially in its Freudian formulation, was unscientific, and I was startled when he insisted that it was. It seemed clear to me that on his own criteria of scientific verification that psychoanalysis was no more scientific than Christian science. In passing, I should note the curious fact that all the logical empiricists or positivists I have known were quite vehement in defending the scientific validity of Freud's basic views—something which in my obtuseness I could never square with their professed philosophy of science.[4]

Doubtless, if both groups of logical empiricists, the Central European and the New York, were aiming to apply the same principles of scientific logic to their evaluation of psychoanalysis, then genetic considerations would have to be invoked to explain how at least one of them was constrained by emotional, perhaps unconscious, factors from adhering rigorously to those scientific principles. If so, which one?

To my question as to how this divergence in philosophy could be explained, Professor Hook kindly responded:

> I do not know how to explain the partiality of the logical positivists toward psychoanalysts—and, as I wrote you, not only of them but of people like Kurt Lewin and Lazarsfeld who agreed with all of Ernest Nagel's methodological strictures but nonetheless said "but it's true."
>
> I don't recall whether it was you or Meyer Schapiro who explained my position and Ernest's as due to the influences of Morris Cohen, but we broke away from

> Cohen on so many points it is hard to explain why we remained critical. Perhaps it was the presence of too much shoddy in the popularizers. I really read and *enjoyed* Freud but couldn't take *Civilization and Its Discontents* seriously or accept *Moses and Monotheism* . . . I greatly admired A. A. Goldenweiser, whose criticism of Freud's excursions in anthropology were more persuasive to me than his defence of his concepts in Personal psychology . . .[5]

Now it is noteworthy that, quite apart from its theoretical explanation, there does then exist a verified empirical correlation, or empirical law, that the Viennese and Berlin empiricists were philo-psychoanalytical while the New York circle was anti-psychoanalytical. Genetic correlations of this kind stand quite apart from their theoretical explanation, but only the latter will help enlighten us as to at least which of the two (or possibly both, or neither) is irrationally motivated. If, however, on presumably independent grounds, we already knew the truth of psychoanalytic hypotheses and were then to explain the irrational opposition of the anti-psychoanalysts, "genetic considerations" might reinforce the validity of the psychoanalytic hypotheses. On the other hand, if such independent evidence for the validity of psychoanalytical ideas is regarded as unconvincing, we would appear to be begging the question in assuming their validity in determining which cultural circle of philosophers was contra-rational.

Actually, however, it is precisely through genetic investigations of this kind that the truth or falsity of psychoanalytical theories is established. We undertake to explain the divergence between the two rival schools of thought concerning the truth or falsehood of basic psychoanalytic hypotheses. If we assume the latter's truth, we shall then have to advance subsidiary hypotheses as to what elements in their social environment exerted an emotionalizing impact on the New York school. Were they shaped by the transmitted American Puritan culture, or the New York lower-class, Jewish, anti-sexual immigrant culture, or a City College philosophical milieu dominated by a would-be father figure who was himself tormented by emotional resistances,[6] or the influence of their next father figure, John Dewey, who retained a Vermonter's distaste for sexual motives, or by the impact of Marxism after the First World War that placed economic motivations higher in the historical hierarchy than the more bourgeois sexual ones.

We may, on the contrary, begin by assuming that the basic psychoanalytic theories are false. In that case, we shall invent all sorts of subsidiary hypotheses to try to explain why the Central European empiricists were misled into accepting them. Experiencing the full impact of defeat and disillusionment after the First World War, they were attracted perhaps toward a psychological theory that might unmask the bogus ideals of militaristic tradition, dedication to the emperor, and battlefield romanticism, as stemming not from some higher metaphysical source but from phenomena as lowly as toilet-training or a domineering father. Perhaps, moreover, as children of the Viennese middle and upper classes, especially its Jewish sector, growing up in an unwonted economic security but with their marriages delayed because of professional exigencies, and their traditional religious restraints obsolete, they gave themselves to uninhibited sexual activities; their sexual overemphasis, in theory as well as practice, was then class-biased and historically strained.

Such genetic considerations would themselves help decide which rival alternative concerning the Central European and New York empiricists was true. If the subsidiary hypotheses concerning the New York school were validated, then weight would be lent to the notion that their scientific judgment had been perturbed or warped by psychoanalytical fixations. Or, if the subsidiary hypotheses concerning the Central European school were confirmed, then we would accept the proposition that class-originated aberrations rendered them susceptible to an acceptance of psychoanalytical ideas contravening their own scientific criteria.

From this standpoint, genetic analysis is comparable to the astronomer's concern with the purity and resolving power of his telescopic lenses. An unusual photographic image may trouble him: Is it the image of a distant star or the consequence of a speck of dust on the great lens or mirror? Is the bit of color on the photograph a record of a "red-shift" in the light from a distant star or possibly a chromatic aberration grounded in the physical structure of the telescopic lens itself? No astronomer would reject such considerations as examples of an "optical fallacy"; why then should a philosopher invoke a "genetic fallacy" to exclude analogous considerations concerning philosophical theses? There is then no "genetic fallacy" as a class of fallacies; rather there are genetic considerations that are well-founded, strong genetic evidence, just as there are cases of purported genetic considerations that are weak, insubstantial.

Leibniz, in his philosophic debate with John Locke over the existence of "innate ideas," thus refused to invoke a "genetic fallacy." Locke had argued that the purported innate ideas were rather the effects of a parent's or nursemaid's admonitions in our early childhood.[7] Leibniz rebutted Locke by saying that the latter hadn't carried his genetic analysis far enough. He agreed that Locke had done well in showing how "under the name of innate principles one often maintains his prejudices," that "under the specious pretext of innate ideas" people avoid "investigating . . . the sources . . . and the certainty of this knowledge. In that I am entirely agreed with him," wrote Leibniz, but he added, "I go even farther." He argued that the human mind is not a blank tablet, that if one probes more deeply one finds ideas that are, as it were, "engraved" in us "by a species of instinct," ideas indeed "of which we are not always actually conscious," truths that "are for us as innate as inclinations, dispositions . . ."[8] Of course, from this Leibnizian standpoint, much further investigation was needed to establish what unconscious ideas were actually "engraved" in the human mind. Leibniz nonetheless calls for a genetic analysis that reaches indeed to origins in a way that Locke's reportage does not. For the rationalist Leibniz, there is thus no "genetic fallacy," only genetic considerations that are relevant, and those that might be irrelevant, that is, perhaps not truly genetic.

Genetic Privilege and Genetic Handicap

An "antinomy of genetic method" (as I might call it) arises, however, in practice to haunt both Leibnizian intuitionist and Lockean empiricist, both Marxist historical materialist and "bourgeois," indeed all genetic analysts. In actual use, the genetic facts often seem insufficient to determine whether the given thinker or school of thinkers was "genetically handicapped," or, on the contrary, "genetically privileged." A genetic privilege exists for a person, with respect to some domain of truths, when his character, configuration of emotions, social circumstances, and historic experiences are such as to place him in a favorable situation for perceiving, receiving, or imagining those truths that will not be thus experienced by people not so constituted. A genetic handicap, on the contrary, exists when a person's character, emotions, and social and historic background are such as to render him impervious to, or desirous of repressing, the existence of certain phenomena that he otherwise might have perceived, received, or imagined.

Genetic analysts, from Marxist to Bloomsburyan, tend arbitrarily to regard their favored class or circle as privileged for access to knowledge, as not being handicapped. The homosexual circle, for instance, that constituted the Cambridge Apostles gave great support to the formation of the intuitionist doctrine of G. E. Moore, to his notion that all utilitarian or scientific definitions of "good" were examples of a "naturalistic fallacy." Lytton Strachey "saw in *Principia Ethica* a justification for homosexuality," and Moore "was not inclined to correct Strachey."[9] G. Lowes Dickinson, a sensitive Platonist and an idealistic advocate for the League of Nations, had together with John McTaggart transmitted to Moore "the romantic idea of Apostolic homosexuality";[10] McTaggart, a sharer of homosexual and philosophical interests with Dickinson, contributed as many as eighteen papers to the Society, one of which, "Violets or Orange Blossoms?" was an explicit "defence of homosexual love."[11] Not only did sexuality by 1894 become "the Apostles' chief interest," but Moore's paper "Achilles or Patroclus?" suggested the underlying homosexual tone: "It was obligatory to make the humorous assumption that all sexual relations were homosexual ones, so that even heterosexual love had to be treated as only a special case of the Higher Sodomy."[12] Moore even ventured the cosmological speculation that the active-passive structure of homosexual relations was connected with the structure of the universe. In a more terrestrial vein, he asserted that copulation was a disagreeable activity, a trial and a trouble, so that a man and a woman who loved one another engaged in sexual relations only because it was necessary for the begetting of children.[13] Moore's candor in talking about his own sexual shame and pruriency evidently had a profound effect on his fellow Apostles.[14] Ethical intuitionism, ethical anti-naturalism, became the ideology of this homosexual circle.

Does its homosexual origin, however, discredit the philosophy of *Principia Ethica?* Does its origin bear upon its validity? Moreover, might one indeed affirm that, far from having handicapped their philosophical vision, the homosexuality of the Apostles made them more perceptive of underlying realities? The homosexual experience, according to this argument, makes its practitioners more spiritual, liberated from materialistic constraints in much the same way that the mathematical imagination liberates one to conceive pure possibilities. The homosexual thinker, from Plato and Socrates on, it is argued, has been genetically privileged, a metaphysical aristocrat, with an access to a pure idea that is not vouchsafed to the vulgarian, the unspiritual heterosexual. By contrast, however, to the claim for homosexual "epistemological privilege," the famed English geneticist and then Marxist J. B. S. Haldane argued that, apart from slavery, it was sodomy that killed Greek science: "Plato and other idealists were extremely tolerant of this latter aberration . . . Today it is common in literary circles, and rare among scientists. For it is a second-rate imitation, like the substitution of words for things, with which it is associated today as it was in Athens."[15] Scientific biologists who have written on questions of ethics have evidently found G. E. Moore's argument on the "naturalistic fallacy" a feeble one. Such scientists as C. H. Waddington, Haldane, and Julian Huxley, who continued to probe the evolutionary and psychological origins of "good," have regarded their work as contributing to a scientific analysis of "good."[16]

Some philosophers likewise have found insubstantial Moore's argument that "good" is unanalyzable;[17] to psychological naturalists, indeed, the doctrine of the "naturalistic fallacy" seems rather to be the functional formula of a resistance mechanism to scientific, reductive analysis. From the psychoanalytical standpoint, the word "good" carries all the overtones of our early conditioning by our parents; to do what is "good" is to enjoy the unconscious reassurance of parental love; to do contrary to "good" ignites feelings of guilt, the threat of withdrawn parental affection and its introjected consequence, self-hatred.[18] From the genetic standpoint, it was bound to seem clear to Moore that "pleasure is good" is not at all equivalent to "pleasure is pleasure," because the word "good" signalizes as well the associations superadded by our moral education that endure recessively in our unconscious. Presumably, these unconscious meanings can be brought to consciousness by psychogenetic analysis. What then remains of the "naturalistic fallacy"? Evidently it expressed Moore's lifelong repugnance toward genetic analysis. He had objected very early to Santayana's ethical naturalism, because Santayana's arguments were so largely founded on psychological and biological facts. "The primitive origins of the various human activities," wrote Moore, have no "direct bearing on . . . the enumeration of goods."[19] Then, many years later, Moore told the eminent genetic psychologist Jean Piaget that the latter's studies were "of no interest at all, in substance, because the philosopher is concerned with true ideas, while the psychologist feels a sort of vicious and incomprehensible attraction for the study of false ideas."[20] This is like saying that the laws of optics are of interest to an astronomer only to inquire into the aberrations of his telescope, but not to explain its effective range and resolving power. Obviously, genetic considerations explain the possibility of knowledge, as Bacon saw, even as they do that of ignorance. The discoveries of great scientific truths—Darwin's theory of natural selection, Einstein's theory of relativity—were also based on genetic conditions, psychological and social, which provided their perhaps indispensable matrix. Truth has no immaculate conception, as most every historian of science knows.

Generally speaking, when a philosopher makes a basic change in standpoint, it is highly likely that that change is grounded in some basic emotional alteration. Bertrand Russell, for instance, after having been an intense intuitionist of objective ethical absolutes found himself persuaded, he writes, to relinquish such absolutes in part by Santayana's arguments on behalf of ethical relativism.[21] Now Santayana's arguments were the ones well known since Locke's time to the English-speaking world. Why did these arguments have a valence for Russell at one time and not earlier? The genetic analyst cannot but observe that Russell's years of intuitionism coincided approximately with his "years of tense self-denial," of sexual abstinence. Santayana's criticisms of Russell's "hypostatic ethics" were published in August 1911.[22] Russell's persuasion to ethical relativism was evidently prepared when, in March 1911, his years of sexual repression ended. To which one still queries: Was sexual abstinence perhaps a genetic privilege that enabled its possessor to intuit higher, spiritual truths?

When philosophies are discarded as obsolete, a kind of unwritten psychogenetic and sociogenetic analysis has usually shown the attraction of those philosophies to have been based on emotions that have in the meanwhile dissolved. John Dewey once observed: "We do not solve them [philosophic questions]; we get over them. Old questions are solved by disappearing . . ."[23] Genetic analysis is precisely the intellectual articulation of this process of "getting over" a question. Many American intellectuals, for example, after the Second World War, abandoned historical materialism for existentialism; with the destruction of the Nazi incubus and the resurgence of the American democracy, they were disenthralled from anxieties that had sombered their thought. Otherwise, virtually no novel argument or evidence had been presented.

Does genetic analysis then terminate in a genetic indeterminacy, that is, an inability to infer from any genetic analysis whether the conditions it analyzes constitute a genetic privilege or a genetic handicap? In that case, a form of the "genetic fallacy" could be affirmed, namely, that we can infer nothing concerning the validity of a proposition by knowing its origins so long as we have no way of knowing independently whether the latter brought privilege or handicap. In which case, the truth or falsehood of any given proposition would have been decided antecedently quite apart from any genetic analysis.

This negative inference concerning genetic analysis, however, is unwarranted. What it overlooks is that genetic analysis is the most searching instrument for studying the epistemic bases for the truth claims of basic philosophical propositions. Descartes's philosophy, for instance, rested on a claim that certain innate ideas existed: the truth of this claim could be validated only by its being able to withstand the kinds of genetic criticisms that Locke made. The epistemic claim of Sidgwick and Moore that an ethical intuition is part of our mental equipment would similarly have to withstand the genetic analysis that aimed to dissolve that "intuition," by making the intuitionist aware that it blended unconscious residues with a resistance-formation. Kant's claim that the Euclidean space and time were a priori forms of intuition was thus shaken by Ernst Mach's inquiry into the origins of the ideas of absolute space and time. The notion that the idea of God is inherent in our consciousness would have to stand unfalsified by Freud's genetic evidence that the idea of God is a projection of the child's fear of its father. The notion of a "genetic fallacy' would simply exclude arbitrarily any empirical testing of the sources of knowledge claims.

Whether a given social or psychological base provides a genetic privilege or a genetic handicap is determined in accordance with the customary scientific criteria. The claim that homosexuality confers an epistemic privilege is sorely contravened by the evidence that homosexuality is virtually unknown among the outstanding contemporary physicists.[24] The ideas of an intuited absolute space and time which were dominant in the homosexual culture-circle of the Cambridge Apostles, and the verbal arguments with which they were demonstrated, scarcely proved as heuristic as those of Einstein and his comrades in the genetically minded, Machian circle of Zurich-Berne students.

Auxiliary hypotheses can, however, be invoked in such disputes. The homosexualists might claim that in the spiritual sciences, involving ethical and aesthetic insight, their genetic status conferred an epistemic privilege.[25] They might claim that sexual energies diverted from female objects rendered one more sensitive to spiritual entities, which coarsened, vulgarized personalities could never apprehend. The discussion and weighing of epistemic claims would thus continue. However, a claim that can be sustained only by multiplying auxiliary hypotheses would run dead against the principle of parsimony, Occam's Law.[26]

Under what logical conditions, then, would genetic analysis invalidate a claim to a metaphysical intuition, that is, a direct, immediate experience of a transempirical reality? The controlling principle is indeed that of Occam's Law: thus, if the genetic analysis establishes a causal chain all members of which are natural occurrences and that terminates, in accordance with causal laws, in an event characterized by all the introspective traits of the purported intuition, then the claim of that intuition to have been generated by some transempirical interrelation, as by some influx of divine energy, or some spiritual interaction, would have to be judged as otiose, as a superfluous, unnecessary hypothesis, to be ruled as contravening Occam's Law.[27] Genetic analysis, for instance, as John Stuart Mill practiced it, thus tried to undermine the intuitionist epistemic claim by using what he called "Psychological Theory." Mill, in effect, defined the role of genetic analysis as showing that such concepts as "intuition" were unnecessary because a reductive psychogenetic explanation could be given of every trait of the alleged intuition;[28] to assume that such a faculty of intuition existed was as scientifically redundant as to hold today to the hypothesis of the physical ether.

The Incompleteness of Genetic Analyses

The aim of genetic analysis in the history of empiricism has thus been founded on a valid principle of scientific logic; if one can take the propositions concerning such metaphysical ideas as God, intuition, freedom, natural right, and show how their origins, their formulation, and the belief in their existence are explicable on the basis of observable psychological facts; if, in other words, a causal line can be established in which all that is asserted concerning these alleged metaphysical entities can be shown to be the consequence of psychological data and laws, with no supervenient fact concerning the character and existence of these metaphysical entities remaining that cannot be explained through such causal lines, then that entity or belief is logically unnecessary. Thus, Arthur O. Lovejoy, America's foremost historian of ideas, though himself devoted to a purely formal account of the filiation of ideas, declared that the genetic analysis of the "immediate certitude of religious experience" had undermined its claims: "The destructive effect of this criticism of the ordinary argument from subjective religious experience to objective theological entities cannot, it seems to me, be denied by a serious and honest reasoner...." How many persons who have believed themselves to possess religious "instructions, even visions" have subsequently come to doubt, noted Hastings Rashdall, whether they were "anything but the outcome of subjective wishes or a disordered brain."[29]

The fact of the matter, however, is that no genetic analysis has really fulfilled the strict scientific canons to an extent that would justify affirming that "God" and "intuition," for instance, can be eliminated on Occamite grounds. In other words, empiricists have never met Leibniz's challenge to Locke. Freud never actually showed that the predisposition to believe in God was simply the outcome of parental scares and fears; the recrudescence of belief in God among Soviet intellectuals, despite all the contrary forces of societal and parental conditioning, might even suggest the existence of an underlying intuition.

Genetic analysis, at the hands of even the most remarkable of its practitioners, tends to claim a completeness that the actual observational facts are too meager to support. Jean Piaget, for instance, declared that "the theory of the filial origin of the religious sense seems to us singularly convincing...";[30] according to his account, moral commands seem to a child to be endowed with a transcendental origin only because very young children have difficulty in distinguishing between what they have invented and what is imposed upon them from without by adults; a superpersonal transcendence thus apparently absorbs what were originally adult-ordained mandates. How then, one asks, do children advance upon the moralities inculcated in them by their parents and their societies? By what criterion does criticism of the received parental morality become possible? The geneticist responds: The very conflict among rival adult moralities helps to develop the young person's autonomous moral conscience. But we ask further: Has such a causally determinist explanation of the genesis of the moral conscience ever been verified? Has it been shown that its developing autonomy is not guided by an implanted intuition, possibly linked to a biogenetic base? Has the psychogenetic causal line been traced in its sequential members so as to enable us to exclude altogether any suggestion that a transcendental component has been operative in the human ethical evolution? Piaget has given a fascinating account of his own "philosophical crises of adolescence"—his devout Protestant mother; their minister, with his weak proofs for God's existence; his father, a scholarly, unchurched historian; his godfather, an enthusiast for Bergsonian creative evolution; and himself, devoted to collecting mollusks.[31] Was Piaget's own intellectual route to his positivist inference the outcome of an overdetermination in emotional rebellion that led him to conclusions that far exceeded the bounds of his observational support?

The notion that, independently of sociogenetic circumstances, a marked propensity to believe in the existence of a divine being is implanted in men still remains consistent with the evidence. Perhaps such a theistic propensity has a biogenetic basis; possibly an intuitive ingredient has acted autonomously.

Genetic analysis, as pursued by its proponents, even Mill and Marx, has indeed been woefully lacking in rigor. Mill, for instance, in justifying his concern with inductive logic, said that he wanted to undermine the belief in intuited political principles, because the latter served the purposes of the conservative party. Whether Mill's *Logic* showed that such principles can altogether be dropped as unnecessary is a question beyond our purview.[32] What is clear, however, is that intuitionism has probably animated liberal political creeds as frequently as it has conservative ones. Jefferson, writing in the Declaration of Independence of truths held to be "self-evident," was appealing to the laws of reason, instilled by God and confirmed in the "common sense" of mankind.[33] Socialists, such as R. H. Tawney, assailing the "acquisitive society," grounded themselves on purported ultimate religious intuitions of human equality and fellowship; and more among the founders of the British Labor Party were perhaps intuitionists than were reductive, genetic utilitarians.[34] Bernard Shaw, for instance, was more of an evolutionary institutionist than a utilitarian.

Stratogenetic analysis, indeed—that is, one founded on the assumption that the class, social origins, or affiliations of thinkers are the primary extralogical factor in determining their philosophic beliefs—is apt to eventuate in far-fetched, caricatural genetic accounts. Thorstein Veblen, for instance, argued that the positivist philosophy was especially congenial to members of the French commercial middle-class who were divorced from physical contact with mechanical productive processes.[35] All of which sounds bizarre when we recall that the greatest proponent of positivism at the end of the nineteenth century, Ernst Mach, was an experimentalist of a high order, who helped inspire the formation in Einstein's mind of the theory of relativity. John Dewey argued that the Greeks' theory of knowledge reflected the separation of social classes; according to Dewey, the "disesteem entertained for the manual worker" expressed itself in a corresponding low esteem for the "prosaic matter of fact knowledge" and the corollary that Pure Ideas, unsullied by empirical impurities, and given only to the aristocratic nonworker, constituted Truth. It followed, from Dewey's point of view, that the genesis of modern science was founded on the growth of a democratic polity that brought "the substitution of a democracy of individual facts equal in rank for the feudal system of an ordered gradation of general classes of unequal rank."[36] The historical facts, however, contravene Dewey's stratogenetic account; actually, British aristocrats, together with King Charles II, were the social mainstay in founding the Royal Society in the seventeenth century; furthermore, the chief centers for scientific research in the first part of the twentieth century were part of the absolutist, undemocratic German Empire, whereas the American democratic states were for a long time quite averse to helping encourage projects of scientific research. Perhaps a genetic analysis of Dewey's rendition of intellectual history would indicate that his own democratic, emotive a priori imposed a pattern that did violence to the historic facts;[37] the democratic outlook might constitute with respect to some questions a genetic handicap rather than the privilege that Dewey took it to be.

Of all the philosophers in the modern era, Leibniz, curiously, was the one most concerned with industrial processes; nevertheless, he, like Spinoza the lens grinder, also believed in pure ideas and the givenness to rational intuition. We might note parenthetically, however, that Leibniz was brought to grief by his industrial interests, Wishing to bring the labor of workingmen under the rational direction of philosophers, he contrived a scheme in 1679 that aimed to enable the Harz state silver-mines to use wind as well as water for keeping the pumps operating continuously. The windmill, however, failed; much worse, the miners threatened to strike unless Leibniz was removed from the premises; he was, they said, "a dangerous man with whom it bodes ill to have any dealings." This was the first recorded rift in history between philosophy and the proletariat, between the intellectual and the working classes. For several years, Leibniz persisted in his project, though his "cost overrun" by 1683 came to more than sevenfold. Finally, the Duke of Hanover, troubled by the miners' unrest, terminated the experiment; Leibniz was deeply disheartened.[38] But his philosophical idealism and intuitionism was thoroughly consonant with his desire to bring science to bear on industrial life.

The Modes of Genetic Analysis

Psychogenetic analysis has been important in the history of philosophic thought; stratogenetic analysis, on the other hand, has had little significance. As far as political ideologies are concerned, however, the stratogenetic determination has been the more primary variable. The various modes of genetic analysis that have been used, and not infrequently confused, need be distinguished; for the "genetic fallacy" is often directed against some irrelevant mode of genetic analysis. The psychogenetic approach itself consists of several varieties. First, there is what we might call the "protogenetic" kind that is concerned with the influence of family and upbringing on one's early years, of the sort that John Locke emphasized. Underlying that level, there is the biogenetic substructure; if diversities in temperament do have a basis in the variant structures of genes, then the character-types underlying the rival philosophies would be denominated as "biogenetically founded." When William James classified philosophies as "tender-minded" and "tough-minded," he was delineating a spectrum of philosophies that might well have its source in biogenetic variations. Such a spectrum might repeat itself throughout the range of different social structures. Under different social systems—capitalist, communist, feudal, primitive—the spectrum of biogenetic characters and their corresponding philosophic propensities would tend to reappear.

Superimposed are the impacts on both personal character and the social climate that are exerted by the events in one's historical placetime. The steady progress of Britain during the nineteenth century, for instance, evidently imparted to Herbert Spencer's thought an optimistic overtone that was not altogether in keeping with his nervous propensity to depression.[39] The events of the First World War and the recurrence of pogroms, by contrast, may have elicited in Sigmund

Freud his theory of the death wish, projecting itself far beyond what either the evidence or his own cultural background of Jewish optimism would have endorsed. We might call this mode of genetic analysis "historico-genetic."

Successive generations moreover tend to revolt against the dominant philosophies of their predecessors. Rough cyclical rhythms are defined in which materialists rebel against idealistic forebears, and vice versa, in which disbelievers rebel against theists, and vice versa. James conjectured that each philosophy expressed some ingredients in our underlying emotional character, and repressed others. The waves in philosophic ideas, their alternation, might then be due to the returns of the successively repressed and incompatible elements of longing.[40] I would call this mode of genetic analysis "generational-genetic."

If there are generational waves underlying the cyclical patterns of philosophic ideas, it is also true that the progressive evolution of technology has given new analogies, fresh models, to those successive generations. Veblen especially was a practitioner of what we might call "technogenetic analysis."[41] Such inventions as the bow and arrow, the weighing scale, the lyre, the wheel, the mirror, the clock have provided a new fulcrum around which philosophic alternatives could pose themselves; cosmologists could ponder whether the world with its recurrent movements was propelled by a wheel of fate, even as they later argued whether the universal clock presupposed a clock-maker, or even a deity who synchronized the mental and physical chronometers. Leibniz and Robert Boyle both responded to the technological inspiration of the clock. Veblen carried technogenetic analysis to the point of unlikelihood; he argued, for instance, that the workingmen in machine factories were shaped in their thinking to an impersonal, causal standpoint, and consequently led to revolt against a system of business enterprise that they regarded as committed to personal, teleological categories. The ideologies of, on the one hand, trade unionists seeking higher wages, security, and better conditions on the job, and, on the other, of the revolutionary unionists, articulating the feelings of the footloose, restless, usually unattached single men, wandering through mines, prairies, fields, and grouping themselves loosely in the Industrial Workers of the World, had nothing to do with Veblen's presumed causal psychology of the machine worker.[42] Veblen's chief disciple, Robert Hoxie, struggled to fit the facts to Veblen's technogenetic analysis; deciding finally that it had all been "bunk," and depressed with his "deconversion from Veblen," he committed suicide.[43]

One could multiply the varieties of genetic analysis, some altogether spurious, that have multiplied.[44] The Nazi claims for a racial-genetic repudiation of the theory of relativity as "Jewish physics" provide the ugliest chapter in the degradation of scientific categories; their counterparts have been the Soviet theses that held the theory of relativity was genetically handicapped by Einstein's bourgeois origin (today it might be "Zionist') and that Mendel's genetics was the distorted intellectual emanation of a Catholic monk.

Every school of philosophy, indeed, tends to be friendly to that mode of genetic analysis which reinforces what it regards to be the proper method of philosophy. Thus, the vogue of linguistic philosophers has been associated with a linguo-genetic thesis, namely, that the structure of a given language tends to determine the basic character of its users' philosophy. The linguo-genetic approach merges partially with the psychogenetic; for, as Franz Boas long ago noted, "linguistic notions never rise into the consciousness of primitive man, and the basic categories are unconsciously formed."[45] The logical-positivist school, however, holding that the proper office of philosophy was the explication of the methods of science, inclined toward a scientogenetic account of philosophic ideas; thus, the dominant scientific advance of a given era, whether Newton's physics or Darwin's theory of natural selection, or Einstein's theory of relativity, or Heisenberg's principle of indeterminacy, was held to have originated a corresponding set of novel philosophies. Religious myths of creation were, from their standpoint, the proto-scientific theories of primitive man which, for reasons of cultural lag, still survive among the backward or unintelligent parts of the population.

The linguo-genetic theory, however, has little evidence in its support. Bertrand Russell used to argue that the problems of being arose because Latin (or Greek) had the verb "to be';[46] yet the same problems preoccupied the medieval Jewish philosophers, though the Hebrew language in its classic form lacked the verb "to be." And the spectrum of diverse philosophical standpoints seems a near-invariant; how much would the distribution among British philosophers of empiricists, idealists, realists, materialists, and commonsensists in 1910 have differed from what it was two hundred years previously, in 1710? Moreover, every scientific advance is shortly accommodated to the diverse philosophical standpoints. William James, Henri Bergson, and C. Lloyd Morgan assimilated the discoveries concerning biological evolution to their respective theistic standpoints; absolute idealists recognized a corresponding insight of their own eternalistic nontemporalism in Einstein's view that the distinction between past and future grows unreal with the advance of physical theory.

Of all the modes of genetic analysis, the psychogenetic, beginning with Locke, has been the most effective critical instrument in the history of empiricism. And the history of the idea of a "genetic fallacy" itself deserves to be studied psychogenetically. The "genetic fallacy" seems to have been invented as a notion for repressing all one's own inner doubts concerning the origins of one's intuitions or one's a priori affirmations concerning ethical, metaphysical, and religious notions; the notion of a "genetic fallacy" serves as a device for maintaining intact the repression of genetic facts in one's unconscious; lastly, it authorizes excluding from the philosophic community whosoever would import genetic considerations into philosophic discussion.[47]

Within the social sciences, sociogenetic and psychogenetic considerations have contributed in recent years to the most basic critique of the Marxist social doctrines. The genetic analysis of the aims of intellectuals was first opened by such writers as Waclaw Machajski and Robert Michels, who noted that a latent motive of the Marxist critique was the desire of the intellectuals to emerge as the ruling class or elite of society. John Stuart Mill had once observed that when he analyzed the social motives underlying the schemes of revolutionists, he invariably found that they housed a project of "liberticide."[48] Schumpeter's writings on the anti-capitalist mentality of the intellectual class further illumined their unconscious aims, helping explain the appeal to intellectuals of Lenin's call in *What Is to Be Done?* in disregard of its possible regressive social consequences. Marxist ideologists have shown the same hostility to genetic analysis when it is applied to themselves as the intuitionists have shown in philosophy; Marxists themselves do not hesitate to use a crude stratogenetic analysis against "bourgeois" thinkers; they experience an anxiety, however, when such considerations are directed against themselves, because they wish to preserve the Marxist unconscious intact, to maintain the "façade" that they speak with the genetic privilege of the proletariat when theirs perhaps frequently is the genetic handicap of the semi-intellectuals, afflicted by jealousy and a compensatory drive for power, or of "high intellectuals," yearning to be philosopher-kings.

Nonetheless, even the most elevated intuitionists have similarly not been unready to avail themselves of genetic considerations to discredit, or ridicule, theories they disliked. John Maynard Keynes, for instance, derided what we might call the "Jevonian anxiety";[49] its exemplar, the economist William Stanley Jevons, feared (as Keynes narrated) that Britain's coal reserves might within a few generations be depleted. Jevons's melancholy conclusions were influenced, Keynes wrote, "by a psychological trait, unusually strong in him, which many people share, a certain hoarding instinct, a readiness to be alarmed and excited by the idea of the exhaustion of the resources." Jevons indeed did so much apprehend an approaching scarcity of paper that he accumulated large reserves of writing paper; a half-century later his children were still using them.

The "Jevonian anxiety," however, it transpires, conferred a genetic privilege on its bearer, not a genetic handicap. Jevons, struggling with poverty, and observing at first hand the brutalities of the gold mining rush in Australia, had grasped the tenuousness of the equilibrium between the human species and its available natural resources.[50] Perhaps the homosexual outlook of the Cambridge Apostles tended in their economic thinking to repress the crude reality of material resources even as in their metaphysical thinking they deprecated naturalistic realities.

In short, genetic analysis opens a path to all sorts of hypotheses for investigation, vertification, contravention. It becomes a matter for empirical investigation to determine whether a particular standpoint constitutes a genetic handicap or a genetic privilege. As applied in philosophical arguments, however, genetic analysis is more likely to terminate in decisions "genetically indeterminate" than is the case with regard to economic or physical problems. Quite apart from their diversities in psychological traits, Jevons and Keynes would have tested their rival hypotheses in terms of their contrary, approximate predictions concerning the depletion of Britain's coal reserves. Even such approximate predictions are not available to philosophers, who may dispute, for instance, with no available crucial test, as to whether a given set of causal laws and conditions are indeed sufficient for explaining the advent of an intuition. In the social sciences, however, the genetic analyses of Marxist ideologies have made a significant contribution; by showing how intellectuals project their own ambitions, language, and even neuroses upon their theories of "historical mission" and "the proletariat," they have tended indeed to undermine the claims to "ideological truth." The intrinsically indeterminate character of philosophical questions, on the other hand, reflects itself in a corresponding indeterminacy with respect to determinations of genetic privilege or handicap.

Curiously, moreover, genetic analysts themselves seem to recognize exceptional processes in which the person "transcends" his social origins and influences. Are there such epigenetic intellectual events that would indeed elude all genetic explanation?[51] Karl Marx wrote of intellectuals ("ideologists"), who, having grasped the direction of history, thereupon decided to renounce their bourgeois background and to commit themselves to the proletarian cause. If this account, presumably a projection of Marx's own personal history, were accurate, then social acts have occurred that "transcended" the laws of historical materialism. Likewise, the English factory inspectors, though presumably agents of the British capitalist system, had, according to Marx, a vocation for truth that was higher than their economic interest. Their Blue Books were a mine of facts for Marx's *Capital*. And Freud too named Popper-Lynkeus as one whose pure character transcended the trammels of neurosis. Genetic analysts themselves thus seem to wish for exceptions to their deterministic genetic causal lines.

Lastly, genetic analysis has sometimes been likened to the cross-examining of a witness in court, where the aim, for instance, is to discredit that witness as a habitual liar, or as a previously convicted perjurer. The genetic analyst, however, is only superficially like the cross-examining lawyer. For the genetic methods are, above all, concerned with establishing *unconscious* influences and *unconscious* determinants of ideas; the suborned witness, on the contrary, is usually a conscious liar. Nor is the genetic analyst, if scientific in his calling, a polysyllabic enlarger of ad hominem argument. The user of an ad hominem argument appeals to a popular prejudice; he does not investigate whether the given genetic conditions constitute a privilege or a handicap; he is not concerned with tracing genetic causal lines. Hitler hated the theory of relativity as begotten by a Jew; Einstein, by contrast, was prepared humorously to accept his theory as "Jewish physics" but would then have inquired as to what elements in the Jewish character or its circumstances had made it sympathetic to the notion of rational, invariant principles.[52]

The probative significance of genetic analysis thus varies greatly as one moves from philosophy to the social sciences and, lastly, to the natural and physical ones. In the social sciences, the genetic hypotheses concerning the rival schools are tested by independent evidence and become part of the observational consequences relevant to the accepting or rejecting of, for example, psychoanalytical ideas. In such cases, the origin of the ideas is relevant to the judgment of their validity, and the "genetic fallacy" is little more than an inflated resistance mechanism. When it comes to metaphysical questions, however, such as that concerning a Jamesian divine inspiration, or determinism and free will, genetic analysis, to be decisive, would have to approximate to the limit of possible explanation; then it might judge whether any lacunae in the causal lines still remained through which a divine ingredient, for instance, or a free decision might operate. Genetic analysis of metaphysical ideas therefore tends to culminate in a domain of indeterminacy, for it cannot usually claim such explanatory completeness. The physical sciences, happily, work with hypotheses which, through their predictive consequences, are usually contravened or verified; herein, the role of genetic considerations approaches a minimum. The sociology of scientific ideas, however, has meanwhile emerged as a science concerned with the extralogical and, indeed, nonlogical genetic factors in their discovery. Einstein's emotional adherence to determinism, in a Spinozist form, had much to do with his reaction to German anti-Semitism, but his scientific efforts at a field theory would usually not have involved a reference to that premise. If Einstein had evidently been resolved to adhere to that standpoint, despite an accumulation of contrary evidence, his fellow physicists might well have deemed it relevant to note the psychogenetic and sociogenetic grounds from which his anti-indeterminist outlook derived.

Notes

1. Henry Sidgwick, *The Methods of Ethics,* 4th ed. (London: 1890), p. 212; J. B. Schneewind, *Sidgwick's Ethics and Victorian Moral Philosophy* (Oxford: 1977), p. 206; cf. William F. Quillian, Jr., *The Moral Theory of Evolutionary Naturalism* (New Haven, 1945), p. 85 ff.
2. Wesley C. Salmon and Maria Reichenbach, in *Hans Reichenbach: Selected Writings, 1904–1953,* vol. 1, ed. Maria Reichenbach and Robert Cohen (Dordrecht: 1978), pp. 76–77, 80.

3. Richard von Mises, *Positivism: A Study in Human Understanding* (New York: 1956), p. 327; Philipp Frank, "Psychoanalysis and Logical Positivism," in *Psychoanalysis, Scientific Method and Philosophy* (New York: 1959), pp. 308–09: "It is a matter of fact that among the founders of Logical Positivism, the members of the Vienna Circle, there have been quite a few scientists who exhibited a certain sympathy with the teachings of psychoanalysis . . . This has been the opinion of Rudolf Carnap and other positivists."
4. Sidney Hook, in *Hans Reichenbach: Selected Writings, 1904–1953*, vol. 1, ed. Reichenbach and Cohen, pp. 33–34.
5. Letter of Sidney Hook to the writer, February 6, 1981; cf. Lewis S. Feuer, rev. of *Psychoanalysis, Scientific Method and Philosophy,* ed. Sidney Hook, *Philosophy and Phenomenological Research,* 20 (1960), 550–52.
6. Heinz Eulau, "Cohen and Freud," *Antioch Review,* 9 (1949), 414–19.
7. John Locke, *An Essay Concerning Human Understanding,* Book I (London: George Routledge and Sons), p. 39.
8. Gottfried Wilhelm Leibnitz, *New Essays Concerning Human Understanding,* trans. Alfred Gideon Langley (Chicago, 1916), pp. 71, 72, 46.
9. Paul Levy, *Moore: G. E. Moore and the Cambridge Apostles* (London: Oxford University Press, 1979), p. 238.
10. Ibid., p. 98.
11. Ibid., p. 103.
12. Ibid., pp. 139, 140.
13. Ibid., p. 142.
14. Ibid., pp. 144–45.
15. J. B. S. Haldane, "How Slavery Killed Greek Science," *Labour Monthly,* 27 (June 1945), 190. Haldane maintained that sodomy (like celibacy and prostitution) was "undesirable and that propaganda in favor of all of them should be discouraged." See also J. B. S. Haldane, *Science and Life: Essays of a Rationalist* (London, 1968), p. 68.
16. C. H. Waddington, *The Ethical Animal* (Chicago: University of Chicago Press, 1967), pp. 6, 54; T. H. Huxley and Julian Huxley, *Touchstone for Ethics, 1893–1943* (New York: Arno, 1947), pp. 157–63.
17. Anthony Quinton writes, "Moore's technique for proving the anti-naturalist principle was crude and unsatisfactory. His successors have discreetly drawn a veil over it . . ." in *Biology and Personality,* ed. I. T. Ramsey (Oxford: 1965), p. 108.
18. "The molding of the newborn infant . . . involves a most surprising process of projection and re-introjection of certain of his own impulses . . . and a whole peculiar mechanism which is described in terms of such concepts as the superego . . . At first sight, the story of the psychoanalysts may seem unlikely, but it seems to me they have now produced enough evidence to make it rather plausible . . .": C. H. Waddington, *The Evolution of an Evolutionist* (Ithaca: 1975), p. 227. Through "socio-genetic transmission," ethical beliefs emerge as part of a new evolutionary mechanism: ibid., p. 278. C. H. Waddington, *Science and Ethics,* London, 1942, p. 137. Such grouplets as the Bloomsbury Circle are useful to mankind in general. By confining their homosexual advocacy and practices to small "endogamous" circles, who then eliminate themselves through a low number of descendants, they are advantageous to society as a whole. Cf. J. B. S. Haldane, "Science and Ethics," in *Science and Human Life* (New York: Arnot, 1933), pp. 98 ff.
19. G. E. Moore, "George Santayana: The Life of Reason," *International Journal of Ethics,* 17 (1907), 253.

20. Jean Piaget, *Insights and Illusions of Philosophy*, trans. Wolfe Mays (New York: 1971), p. 25; Margaret A. Boden, *Jean Piaget* (New York: 1979), p. 97: "Anglo-Americans likewise regard psychological facts about the history of a concept or belief as in principle irrelevant to its philosophical justification or epistemological worth. As a result, such philosophers feel that there is no need for them to read Piaget—and if they do, they accuse him of systematically committing the genetic fallacy."
21. George Santayana, "Russell's Philosophical Essays: III. Hypostatic Ethics," *Journal of Philosophy, Psychology and Scientific Methods*, 8 (1911), 421–32; Russell wrote: "It was Santayana who first led me to disbelieve in the objectivity of good and evil . . ." (Bertrand Russell, *Mysticism and Logic* [New York: 1929], p. v).
22. *The Autobiography of Bertrand Russell*, vol. 1, pp. 200, 274.
23. John Dewey, *The Influence of Darwin on Philosophy, and Other Essays in Contemporary Thought* (New York: 1910), p. 19.
24. A study of "life-styles" of forty research scientists at a West Coast university was done by a clinical psychologist, Bernice T. Eiduson. Although, as she acknowledges, her interviews "were not so sustained or so intensive that extremely intimate material was proffered," none of the scientists indicated any homosexual propensity. Their sexual behavior was generally conventional, and she noted: "The scientific community as a group, however, seems to uphold these conventions." Cf. Bernice T. Eiduson, *Scientists: Their Psychological World* (New York: 1962), pp. 204–26.
25. Benjamin De Mott, *Supergrow: Essays and Reports on Imagination in America* (New York, 1969), pp. 24–25.
26. Lewis S. Feuer, "The Principle of Simplicity," *Philosophy of Science*, 24 (1957), 113–14.
27. Lewis S. Feuer, "The Bearing of Psychoanalysis Upon Philosophy," *Philosophy and Phenomenological Research*, 19 (1959), 333–34; Lewis S. Feuer, *Psychoanalysis and Ethics* (Springfield: 1955), p. 7. Many years ago the writer presented this standpoint to the last Unity of Science Congress at the University of Chicago, in 1941. The paper was absorbed into one published as "Ethical Theories and Historical Materialism," in *Science and Society*, 6 (1942), 242–72. Its use of psychoanalytical ideas elicited a collective rejection by several Marxist writers. Evidently the application of the genetic method to the Marxist ideology itself aroused anxieties in Marxists' unconscious.
28. Thus Mill tried to refute Hamilton's view that we have an "intuitive knowledge" of an external world by showing that "the Psychological Theory" could explain all the traits of the alleged intuition without going beyond "the order of our sensations coupled with memories"; one could further explain how the belief itself in an intuition of the external world arose from these verifiable psychological facts. See John Stuart Mill, *An Examination of Sir William Hamilton's Philosophy* (New York: 1873), vol. 1, pp. 203, 236, 243.
29. Arthur O. Lovejoy, "James H. Leuba, A Psychological Study of Religion: Its Origins, Functions and future," *International Journal of Ethics*, 24 (1914), 218; H. Rashdall, rev. of *Varieties of Religious Experience*, by William James, *Mind*, 12 (1903), 250; Herbert Feigl, "Critique of Intuition According to Scientific Empiricism," *Philosophy East and West*, 8 (1958) 14–15.
30. Jean Piaget, *The Moral Judgment of the Child*, trans. Marjorie Gabain (New York: 1932), pp. 88, 385–86.
31. Edwin G. Boring, et al., *A History of Psychology in Autobiography*, vol. 4 (Worcester: 1952), pp. 239–40.

32. *Autobiography of John Stuart Mill* (New York: 1924), pp. 192–93.
33. Carl Becker, *The Declaration of Independence: A Study in the History of Political Ideas,* 2nd ed. (New York: 1942), rpt., n.d., p. 26.
34. Cf. my *Ideology and the Ideologists* (New York: 1975), pp. 64, 118.
35. Thorstein Veblen, *Theory of Business Enterprise* (New York: 1904), p. 367.
36. John Dewey, *Reconstruction in Philosophy,* pp. 12, 13, 65, 66. Similarly, according to the Bolshevik sociologist Nikolai Bukharin, the *Summa Theologica* of Thomas Aquinas "clearly reflects the feudal conditions of his philosophy." *Historical Materialism: A System of Sociology,* tr. (New York: 1924), p. 186. N.I. Bukharin, "Theory and Practice from the Standpoint of Dialectical Materialism," in *Science at the Crossroads* (London: 1931), pp. 14–18.
37. A. Hunter Dupree, *Science in the Federal Government: A History of Policies and Activities to 1940,* (1957; rpt. New York: 1964), pp. 19, 21, 24, 40.
38. Cf. R. W. Meyer, *Leibnitz and the Seventeenth-Century Revolution,* trans. J. P. Stern (Chicago: 1952), pp. 108–09; Ronald Calinger, *Gottfried Wilhelm Leibniz* (Troy: 1976), p. 18.
39. Cf. Richard L. Schoenwald's psychogenetic study of Spencer, "Town Guano and 'Social Statics,'" *Victorian Studies,* 11 (1968), 699 ff.; Herbert Spencer, *An Autobiography* (New York: 1904), vol. 1, pp. 553–55, 579–80.
40. See my *Ideology and the Ideologists,* p. 59. The phenomenon of fashions in ideas is related to the generational wave. The philosopher-sociologist L. T. Hobhouse observed of his students at the London School of Economics: "These generations are extraordinarily short-lived. I can count up the intellectual fashions that have taken and held my students for a brief space. When I began in 1907 there was a wave of social idealism. Then very soon came suffrage, then syndicalism, then the war, then guild socialism, then Freud. Freud, nothing but Freud, for three or four years; now, thank goodness, that is going out, and we have mostly Elliot Smith and the Diffusion Theory . . . Each of these waves absolutely submerges everything for the time being . . . It's lost labour to refute these things. They just die out in time." See J. A. Hobson and Morris Ginsberg, *L. T. Hobhouse* (London: 1931), p. 68.
41. Thorstein Veblen, *The Theory of Business Enterprise,* p. 367.
42. Cf. Cornelia Stratton Parker, *An American Idyll: The Life of Carleton H. Parker* (Boston: 1919), pp. 144–48; cf. Selig Perlman's notion of "job consciousness," in *A Theory of the Labor Movement* (New York: 1928), p. 169.
43. Alvin Johnson, *Pioneer's Progress: An Autobiography* (New York: 1952), pp. 206–07; Robert Franklin Hoxie, *Trade Unionism in the United States,* 2nd ed. (New York: 1926), pp. 365–67; Joseph Dorfman, *Thorstein Veblen and His America* (New York: 1937), p. 311.
44. Thelma Lavine, in a keen essay, applied what she called the "four basic laws of situational logic" to give a rationale for genetic analysis. Basically, she founds it on the notion that beliefs that were designed to solve the problems of one period may be presumed inadequate for those of a later one. Is this "law" helpful? The Protestant virtues, thrift, frugality, industry, may have had in mind the problems of early capitalist accumulation, but all societies have found them valuable. Genetic analysis is at its best in confuting solipsism, though the recurring doctrine hardly provides an answer to some previous "problematic situation" (T. Z. Lavine, "Reflections of the Genetic Fallacy," *Social Research,* 29 [1962], p. 333).
45. Franz Boas, Introduction to *Handbook of American Indian Language* (Lincoln, Nebr.: 1966), pp. 63–64, 66.
46. Lewis S. Feuer, "Sociological Aspects of the Relation Between Language and

Philosophy," *Philosophy of Science,* 20 (1953), pp. 85–100; rpt. Robert A. Manners and David Kaplan, eds., *Theory in Anthropology* (Chicago: 1968), pp. 411–421.
47. Cf. J. B. Schneewind, *Sidgwick's Ethics and Victorian Moral Philosophy* (Oxford: 1977), pp. 205–206.
48. Lewis S. Feuer, "John Stuart Mill as a Sociologist," in *James and John Stuart Mill: Papers of the Centenary Conference,* ed. John M. Robson and Michael Laine (Toronto: 1976), pp. 89, 104–05; F.A. Hayek, ed., *John Stuart Mill and Harriet Taylor: Their Correspondence and Subsequent Marriage* (Chicago: 1951), p. 216.
49. John Maynard Keynes, *Essays in Biography,* new ed. (London: 1951), p. 266. Bertrand Russell much annoyed John Dewey by characterizing the pragmatic theory of knowledge as an expression not of the scientific mentality but of American industrial motives; cf. P. A. Schilpp, ed. *The Philosophy of John Dewey* (Chicago: 1939), pp. 156.
50. *Letters and Journal of W. Stanley Jevons,* ed. H. A. Jevons (London: 1886), pp. 66, 101, 123, 127, 129; R. D. Collison and Rosamond Konskamp, *Papers and Correspondence of William Stanley Jevons* (London: 1972), vol. 1, p. 38.
51. Lovejoy once noted: "If the thing to be accounted for is truly something new, an 'emergent' or pure 'mutation,' then, though the theory may correctly describe the circumstances preceding, or attending its emergence, it cannot deduce the necessity of its emerging . . ." In this sense, "it is of the nature of really genetic theories *not* to explain" (Athur O. Lovejoy, *Reflections on Human Nature* [Baltimore: 1961], pp. 85–86. A fully developed genetic explanation would, however, take the form of a completed psycho-social intellectual law.
52. Cf. Raphael Patai, *The Jewish Mind* (New York: 1977), pp. 352–53.

7

The Reasoning of Holocaust Theology

I

The Holocaust, the destruction of most of Europe's Jews, was a phenomenon that seemed to sunder the ordinary arithmetic of human evil. Therefore, thinkers of high intellect and character have felt that it carried for mankind a profound philosophical message. Beside it, the verbalisms of philosophers about the "problem of evil" seemed trivial and obsolete. The "terms of philosophizing" (as we might call it) were presumably permanently altered. If the theory of relativity, when it was experimentally verified, produced such a basic change in the "terms of philosophizing" about science, then the Holocaust might well involve similar consequences for our religious beliefs and philosophy. Does the Holocaust have such a philosophical bearing, and if so, in what fashion?

In the eyes of many, the Holocaust was a unique event, perhaps the most unparalleled case of absolute evil in human history, a radically novel phenomenon. Does this alleged uniqueness establish, as some thinkers hold, the unique philosophical significance of the Holocaust?

Let us note, to begin with, that there is a great deal of inconsistency in the discussion about the "uniqueness" of the Holocaust. A unique event is one that is unrepeatable, that cannot happen again, that is, of necessity, a class with one member. Yet the aim of the discussion concerning the Holocaust is largely (and appropriately) to ensure that the like of it shall not happen again. In other words, given the nature of men and historical conditions, other Holocausts could take place, and it requires all our efforts at enlightenment and prudential measures to ensure that they shall not recur. We must endeavor that the Holocaust remain an isolated occurrence, but if it were metaphysically unique, we should have no troubles on that score. Soviet society, during Stalin's final year, for instance, was fast acquiring the kind of mentality that precedes recourse to holocaust.[1] An intent of holocaust on the part of its neighbor-states in the Middle East is always feared in Israel; indeed, under certain conditions, such an intent might be strengthened by Soviet technology and ideological condonement. The occurrence, actual or possible, of a class of holocausts is clear evidence that sociological laws underlie them; a unique event, on the contrary, would constitute a surd with respect to sociological law.

Now the Holocaust was not so much unique as rather unprecedented. Hitherto, the mass destructions of peoples have been either those of backward tribes or nations destroying other backward ones, or those of a backward society destroying a more civilized one, as when a left-wing Turkish revolutionary government encouraged the massacre of its Armenian minority, or the Tartars of Genghis Khan massacred the townsmen of Central Asia. Occasionally, a more civilized people has desired to eliminate a backward one, as when early American colonists sought to extirpate Indian tribes. The Holocaust, apart from the vastness of its scale and its chemical technology, was the first case of one advanced people destroying another advanced one. This was unprecedented, though not necessarily unique.

To many persons, the Holocaust seems unique in the sense of what we might call a "negative miracle." A positive miracle is one such as the alleged parting of the Red Sea that would have contravened the laws of nature in a fashion so ethically purposive as to testify presumably to God's intervention in history. A negative miracle, on the other hand, would constitute evidence for God's absence or withdrawal from history, if not the actual intrusion of a dualistic anti-God, the Devil. The existence of God has always been argued for in contrary ways. On the one hand, it is inferred from the fact that great comprehensive laws of the physical order exist; on the other, alleged impressive exceptions to the physical laws, evidently advancing goodness in the world, are also taken as evidence for God's existence.

Now the Holocaust cannot be regarded as an event that contravened the laws of either sociology or physics. If it involved an unprecedented number of victims, that by itself was not a fact that contravened the laws of sociology.[2] The possibly several millions of Hindus and Moslems who were massacred, often in most bestial fashion, with the achievement of their independence in 1947, did not testify by their numbers to either the violation of sociological law or the presence of God.[3] Nor does the fact that the massacre of possibly upwards of a million Armenians was planned and directed by the Turkish government itself give it the aspect of a metaphysical uniqueness.

Long ago, such thinkers as Freud, Einstein, and Heinrich Heine doubted whether the ethics of civilization had more than superficially replaced the drives of savagery. Freud thought that the brutalities of the First World War had made evident the workings of an aggressive drive, the violent energies of the id, that were usually curbed in unstable equilibrium with the institutions of civilized life. He had reluctantly accepted the view of a deviating pupil, Alfred Adler, that the aggressive instinct was an independent force in the human psyche not subsumable as a mode of man's sexual energy. William James, America's greatest psychologist, believed, despite his pacifist aims, that man was ultimately a bellicose animal. Justice Oliver Wendell Holmes, Jr., a great jurist, regarded human morality as a thin veneer of politeness spread upon a bundle of cruel drives.[4] The Holocaust might, indeed, be regarded as a macro-social confirmation of this aspect of Freud's and James's conceptions of human nature. When Einstein, in a celebrated correspondence with Freud in 1932, called upon the penetrating, life-long inquirer to analyze how the human races might ward off the outburst of a Second World War that was already threatening, Freud could do little except recognize the power of the universal drive toward aggression, and hope that it might be partially contained by the Eros in man. Five years later, in 1937, Bertrand Russell thought that Europe, careening toward disaster, was verifying Freud's theory of the death-wish.[5]

Whatever the character of the universal propensity to aggression, specific social circumstances have given to that drive the specific social form of genocide. It is a melancholy generalization that, in the twentieth century, the phenomena of genocide have occurred exclusively through the workings of socialist or leftist-minded dictatorships. The Young Turks' regime which massacred the Armenians in 1916 was suffused with presumably high aims for social betterment; the German Nazis had removed the blight of unemployment from the lives of its workers by a bold socialistic policy that evoked the praise of the dramatist, Bernard Shaw and the British economist, John Maynard

Keynes.[6] The genocide executed by the Cambodian Communist regime against a section of its people, the forced exodus and genocide of Chinese "boat people" by the Vietnamese Communist regime, the massacre of Arabs by the black leftist leaders of Zanzibar, the tribal genocides under several African socialist rulers, and the extensive Soviet extirpation of several of its Asian minorities[7] suggest that there is a high holocaustal potential in socialist totalitarian societies. The continuing spread of such social regimes bodes an ominous recurrence of genocides; the Holocaust, far from having been proved to be a unique phenomenon, may have possibly proved the precursor of such multiple phenomena.

II

Does the Spirit of God make of the genocidal centers its especial sanctuary, its *Shechinah,* where its exceptional presence can impart the Divine Message most effectively? Were the burning bodies in Auschwitz our modern analogue of the burning bush that inspired Moses in the Midianite desert? The burning bush heightened the belief in God; the furnaces of Auschwitz diminished it. A majority of the survivors may have believed in the existence of God before the Holocaust; that was not the case afterward when the proportion of believers (in a studied sample) had declined to 38 percent.[8]

> Few survivors felt that if the voice of God is to be heard at all it would be heard through the bitter struggle of a painfully tormenting substance.

The belief that the Jews were a chosen people also evidently subsided among survivors; the 41 percent who had held such a view before the Holocaust were reduced to one-third thereafter. Other creeds were even more strenuously repudiated for having failed during the crisis, and survivors often characterized them as damnable.[9]

Was the Holocaust, then, the instrumentality through which God's will achieved the founding of a state of Israel? It is probably true as a matter of historical causation that, had there not been a Holocaust, a Jewish state might not have been created in 1948. Nevertheless, only a small number of the survivors, six percent of them, dared assert that the founding of Israel was worth "the sacrifice of the Six Millions."[10]

Perhaps, however, it should be borne in mind that ideas of the Holocaustal survivors are naturally imbued with the kind of transient bitterness that affected the patriarch Job. In his miseries, Job, calling God to account, would have challenged Him before an impartial tribunal; doubtless the survivors would have challenged God as well. Despite the survivors' predominant standpoint, distinguished theologians and many young Jews continue to be drawn to a Holocaust Theology. From Auschwitz, they affirm, there emanates a new commandment, a six hundred and fourteenth, to add to those enumerated in the Pentateuch: that the authentic Jew today is forbidden to assist Hitler to a posthumous victory—namely, the obliteration of the Jewish identity through assimilation, inattention, or indifference.[11]

But what does it mean to grant Hitler a posthumous victory? Hitler's program was the physical extermination of the Jews; he was as violently opposed to accepting any assimilation of the Jews as he was to their ethnic persistence. What Hitler hated was the Jews' physical being in any social form, the sheer Jewish genetic continuance in the human species. Those who condemn Jewish assimilation, it might consistently be argued, are the ones who are handing Hitler a "posthumous victory." Indeed, whenever one reverts, or regresses, to patterns of tribalist-shaped thought, whenever one abandons reason and mocks against the cosmopolitan, eighteenth century rationalism, one is, to that extent, granting to Hitler a fraction of the victory which he sought. A Jewish intelligence, battered into a tribal irrationalism would, indeed, constitute a partial victory for Hitler; a Jewish survival in the biological sense in humanity's genetic basis, one might argue, might, by contrast, represent a big defeat for him. Would, however, the dispersal of Jewish genes throughout the "gene-pool" of the world's populations involve a tremendous loss of the most concentrated intellectual powers of mankind, a loss that would super-add to the Holocaust that was the greatest genetic catastrophe in human history? In any case, the alternatives for social decision will be considered most wisely if we disregard asking what Adolf Hitler's preferences would have been; we should decide the terms of our partnership with future generations (to adapt Edmund Burke's phrase) as rationally as we can in terms of our own desires and feelings, and not shape our philosophy by asking ourselves what an evil Nazi madman would have wished.

For Hitler's basic attack was against reason, the one common bond of humanity; to the extent that we capitulate to the irrational, we are, indeed, giving Hitler an element of victory. If, on the other hand, we were to succeed in restoring full employment, we would scarcely be conferring upon Hitler a posthumous triumph even though he, too, aimed to eliminate unemployment. Although tribal feelings and identifications have their just place among the emotional ties that we would preserve, tribalist ideology however, like all ideology, is a species of assault against reason.

III

Catastrophes to life and the human spirit are a recurrent theme in biological history; every destroyed species, variety, tribe, people, and race has been an example of radical evil, which is hardly a new phenomenon under our sun. During the Crusades, when Christians and Moslem warriors alike were venting their military frustrations upon the unarmed Jews, Maimonides, with magnificent serenity, elaborated his proofs for the existence of God. The fluctuations in social outlook, of pessimism and optimism, do, indeed, have a law of their own that is relatively independent of the traumas and triumphs that we sustain.

The Lisbon Earthquake of November 1, 1755 is often said to have provoked a widespread revulsion against Leibnizian optimism, and scholars have expected that a similar reaction would be evoked in our time by the Holocaust. The Lisbon Earthquake is said to have been the "disaster that had shocked western civilisation more than any other event since the fall of Rome in the twelfth century."[12] "What crime, what misdeed have the children done / Who lie crushed and bleeding on their mothers' bosoms?," cried Voltaire in his *Poem on the Disaster of Lisbon*. And the lesson he drew was: "The elements, animals, the humans, all are at war / We must recognize it, evil rests upon the earth." Yet Voltaire had surrendered his optimism long before the Earthquake, which only added its dramatic shattering of the "all's well" doctrine, whose shallow cheerfulness permeates Alexander Pope's *Essay on Man*.

Now the Holocaust was evil in multiple dimensions compared to the simple geometry of the Lisbon Earthquake. The latter was a natural accident, a consequence of physical forces, involving no act of human will, no decision to annihilate a branch of the human race, no indictment of humankind;[13] even the Portuguese government, under the Marquis of Pombal, exerted itself at once to bring succor to the injured and orphaned. If God, the Author of the laws of seismology, were nonetheless to be held accountable, then the Holocaust, by contrast, provides a magnified bill of particulars against both God and His human creation. For the Holocaust, a deliberately planned and organized political act, enlisted in its execution many thousands of executioners and their assistants. Its motive: to obliterate a whole people because they were too intelligent, too creative, too venturesome, or too proud to have allowed themselves, during two thousand years, to vanish in the surrounding population.

The Holocaust cannot be justified as an incident in God's plan for a higher perfection on the earth. For the Holocaust was, indeed, a case of Radical Evil. What is Radical Evil? I should define it as an evil that is so unalloyed, so infinitely debasing and cruel, that it cannot possibly be transfigured, or compensated for, through some purportedly higher historic good. True, without the Holocaust, there probably would have been no state of Israel; without the Holocaust, the Israeli soldiers might not have been imbued with the moral strength to reject as unthinkable any acquiescence to defeat. Nonetheless, who would dare say that the creation of the state of Israel was worth an Auschwitz, a Treblinka, a Maidanek? No calculus of martyrdoms can provide us with such an equation, no Benthamite solution for a maximum of pleasures and a minimum of pains. If some transcendental mathematician had thus solved the equation, he would have had to repress the victims themselves, who were given no choice as coefficients of extinction.

For among the grimmest aspects of Hitler's Holocaust was that he contrived to deny to the victims the status of martyrdom itself. A martyr is a person who has been offered a choice. If he accepts one alternative, he is spared; if he refuses, he pays with his life. Until the Holocaust, the massacres of Jews in Europe were much contingent upon their refusal to be converted to Christianity. The Cossacks in 1648, led by Chmielnicki, massacred Polish Jews on a scale that was genocidal, yet, as Heinrich Graetz wrote concerning the murders at Nesterov:

> After they had robbed the Jews of all their possessions, the Cossacks gave them the choice between death and baptism. The Jews, however, scorned the offer, . . . (The following year, in 1649, when a peace was temporarily concluded) Jews who had submitted to baptism under compulsion were given permission by the king [of Poland] to return to their ancestral faith.[14]

The Jewish victims of Hitler were not accorded the privilege of martyrdom, the vestigial residue of their human status. Again this was unprecedented in the annals of persecution; not even Torquemada of the Spanish Inquisition would have dared suppress the last minimal choice. Such cruelty, again not necessarily unique, from a metaphysical standpoint, remains grounded in the nature of man and his multiform possibilities.

European thought recovered quite quickly from the impact of the Lisbon Earthquake. An optimist postulate was soon reinstated by Western philosophers. Immanuel Kant, a few months after the catastrophe, wrote essays on its various aspects that held up the American philosopher and inventor, Benjamin Franklin as "the Prometheus of the New Age," the exemplar of those forces that would direct nature's energies toward satisfying human needs.[15] Soon the French revolutionaries, with high optimism, were to undertake their purported road to human perfection and Condorcet, the friend of Voltaire, was to write—in the very shadow of the guillotine—his celebrated sketch of the history of human progress.

IV

Oddly enough, Holocaust theology appeals most to those who still wish to translate, or sublimate their longing for a supernatural redemption into an axiom of natural earthly progress. It still nourishes the conviction that the Holocaust miseries were an agency in the Divine Plan for Human History for a higher happiness.

The Holocaust itself did plainly tend to contravene the notion that God had underwritten a guarantee for a law for earthly progress. The Holocaust did confute crypto-theology, that is, the ideology of history, with its tenet that high technology makes for high-mindedness in ethics. Moreover, it would violate common sense to maintain that God's will had made Auschwitz into his indwelling-place, his *Shechinah*. If the prospect of earthly progress rested on God's guarantee, then the confutation of progress would show that God does not exist, and the Holocaust would serve as the Atheist's Empirical Argument.

The existence of God, however, cannot be linked integrally with the notion of indefinite human progress. When all theodicies collapse, the proposition that alone might withstand criticism is that God's will could be fully realized only in an Eternal Noumenal Reality. Only in a divine domain might the sacrificed children and the wasted, tormented lives of a Lisbon and an Auschwitz experience a re-significance. Like the physical mass that approaches infinity as its speed nears that of light, perhaps the truncated human lives in an Eternal Noumenal Setting are enhanced to realize potentialities that were annihilated in transient material existence. Radical evils are not transfigured in any sequence of earthly time; only if a noumenal reality exists, co-involved with, yet transcending, its polar phenomenal contrary, could Job's challenge be met; otherwise, the existence of that evil would invalidate that of God. No law of history, or earthly theodicy, could possibly justify the ways of God; His aims, if He exists, are not fulfilled in our space-time. A successor world in time, even a beautiful one, could not undo the pain inflicted upon a destroyed species or people that has been excluded from a share in that successor world.

The universe of time, with all of its evils, horrors, and goods, we might extrapolate, is, through unknown "transformation equations," wrought into an External Noumenal Reality in which the divine ingredient is pervasive and enlarged. How this transformation is done can be characterized only in metaphorical terms. The English idealist philosopher, F.H. Bradley, once struggled with all of his verbal resources to suggest how temporal facts might be re-cast into the reality of an Absolute Experience: "the finites blend and are resolved," he wrote; they are "transmuted;" their "divergences" are "absorbed," their "discordance ... overruled", a "consummation" in which all distinctions are "suppressed."[16] The Eternal Noumenal Reality is not, however, an Absolute which, perforce, sponge-like absorbs all phenomenal realities, for much of that phenomenal world might utterly vanish, as most of its "values," its evils, would approximate to zero, and we might say mathematically become cancelled terms. The Noumenal Reality is not approached through a temporal series as the limit of a historic process of progressively diminishing cruel events; it is not the end-term of a progressive series of realizations. Yet even if there be no progressive series, nevertheless, such goods as have been actualized among all its members might be trans-realized, that is, their distinctive values would be selected and intensified in what we may call a transjective reality. Perhaps the hatred-terms of existence vanish and men become ends only, never means. If cruelties, killings and tortures do not become the locus of a divine message when their numbers are increased manifold, then, presumably, in the Noumenal Reality they are filtered out to have no part in the divine character of things.

According to Jewish theologians in the massacre-laden centuries of the Middle Ages, the coming of a Messiah would not abolish human imperfections and evils; his advent was a political forecast, not a religious principle. The rule of the Messiah was to be a temporal event bringing more peace and good health but not altering the class structure of society, said Maimonides. The era of the Messiah's rule was not to be confounded with the eternity of the human mind or of active intellect. Joseph Albo noted that one could disbelieve altogether in the occurrence of physical resurrection, and still accept the Law of Moses.[17] Realizations in historical time can, indeed, never obliterate evil or transform it completely into a higher good; only in the eternal realm would evil be deprived of whatever reality it might have possessed. Though the devices of idealistic philosophy for rectifying the evil appearances of phenomenal existence were scarcely describable, that was no adequate reason for dismissing them. Every philosophy seems, indeed, to terminate in some surd proposition, as it endeavors to approximate, or to bend, realities to a plausible fitness with its conceived scheme. Idealistic philosophy remains unconvincing, however, because it seems impelled to hide from its sight the irreducible character of radical evil. We don't quite believe that the phenomenal bestialities of the Holocaust can be "transmuted," "blended," "resolved," or "absorbed." We can think of them as excluded but never elevated into a constituent, through some metaphysical chemistry, of a higher good. Perhaps the eternal domain might be purified of them but could not incorporate, sublate, or assimilate them.

Perhaps in an eternalistic supra-world, or anti-world, prefigured in the daring relativist speculations of the logician, Kurt Gödel, while the world's goods are trans-realized, the evils, on the other hand, are nullified. In such speculation, we can be guided by the fact that a rare anomaly in experiment or observation can set the direction for a basic reconstruction of scientific theory. Some minor experimental effects, valid only for unusual speeds, provided the background for Einstein's theoretical reconstruction. The lodestone and the lightning, marginal phenomena compared to the everyday mechanical pushes and pulls, proved, despite their marginality, to provide the penetrative access into underlying physical reality. Similarly, the comparatively rare manifestations of the nisus toward ethical good in this universe, not to mention the rare occurrence of conscious, living existence against the background of the bleak, impercipient activity of the physical world, are like observed "effects," outcroppings that seem inexplicable without the assumption that they arise from an underlying, or co-involved, polar spiritual reality. Perhaps the dross of history, the pain and sacrifice of millions of species rendered extinct, the tortured fears of primitive humans, the never-ending anxieties of persons who, though advantaged by consciousness in the struggle for existence, are penalized by being the only animals having a foreknowledge of death,—perhaps all that is separated and distilled away from Noumenal Reality. It is not solely a will-to-believe that projects the supra-natural order, for its outcroppings in the material world seem real enough. The conviction that beautiful simple laws governed natural phenomena was, from Newton and Leibniz to Einstein and Eddington, animated by belief in a reason-desirous God. With nature thus rendered rational, may we venture a counterpart "great extrapolation" to an underlying beauty in the living and mental domains, that otherwise seem so ugly and repugnant? Higher noumenal laws, as yet unknown to us, might embrace both those of the material world and an underlying or counterpart spiritual reality, of which we perceive only the supervenient effects.

It is the rare advent of good, not the frequent occurrence of evil, that poses the anomalous phenomenon,—the genuine problem. According to the law of entropy, our universe is one in which all moves toward "degradation," with elegant order and beauty disappearing. That the good interposes itself and manages to survive, for a while, against such overwhelming odds, suggests that its source is a noumenal one, whose expression in the physical world is forever inadequate, even distorted.

V

Apart from any putative religious commandment associated with Auschwitz, for this historical era the Holocaust's events have probably undermined whatever plausibility the temporalist theology of such men as William James and Bernard Shaw might once have had. James had conceived a pragmatic theology according to which the salvation of the world was potential in human choices, "the condition being that each several agent does its own 'level best'...." God and men were to cooperate in meeting the world's dangers:

> The world's safety is unwarranted. It is a real adventure, with real danger, yet it may win through. It is a social scheme of cooperative work genuinely to be done.[18]

As a pragmatist, James made an immense leap of faith through affirming the pragmatic postulate: that, much as in life, whether in courtships or commercial enterprises, the victory goes to the man with confidence in his cause, so, likewise, by faith in God and His adventure we would provide the infantry for His generalship.

This pragmatic postulate of faith was, for many persons, shattered, perhaps irretrievably, during the Holocaust. Among the Ghetto fighters, Marxists with their long-term faith in the international working class movement waited pitifully for aid from the Polish workers' movement, and looked hopelessly into the skies for some solitary airplane gesturing its support from the advancing Soviet armies, for a single bomb to be dropped on the lethal gas chambers; but their will to believe conjured nothing from the metaphysical "social scheme of cooperative work." In Britain, the Home Secretary, a Laborite in the Wartime Coalition Cabinet, balked at receiving 1,000 Jewish children lest anti-Semitism be provoked;[19] American officials succeeded not in only keeping Jewish wartime refugees from these shores, but, also, in throttling negotiations for saving some small fraction of those scheduled for extinction at Nazi hands. The Vichy French regime bartered Jews to preserve a shameless French parody of autonomy; the republican conscience which once proclaimed the Rights of Man reached a nadir of degradation. As long as the Holocaust is a living memory, it is doubtful that the will-to-believe of pragmatic theology will ever have the resonance of truthfulness. Henceforth, the will-to-believe smacks of make-believe. And Shaw's Life Force seems a mis-nomer for a Death Drive in which the worst not infrequently unite to destroy the best. "Extermination" became a central theme in his later drama.[20]

The pragmatist postulate was essential if the realization of the world's possible perfection in time was to have at least a fighting chance. With that postulate gone, however, and our theistic views detached from the prospect of temporal realization, we shall look frankly to Eternal Noumenal Reality, outside time-processes, as that wherein good alone perdures. A book of essays, *The God that Failed,* published more than thirty years ago, was composed of autobiographies of men who had traversed the experience of Marxist ideology and found it wanting. More properly, it was but a sub-species of experiences whose generality deserved a book, *All Gods Fail,* for, indeed, all temporal gods are predestined to defeat and decay, and whatever hopes we have for the enduring significance of human greatness must be founded on an extra-historical noumenal reality.

Why did the Holocaust as an event evoke such little response from avowed professional philosophers or writers on ethics? No figure comparable to Kant arose to discuss the Holocaust as he had the Lisbon Earthquake. Bertrand Russell, in the last years of his life, ignored the subject; logical empiricists, linguistic analysts and scientific philosophers evinced no interest. Probably the Holocaust was a non-problem because Western European thought had predominantly arrived at a pervasive pessimistic standpoint long before the Second World War. In empiricist circles at that time the pursuing of one's "values" in logical arguments and critical studies smacked of a desperate effort to obliterate in one's work the meaninglessness of things; an unwritten understanding prevailed that "ultimate" questions were not in good form. What Thomas Henry Huxley called the "nightmare" that had descended upon European intellectuals with the primacy of deterministic convictions still weighs upon Western thought. The hapless Eddington was pounced upon by positivist reviewers for venturing to link the principle of indeterminacy with the freedom of the will. The intensity with which they did so suggests that they were trying to harden themselves to their own atheism by virulently criticizing a philosopher who externalized their own inner longings or underbeliefs. At the Unity of Science Congress that convened at Harvard University in September, 1939, on the very first day of the Second World War, one had the sense of a last ritual enacted by those who apprehended that human values were a fragile, transient occurrence soon to be overwhelmed by the irresistible powers of unreason. Outstanding scientists, as refugees, already seemed to accept defeat as their lot. As for Americans, all had read "A Free Man's Worship" and agreed with it, except for the title. Men were not "free," nor did they want to worship; the title should rather have been "Why Life Is Not Worth Living, and What One Can Do To Forget About It." Moreover, the question itself was perhaps meaningless, since one had to be alive to ask the question, and the answer itself could not be acted upon dispassionately because, once alive, the imperatives of life,—through a kind of "vitacentric predicament"—cancelled the choice of non-existence.

Holocaust theology will not, however, dispel such pessimism by claiming that Auschwitz, with a mandate for the survival of Jews as Jews, is the source of a new philosophy. Whether the Jews will survive or not depends in large part on whether Jewish culture remains sufficiently attractive to counteract the momentum toward assimilation. The real problem is whether a renaissance of the religious spirit, congruent with the scientific standpoint, is possible, whether Huxley's "nightmare" can be undone without recourse to fantasy, whether the kind of conjunction of science with religion that Descartes, Spinoza, and Leibniz knew can again become part of our philosophy.

Every philosophical system, we must all recognize, eventuates in some surd. The materialist, for example, finds that, try as he may, no law or device avails to persuade him that he can explain the advent of mental events or entities. At such junctures, the philosopher makes what is as much a leap of repression as a leap of faith; the insoluble problem is banished, perhaps as meaningless or illegitimate, by some convention of language or the laboratory. But if the philosopher dares enquire into the sources of his philosophical denial, he finds to his discomfort that some social circumpressure, such non-logical influences as the power of fashions, circles of associates, and social animosities, have been most efficacious.

A philosophical inquiry, pursued with an utmost discounting of genetic circumpressures and intrapressures, may terminate with a heightened sense of the inexplicable. If the categories of the human, scientific systems do not altogether avail for philosophical understanding, perhaps we shall resolve such problems, though not solve them, in terms of what Leibniz called "the principle of the fitness of things," the working postulate according to which a harmony prevails in the underlying realities.[21]

Notes

1. At the time of the alleged Jewish Doctors' plot in Moscow in 1953, as the novelist Ilya Ehrenburg described it: "Everyone said there was chaos in the hospitals, many patients regarding the doctors as dangerous and refusing to take any medicines" (*Post-War Years 1945–1954,* tr. Tatiana Schebunina, [Cleveland, 1967], p. 298).
2. Even the degree of resistance that the Jews showed to the Nazis followed approximately a sociological law: the smaller the given Jewish community was, relative to the surrounding host population, the more vigorous was its resistance. The Dutch Jews thus resisted more energetically than did the Polish Jews (Yehuda Bauer, *The Jewish Emergence from Powerlessness* [Toronto, 1979], p. 38).
3. "An extensive campaign of genocide was conducted against the Muslim population of east Punjab, Deli, Ajmer, and the Indian states of Kapurthala, . . . , etc., etc. The Muslims in these areas were systematically massacred" (Jyoti Bhusan Das Gupta, *Indo-Pakistan Relations 1947–1955* [Amsterdam, 1959], p. 116).
4. *Justice Holmes to Doctor Wu: An Intimate Correspondence, 1921–1932* (New York, 1947), p. 37; *Holmes-Laski Letters: The Correspondence of Mr. Justice

Holmes and Harold J. Laski, 1916–1935, ed. Mark DeWolfe Howe (Cambridge, Mass., 1953), Vol. II, p. 837.
5. Sigmund Freud, "Why War?", in *The Standard Edition of the Complete Psychological Works of Sigmund Freud,* ed. James Strachey (London, 1964), Vol. XXII, p. 211; Bertrand Russell, "Two Statesmen," *New Statesman and Nation,* XIII (1937): 416.
6. Bernard Shaw, *Complete Plays with Prefaces* (New York, 1963), Vol. V, p. 641.
7. Alexander M. Nekrich, *The Punished Peoples: The Deportation and Fate of Soviet Minorities at the End of the Second World War,* tr. George Saunders (New York, 1978), pp. 96, 104; Robert Conquest, *The Nation-Killers: The Soviet Deportation of Nationalities* (New York, 1970), p. 162; Alan Fisher, *The Crimean Tatars* (Stanford, 1978), pp. 145, 149, 154, 159, 161.
8. Reeve Robert Brenner, *The Faith and Doubt of Holocaust Survivors* (New York, 1980), p. 92.
9. Ibid., pp. 186, 84.
10. Ibid., p. 242.
11. Emil L. Fackenheim, *The Jewish Return into History* (New York, 1978), pp. 22, 109; Emil L. Fackenheim, *God's Presence in History: Jewish Affirmations and Philosophical Reflections* (New York, 1970), pp. 71, 83.
12. T.D. Kendrick, *The Lisbon Earthquake* (London, 1956), p. 122.
13. Evidently there was some effort, however, to impute the earthquake to the tolerance that the Portuguese had shown the Jews. Cf. T.D. Kendrick, *Op. cit.,* pp. 86, 106.
14. H. Graetz, *Popular History of the Jews,* tr. A.B. Rhine, ed. Alexander Harkavy (New York, 1935), Vol. V, pp. 89, 94, 92; S.M. Dubnow, *History of the Jews in Russia and Poland,* tr. I. Friedlaender (Philadelphia, 1916), Vol. I, pp. 148, 150, 156–157.
15. Ernst Cassirer *et al., Vorkritische Schriften* (Berlin, 1922), Band 1, p. 484.
16. F.H. Bradley, *Appearance and Reality: A Metaphysical Essay,* sec. ed. (London, 1899), pp. 429, 430, 488, 467, 488, 462.
17. *Moses Maimonides' Treatise on Resurrection,* tr. Fred Rosner (New York, 1982), pp. 78–79. "Maimonides makes a rigid demarcation between the Messianic Age in which events retain their physical context, and the world to come where existence is spiritual only" (p. 17); "Rabbi Hillel did not believe in the coming of the Messiah at all, ... it is not a fundamental principle of the Law of Moses, ..." (Joseph Albo, *Sefer Ha-'Ikkarim, Book of Principles,* tr. Isaac Husik [Philadelphia, 1929]. Vol. One, p. 47); Joseph Klausner, *The Messianic Idea in Israel,* tr. W.F. Stinespring (New York, 1955), p. 415; Rev. A. Cohen, *The Teaching of Maimonides* (New York, 1968), pp. 226–228.
18. Cited in Arthur O. Lovejoy, *The Thirteen Pragmatisms and Other Essays* (Baltimore, 1963), pp. 76–77; William James, *Pragmatism* (New York: Meridian Books, 1955), p. 187.
19. Bernard Wasserstein, *Britain and the Jews of Europe 1939–1945* (Oxford, 1979), pp. 112–113.
20. Bernard Shaw, *Complete Plays with Prefaces* (New York, 1963), Vol. V, p. 479f.
21. Gottfried Wilhelm Leibniz, *Theodicy,* tr. E. M. Huggard, ed. Diogenes Allen (Indianapolis, 1966), pp. 59, 65.

8

Confronting Evil and Its Unreason

Rationalist Reflections on the Holocaust

Almost all his readers will share the feelings that George Steiner has expressed concerning the vast tragedy of the Holocaust.[1] The differences of opinion that some of us may have pertain to the philosophical and sociological judgments that he attaches to what he prefers to call the Shoah.

Professor Steiner doubts not only whether human language is adequate to understand the horrors symbolised by Auschwitz. He asserts in addition that the concepts required are "not amenable to rational analysis". In his view, the catastrophic experience of the Jewish people proves "the essential inadequacy of pragmatic-positivist levels of experience". The Holocaust that has now burdened them with "the inconceivable guilt of God's indifference" arose, he maintains, from "the core of European history and culture". And only "theological-metaphysical analysis" will help explain its phenomena—including the persistence, for instance, of "Jew-hatred" even when and where there are no Jews left. What he calls (mistakenly, I think) "the Auschwitz experiment" has for him contravened all philosophical rationalism, leaving us solely "the haunting tenacity of the metaphysical and the theological presences in our psyche". If Auschwitz cannot be written or spoken of intelligibly, George Steiner seeks none the less for a "rehumanisation of language" that would restore its lost capacities to speak to God.

Does the Holocaust raise a basic problem of language, of "Jewish hermeneutics"? Job, although a non-Jew, and reprobated by some of the ancient rabbis, could still adapt his charge against existence: "Cursed be the era in which we were born, cursed be its peoples and its rulers." He would still deny to God's and history's three yes-men that the Jews were being punished for wrongs they had done, though at the end too, Job would reiterate a faith in God even though He slays us. God was still addressed by the Jews after the massacres in 1648 had killed perhaps half their number in Poland. (Meanwhile, the chieftain of the Cossack barbarian murderers, Chmielnicki, is still extolled in the Soviet annals as "a national hero", and the only American scholarly biography of him filters out the massacres into oblivion.)

Perhaps the adolescent Spinoza in Amsterdam, as he listened to the newly-arrived refugees, decided that it made no sense to say that God was "good". And in 1904, after another in the series of Kishineff massacres, the Hebrew poet Hayyim Nahman Bialik, crying out against the Russian pogroms and defilements, shelved his faith in a God that had abdicated and urged instead the defensive use of weaponry guided by human arms and eyes.[2]

It is not language, or its resources, that fail us. It is, rather, that with the Holocaust the upper limit of the human organism's empathetic tolerance of pain is reached; our pain-receptors become dulled. We have our endurance limits, as metals have their fatigue. And perhaps, in our primitive evolutionary past, such limits helped our ancestral tribes to survive their ordeals of massacre. Why is it that no film based on the Holocaust has affected audiences as much, say, as *Chariots of Fire,* dealing (in part) with an Olympic runner who was a Jew? It is because we respond to hope, effort, achievement; whereas the Holocaust presents predominantly a history of passive suffering, unredeemed by even the martyrdom of some defeated alternative.

The "quantitative" standpoint is regarded as a sacrilegious approach to Jewish passivity during the Holocaust. Nevertheless, its use shows that, as compared to other peoples, the Jewish resistance displayed at least an equal heroism against odds that were rather greater than others confronted. When any guerrilla or partisan resistance is evaluated according to the criteria of military science by its analysts, a series of questions is asked: How much aid could the partisan group have counted on from the surrounding local population? How much moral support did they receive from other nations? What supply of arms was made available to them?

As compared to the French Resistance, also the Norwegian, the

Dutch, the Belgian, the Polish, the answers are starkly simple: the Warsaw Jews received virtually no support from the surrounding local populations. Not a single radio message of the Allies' concern emanated from transmitters in Britain; the arms that came to them from well-supplied Polish sources were pitifully meagre, and usually required payment. By contrast, such European peoples as the French received radio communications all the time, were favoured with British missions, and were regularly provided with military supplies. Despite such contacts, close to a million French workers volunteered to be collaborationist labourers in German factories.

The Jews, in fact, were the one people to whom the eyes, ears, and hands of the world were practically closed. That under such conditions an armed resistance was sustained for the last weeks is exceptional to the laws of military probability. The one organisation—the Polish Communist Party—that had been designed for underground action had been "liquidated" by Moscow in 1938; and Joseph Stalin so much hated the idea of an autonomous Jewish people that in 1942 he executed the leaders of the Jewish Bund, Ehrlich and Alter, as "Nazi spies". (The Russian regime still fails to acknowledge this crime.)

Mathematical language, when it is used dispassionately, does not fail to tell the truth of even such events as the Holocaust. The world's leading authority on the destruction of European Jewry doubts that the entire fire-power of the Warsaw partisans exceeded that of a single US infantry company. But the extraordinary event was that such an improvised resistance appeared at all under overwhelming contrary moral and material obstacles, especially when one bears in mind the relative capitulations of more favourably provisioned peoples.

The Holocaust indeed has not as yet been recorded in the pages of some great work of literary art. Many looked to the supreme Jewish story-teller, Isaac Bashevis Singer, to translate into fiction the reality of the destruction of the Warsaw Ghetto. Singer had lived as a young writer in Warsaw, attended its Writers' Club meetings, and described mordantly the lucubratory debates among its Communist, Trotskyist, and Marxist adherents. He judged, however, that he could not write convincingly of the Battle of the Warsaw Ghetto, in which no miracle intervened; he could not people it with his dybbuks, Cabbalistic devils, and malignant spirits, or draw on his readings in Gurney's *Phantasms of the Living*; he could not profane the battle's reality with his make-believe world. Moreover, notwithstanding his sarcasm for pathetic and pretentious ideologists, the fact is that every standpoint and every party—liberal and conservative, all the varieties of Marxism and anarchism and, most of all, orthodox religion—failed during those testing years to help avert catastrophe for the Jews. Singer with his Cabbalistic resignation found alien the courageous will of his fellow Polish Jew, Marek Edelman, the (still surviving) second commander of the Warsaw detachment, who strangely then chose to remain in Poland (where he practises cardiology). Did I. B. Singer sense that the surrender of the intellect, the recourse to mythical interventions, would have constituted a sacrilege upon the memories of the Ghetto's defenders?

Are the Jews, as Steiner says, "a traumatised, crazed people"? Traumatised, undoubtedly; but "crazed", no. Those who are self-consciously Jews have responded rationally to the truncating gash aimed into their collective consciousness. Rationalism remains ascendant among the Jews; they have generally not regressed into anti-science movements; indeed, in the universities, the Jewish students turn increasingly to the sciences, and away from the socio-humanities whose teachers have too often shown themselves to be ideological and irrational. They take most pride in their Nobel Laureates in the sciences, and in the USA admire those applied scientists who helped save American military and naval power during the War and post-World-War times—Oppenheimer, Teller, Rickover. On a recent journey to Australia, I found that the most admired among Jews is Sir John Monash, the engineer who commanded the Australian Corps during the First World War (and whom such as Lloyd George and Montgomery of Alamein regarded as the outstanding military leader in the Allied Armies).

Probably much of the Holocaust-originated pessimism persists because a basic sense of justice (rationalised retaliation) was largely frustrated among the Allied combat soldiers and the liberated Jews. At one Nazi death-camp that American soldiers captured, the GIs were so horrified by what they saw that they turned the Nazi ex-guards over to the freed Jewish prisoners, who then killed them. Technically, a trial should have taken place, and indeed it was to Churchill's honour that he insisted against Stalin that legal procedures should prevail. But no one has ever dared suggest impeaching the American soldiers who allowed a Hobbesian justice where the Nazi state had made a norm of murder.

Would the world today have felt more "balanced" psychologically

Confronting Evil and Its Unreason   185

if, as originally intended, the Atomic Bomb had been invented a year or two earlier, in 1943 or 1944, and used against the chief Nazi military centers? The spectacle of the murderers and their large aggregates of accomplices who eluded justice in both West and East Germany was scarcely mitigated by the capture of an isolated Eichmann. The combative, militant energies that should have been expressed externally against Nazi targets were then dammed and turned inward among the peoples of Western civilisation. From this process, Jews were to some extent rescued psychologically when the advent of Israel brought the opportunity for a Jewish Army that, without benefit of ideology, fought solely and simply for their own lives and welfare.

Is it, then, true that the "causal dynamics" of the Auschwitz asphyxiation-chambers arose "from inside the core of European history and culture"? A cultural determinism can be every whit as depressingly fatalistic as an economic one, and perhaps as equally fallacious. For one thing, European history was probably more pluralistic and multi-cored than Professor Steiner allows. By and large, the rationalistic-scientific movement worked mainly outside the lines in which anti-Semitism flourished. When Francis Bacon recorded his vision of the new scientific culture in his *New Atlantis*, he made its proponent the Jew Joabin. Probably Bacon had heard tell of the Jewish metallurgist, Joachim Gaunse, whose innovations in the smelting of copper had helped cast the cannon deployed against the Spanish Armada. To German philosophers after Hegel, Bacon was a "vulgar empiricist", though it was the Baconian inspiration (as the great rationalists Descartes, Spinoza, and Leibniz acknowledged) that stirred them in a common enterprise.

In a tautological sense, the Holocaust was an outcome from "the core of European history"; but from the factual standpoint other outcomes, including happier ones, would have been equally, or even more likely. The Kaiser's Imperial regime did not foreshadow Hitler's Holocaust. During the First World War, for instance, when it appeared that the leftist Young Turkish regime was considering a supplement to its massacre of the Armenians with one of the Palestinian Jews, the German Emperor intervened to prevent such an act. The German Army that occupied Polish and Russian territories during the First World War was relatively benevolent toward the Jewish population. Russian-imposed restrictions were abrogated, elementary education became mandatory, and secondary schools and universities were opened to Jewish students; Jews were given a fair representation on municipal councils; and liberal German Jews were brought in to help administer the occupied Jewish areas.

In short, the "core of European history" has had inconsistent strains. One can think of many eventualities that might have cancelled the possible occurrence of a Holocaust. If America had joined the League of Nations, and lent its military and economic power to preventing the Nazi entry into the Rhineland—if Right-wing Bolshevik leaders had been of a tougher character—or had Trotsky not been enthralled by an overcompensating Party loyalty from calling upon the Red Army to remove the gangster Stalin—or if the German Social Democrats, under a more energetic leadership, had directly embattled Hitler's thugs in their Nazi Party strongholds—or if Leon Blum, the first French Socialist Premier, had not shrunk from the idea of a preventive war against the Nazis—or if the British public and Parliament had been less enveloped in self-comforting pacifist garments.

There was a Germany, too, that had been rational, cosmopolitan, and free from "Jew-hatred", and whose working class had welcomed enthusiastically the personal leadership of Ferdinand Lassalle. Kant, Schopenhauer, and Nietzsche found their most encouraging words among Jewish associates, and the German Social Democratic Party was the only one that made Darwin's ideas central to its mode of thought. Eduard Bernstein's revisionism in time became his party's unofficial philosophy; and he was so much beloved by his comrades that they could not expel him when his philosophy was recognised to be a first repudiation of Marxism and ideology. The Austrian working class took for its spokesmen at different times such "non-Aryans" as Victor and Friedrich Adler and Otto Bauer. If their movement was ultimately broken and defeated, it was not historically necessary that it should fail. The mark of the Nazis, like that of Cain, was not the mark of an historically inevitable crime.

In George Steiner's judgment, the anti-Jewish feeling at "the core of European history" essentially derives from Christian writings, like those of St Paul, which assail the Jews for rejecting Jesus. But anti-Jewish sentiments were prevalent among Roman intellectuals before Jesus's mission, and continued afterwards. Cicero, Tacitus, Horace, and Quintilian (philosopher, historian, poet, and rhetorician) all thought that Judaism was superstitious or nonsensical. Cassius, Plutarch, Strabo, Sextus Empiricus (historians, geographer, and sceptic) agreed with this opinion, while Seneca, the Stoic minister of Nero, and Juvenal, the satirist, both thought that the Jews' observance of the Sabbath encouraged people to be lazy. There were anti-Jewish outbreaks in Alexandria on the part of its "Greek" community in the time of Philo Judaeus; and Christianity was not involved.

In short, anti-Semitism has tended to arise whenever a population regards the Jews as an "elite" people, standing apart from the cultural majority in important respects, but nevertheless thinking of themselves as potential candidates for membership in that society's elite. Paul of Tarsus evidently judged shrewdly that, by aligning himself with the anti-Jews, he would make the acceptance of Christianity that much more palatable to the masses. Joseph Stalin reasoned similarly that his Communist regime would gain in popularity with the traditionally anti-Semitic Russian and Ukrainian masses if his "Moscow trials" succeeded in driving virtually all Jews out of Soviet politics. Even today, few people care to consider that this might have been an underlying motive for the "Moscow trials". And Adolf Hitler actually bound the German people closer in loyalty to himself by attacking the Jews. For he made the German people into what might be called "a community of guilt", all sharing to some extent a complicity in the genocide, more active for some than for others. A common guilt guaranteed a common bond to fight unyieldingly, and thereby to postpone or evade the accusations and punishment they otherwise expected.

It is a natural temptation to blame the world's ills upon philosophies we don't like, even though this procedure consists usually of examples of what we might call "the ideological fallacy". Professor Steiner thus holds that "the Auschwitz experiment" somehow discredits the "pragmatic-positivist" philosophical amalgam.

To conjoin the two philosophies is a significant blunder. William James, the greatest of the American pragmatists, wrote page after page to define a God finite in His powers but struggling forever against forces of evil The horrors inflicted upon all creatures, even the lowliest, by the struggle for existence, haunted James; few philosophers have felt so personally what their professional literature calls "the problem of evil", for James would accept no dialectical device for sublimating evil into good. The founder of pragmatism, Charles S. Peirce, similarly thought the "problem of evil" the most profound for human thought, even if insoluble, and prophesied darkly that the latter part of the 20th century would witness the release of massive forces of evil on a scale hitherto not seen. John Dewey, America's leading pragmatic philosopher for half-a-century, was the most philo-Judaic thinker in the history of philosophy. Many were the Jewish students guided in their contributions by Dewey's generous, thoughtful spirit, while Jewish trade-unionists, obscure Zionist teachers, and lonely academic outcasts called upon him. He, without condescension, understood their problems, and acted. All the *pragmatic* philosophers rejected *positivism*. What is gained by joining them rhetorically?

It is noteworthy that after 80 years "pragmatic" is today used more often than any other philosophic adjective to connote a person who, without recourse to ideology, tries to resolve problems in a way that advances the well-being of all. The word "existentialist" has meanwhile joined the limbo of tattered placards.

In any case, can we use the political beliefs of philosophers to verify or falsify their philosophies? Are we to regard Bertrand Russell's logical atomism as verified by his life-long repugnances toward America? Or are we to reject altogether Henri Bergson's intuition of free will because he never published a passage against the injustices in the case of Captain Alfred Dreyfus? And why do hermeneutists who make such linkages usually forbear from citing that it was Heidegger-the-hermeneutist who signed as Rector for his university the decree that deprived all Jewish students of scholarship aid?

Rational analysis has not been overwhelmed, or undermined, or overthrown, by the macro-evils of the Holocaust. Rational philosophers and scientists did indeed fear that such a catastrophe was potential for modern civilisation. Herbert Spencer, the melancholy philosopher of Evolution, was apprehensive in his last years of the "rebarbarisation" that was beginning to take place among civilised peoples, and was profoundly depressed. Charles Darwin himself pondered that "at no very distant date . . . an endless number of the lower races will have been eliminated by the higher civilised races throughout the world". Given his own moral sympathies, he detested the consequences that he acknowledged as foreseeable from his theory of natural selection and the struggle for existence. Freud had only a slender hope that European civilisation could contain the energies of aggression and hatred that were seething in 1932, and he had found no nostrum in Romain Rolland's Asian "oceanic" mysticism. Einstein never tried to give a course of lectures in Germany after 1919 when the German "students" threatened to assassinate him; he found a philosophic answer in Spinoza's determinism, in explaining the recurrence of barbarism with a civilised face.

The Holocaust, though it has contravened no epistemologies or technical philosophic tenets, has nullified the cheerful mood that Samuel Alexander, the Anglo-Jewish philosopher, best expressed when he wrote in 1892: "However much bloodshed is involved, no struggle of nations aims at extermination..." His "natural piety" was founded on the conviction that Love conquers Strife, and that Emergent Evolution finds new culminations in the successive ideals of mankind. That belief is quite gone. Rational reflection tells us that his contemporary, E. Ray Lankester, was more closely correct in perceiving that human history might well illustrate the far more pervasive pattern of degenerative evolution, that regress rather than progress may always entangle the evil choices of mankind. Benjamin Franklin, the wisest American of the Enlightenment, almost abandoned all faith in humanity when a company of Pennsylvania frontiersmen massacred an entire settlement of peaceful, defenceless Indians, men, women, and children. "The Blood of the Innocent Will Cry to Heaven for Vengeance", Franklin wrote. That vengeance never came; there is no law of retribution in earthly history. But Franklin did the next best pragmatic thing; he volunteered to carry a musket in the force that was recruited to march against the massacrists. He could never, however, remove his doubt (as he wrote to his fellow scientist Joseph Priestley) whether "the Species were really worth producing or preserving."

Against such a background, rational men in our time, far from renouncing their rationalism, will in this post-Holocaust era hold all the more steadfastly to a rationalism tempered by a stoic outlook. It has in the past availed to help civilisation withstand barbarian insurgence.

Notes

1. Professor George Steiner had in February, 1987, published his moving essay "The Long Life of the Metaphor," in the British magazine *Encounter* Vol. LXIX, No. 2.

2. See, in *Encounter,* the "Men & Ideas" portrait "On Rereading Bialik", by David Aberbach (June 1981).

9

The Philosophical Method of Arthur O. Lovejoy: Critical Realism and Psychoanalytical Realism

Arthur Lovejoy has for half a century been the most powerful critical intelligence in America. During his life-span, he has seen the rise and fall of philosophical movements, absolute idealism, pragmatism, neo-realism, behaviorism, positivism. His arguments against these various doctrines have, to my mind, been convincing and unanswerable. Their advocates, however, have usually been quite unaffected by Lovejoy's criticisms. The question therefore arises: why does Lovejoy's method of philosophical inquiry fail to detach persons from what presumably are their intellectual confusions? Does it stop short of the full analysis of philosophical ideas? If so, in what ways must it be developed to fulfill its contribution to philosophical understanding? We shall find that Lovejoy's method, when it is made completely explicit in its intent and assumptions, leads from his own "critical realism" to a psychoanalytical, or genetic, realism. Precisely where Lovejoy's method falters, as it encounters the hard core of irrationality which makes most philosophies immune to any criticism, there the use of psychoanalytical considerations as a mode of philosophical demonstration is the next job.

What then is Lovejoy's method of philosophical inquiry? It consists in asking whether the philosopher does, as a matter of fact, hold to the philosophy which he professes. He confronts the philosopher's formal views with his informal actions, opinions, and remarks, and he finds that the philosopher informally acknowledges the truth of beliefs which he denies when he assumes a professional posture. He finds that all philosophers in their everyday action share a common world and common beliefs which they can deny only if they are ready to follow absurdity, even irrationality.

Where the philosopher persists in an irrational view, Lovejoy will elaborate its consequences to show that the philosopher will either have to contradict what he usually takes for granted, whether it is his own existence, that of others, the physical world, or an unalterable past, or to embrace in rigorous consequentialism, some self-destructive or other-destructive commitment.

On the face of it, it would seem that Lovejoy is taking common sense as the final arbiter in philosophical matters. To this, however, Lovejoy would not altogether agree:

" 'Common sense' is, it is true, a notoriously vague term. It presumably refers to the unexamined beliefs or assumptions of the 'plain man'. But there are many degrees of 'plainness' among men; and with respect to the epistemological question, it is probable that the average man's conceptions fluctuate confusedly between naïve realism, *i.e.*, a form of epistemological monism, and epistemological dualism."[1]

Lovejoy might rather indeed be characterized, in Charles Peirce's term, as a "critical common-sensist."[2] He would test all philosophies, including the varieties of conventional common sense, to ascertain whether they are indeed unescapable in human action and existence. The "common sense" statements of a particular culture may thus be belied by the actual practice which enables it to cope with its environment. Human experience, furthermore, includes the findings of science, and these extensions of knowledge are likewise "considerations" for the philosopher. Lovejoy does not turn "common sense" into a legislator of metaphysical truth. As an emergent evolutionist, he would reject, for instance, that bit of mediaeval common sense which holds that "the cause equals the effect."[3] But where the beliefs of common sense are found to rest on an instinctive level, and to be indispensable as the basis for our lives as physical, biological, psychological, and social beings, then Lovejoy regards such beliefs as part of "the natural and spontaneous epistemological creed of mankind," "a part of the unformulated working epistemology of our race."[4] In this inquiry into what he calls "descriptive epistemology," the matrix of study is human action and experience as a whole, not one portion of it, such as its language, at a given time. For the "common sense" congealed in linguistic habits may be in a situation of "cultural lag" with respect, for instance, to our practical working activities. There are levels of common sense, from the peripheral to the more central, as there are levels of consciousness. Those philosophers who would deny the "working epistemological creed" Lovejoy, indeed, regards as abnormal. To thinkers, for instance, who find the notion of "intentionality" a paradox and a mystery, Lovejoy replies: "on the contrary, it is a notion obviously entertained by the plain man and by most philosophers in their normal moments." At such points, Lovejoy's method, as he uses it himself, enters upon a mode of argument which is none other than the psychoanalytical.[5] For what we mean by the "normal moments" of a philosopher will be defined precisely in psychoanalytical terms: they are those moments when anxieties and neuroses have not obtruded to shape our perceptions and actions. Independent evidence, apart from the philosophizing itself, is available for the distortional effect of underlying anxieties. Let us therefore enquire in several examples how Lovejoy's method reaches a stage in its use where genetic and psychoanalytical considerations become required.

(1) The gravamen of Lovejoy' critique of John Dewey's pragmatism is that Dewey projects his moralistic and practical bent into the definition of "knowledge"; Dewey's definition of "knowledge," we might say, is ideological rather than scientific. Dewey, for instance, refuses to regard "the reminiscences of old age" as instances of "knowledge." Knowledge for Dewey must always have an import for future action and control of the environment: "The finished and done-with is of import as affecting the future, not on its own account.... Given a world like that we live in, ... experience is bound to be prospective in import.... To isolate the past, dwelling upon it for its own sake, and giving it the eulogistic name of knowledge, is to substitute the reminiscence of old age for effective intelligence."[6] Lovejoy replies that retrospective judgments are, however, made, and whether they have practical consequences or not, still their reference is to the past, not the future. One may dislike reminiscences as a sentimentality of old age, devoid of practical issue; nevertheless the reminiscences are truthful or false, accurate or inaccurate; they are offered as a documentary record of the past. Dewey's wish to exclude such knowledge of the past from the class of "knowledge" is, says Lovejoy, "a sort of moral appraisal masquerading as a logical analysis."[7]

Now Dewey, of course, was aware that people do reminisce. He, however, disliked reminiscences. "Dewey was very loath to talk about the past except in answer to a direct question," writes Sidney Hook, "and even then reluctantly. All he wanted to talk about was the future! He never in all the years I knew him (25!) reminisced except about his two young children who died in Italy . . ."[8] This was the psychological basis from which Dewey drew the "pragmatic generalization" that "all knowledge is prospective"; pragmatism differed, he wrote, from traditional empiricism in having a futurist significance, "it does not insist upon antecedent phenomena but upon consequent phenomena." For it, "the future is not a mere word"; it "replaces the value of past experience, of what is already given, by the future, by that which is mere possibility."[9] Dewey could subscribe to the view of his fellow New Englander, Nathaniel Hawthorne, in *Old News:* "All philosophy that would abstract mankind from the present is no more than words." To all of which, Lovejoy, although sharing Dewey's concerns as a citizen and reformer, continues to observe that people's memories do refer to the past, not to the future, and not to the present, and that such retrospective judgments cannot by any pragmatic alchemy be transmuted into prospective ones.

What then does Dewey say in reply to Lovejoy's criticism which accords entirely with the plain man's standpoint? Dewey does not abandon his pragmatic definition of knowledge. Rather he simply refuses to call reminiscences "knowledge"; they are only "esthetic affairs," random thoughts which cannot provide statements in a logical sense. "The reminiscences of old age are a compensatory withdrawal," says Dewey, "from the actual scene and its imminent problems, its urgencies for action." Likewise a historian speculating about the past in an antiquarian spirit is merely indulging, says Dewey, an "esthetic fancy," devoid however of any scientific status.[10]

The actual psychological evidence seems to me to sustain Dewey's view that all historical judgments have a "practical" import, provided that "practical" is taken as denoting the satisfaction of some emotion oriented to the well-being of the person. The pure antiquarian is studying the past not solely for its own sake but because escape to the past somehow helps him to endure the anxiety-ridden present. Even Lovejoy's own remarkable studies in the history of ideas have seemed to me less "disinterested" in a psychological sense than he says. I have wondered whether this sturdy realist's devotion to the history of theology and romantic metaphysics is not a kind of re-enactment of trauma of his own intense upbringing by his father, a Reformed Episcopal clergyman, who drilled his young son to read the Bible in Greek and Hebrew. But even though history has a "practical" intent, it does not alter the fact that the scientific content of its statements is retrospective. Moreover, what Dewey really wishes to establish as necessary for historical knowledge is the reformer's kind of "practical" motive, not the antiquarian's, and not the resignationist's; he wishes history to become a help in reconstructing the present, and not a device for resignation to things as they are. And the reformer's intent is certainly not a necessary condition for historical inquiry.

Against the reformist futurism of Dewey, Lovejoy reiterates that we cannot by verbalisms alter the fact that all of us do make retrospective judgments about the past; we cannot banish this experience by flying in the face of it, as Dewey does when he says that "the present or future constitutes the objective or genuine meaning of the judgment about the past," and that "the past occurrence is *not* the meaning of propositions." All of us have memories, and all of us know their pragmatic limitation. As Lovejoy writes: "Memory does not raise the dead nor history rebuild Babylon." But memory, for all that, remains knowledge even though it has no reconstructionist outcome. What, however, if the pragmatist is unshaken in his resolve to call reminiscences "esthetic fancies" or "meaningless affairs"? Then it would no longer avail Lovejoy to appeal to the indispensable commonalty of human belief and practice. He would have to go on to show explicitly what occasionally he suggests, namely, that the pragmatist is not altogether sincere, that he does not really believe what he is saying, that he is trying inconsistently to adhere to an emotional stance. Lovejoy would have to show that underlying the pragmatic animus toward retrospective judgments is an emotional desire to repress the past.

Despite his dislike for genetic considerations, Lovejoy thus finds himself compelled to make brief ventures into genetic analysis.[11] He cannot deal otherwise with resistances to acknowledging universal facts of human experience. Lovejoy begins his analysis by appealing in the first place to the instinctive assumptions which every effective agent must make, and which would constitute the foundation for a "truly pragmatic logic." We all acknowledge the existence of other persons and other objects existentially distinct from us in past and present; we refer to the future as well. When Dewey, however, persists in his desire to overcome such "dualisms" of subject and object, Lovejoy essays what is, indeed, genetic analysis. He sees in Dewey's views the "expressions of an obscure feeling that nothing ought to be treated as 'known' which is not immediately given."[12] In the last analysis, however, unless the character of the "obscure feeling" from which the given philosophy emanates can be made explicit, we shall not have been able to come to terms with that philosophy. In that sense, in Dewey's own words, "we would be rendering a service to the cause of philosophical sincerity if we would openly recognize the motives which inspire us."[13]

Only genetic analysis can show that a philosopher is trying by means of either repressive mis-reports of experience or of ideological, repressive definitions to reduce some realities to nonexistence, and to overdetermine others into a universal existence. A definition is repressive if it proposes as a defining characteristic for the object in question some trait which will limit the application of the term to that sub-class only of the objects which we like, and which will exclude that sub-class which we hate. A converse repressive definition is also possible wherein we construct it to include only that sub-class of objects which we hate. Genetic analysis is the final method for the refutation of philosophies which rest on repressive mis-reports or paradoxical, repressive definitions. If a pragmatic philosopher insists on denying that we have memory-judgments which refer to the past, with virtually no prospective intent, we can appeal to his own occasional reminiscences, when he allows his experience to be spontaneous, and unshackled by his formal epistemological "off-limits" signs. But if our pragmatist insists on mis-reporting his memories as "esthetic fancies" or claims indeed that they always have a reconstructionist intent, we are in a situation where we must show that some strong emotion is actually leading him to mis-report or mis-classify the experience in question. We must show what repressive motives are at work, or what sort of "bad faith" is involved. Lovejoy's own arguments against pragmatism always have, as he recognizes, an "obvious" quality about them. He is the spokesman of the "plain man" against philosophical paradoxes. With this standpoint, we can try to show the repressive philosopher that in his informal moments, in his verbal slips and unguarded speech, he acknowledges realities which he formally denies. But if the philosopher nonetheless persists in what to the plain man are projections of his repressive and expressive motives upon realities, there is no alternative except to enter into the analysis of underlying unconscious motives. This, as I have said, Lovejoy comes near to doing when he characterizes Dewey's pragmatism as "a sort of moral appraisal masquerading as a logical analysis," or as an expression of an "obscure feeling." If we are dealing with logical "masquerades" of obscure feelings, we have entered into the domain of genetic description and analysis. Critical realism merges into a psychoanalytical realism. We might as well face the problems of the psychoanalytical understanding of philosophies honestly and fully.

There is, however, another mode of resistance which many philosophers, including the pragmatic ones, use against "critical" analysis. It consists in saying that philosophy has basically nothing to do with truth, that its office is solely a reflection of the underlying culture. American pragmatism, for instance, Dewey notes, articulates "the practical element which is found in all the phases of American life"; it reflects "the progressive and unstable character" of American civilization by depicting "the world as being in continuous formation, where there is still place for indeterminism, for the new and for a real future"; it therefore assigns to the individual mind "a practical rather than an epistemological function."[14] Another civilization will view the world differently, but according to Dewey (in this mood), a philosophy is not to be scrutinized for its truth or falsehood; rather it is to be taken as expressing the "meanings" which that civilization perceives in things. Civilizations are the measure of all things, and philosophies are the products of so many cultural relativities. Thus, says Dewey: "Meaning is wider in scope as well as more precious in value than is truth, and philosophy is occupied with meaning rather than with truth . . . ; truths are but one class of meanings, namely, those in which a claim to verifiability by their consequences is an intrinsic part of their meaning. Beyond this island of meanings which in their own nature are true or false lies the ocean of meanings to which truth and falsity are irrelevant. We do not inquire whether Greek civilization was true or false, . . . In philosophy we are dealing with something comparable to the meaning of Athenian civilization or of a drama or of a lyric."[15] From this relativist standpoint, all of Lovejoy's criticisms of pragmatism are misconceived. For his base for philosophizing is the plain man, a spokesman not of a particular culture but of the whole evolutionary history of the human race who has laboriously fashioned for himself a "working epistemology" which is invariant with his human status throughout all cultures. The plain man philosophizes with an eye to truth, which he respects as the means for his survival. It is Dewey, curiously enough, who turns philosophy into a matter of "esthetic fancies."

At this juncture, Lovejoy would be obliged to undertake a critique of cultural relativism. This he has indeed partially done in the notable essay, "Present Standpoints and Past History," where he has analyzed the logic which seeks to entrap the historian in a relativist commitment. No working historian, says Lovejoy, could base himself on such a pragmatic philosophy as that of George Herbert Mead according to which there is no "real past," no "set of events which is irrevocably there in the past, to which histories seek a constantly approaching agreement," but rather a past which is in its detail altered with the needs of each emergent present. All knowledge, says Lovejoy, rests "upon the postulate of the reality of past events and of the possibility of now knowing something of what they were." He recognizes: "It *is* a pure postulate; for it can never be empirically verified in any strict sense."[16]

Here again we come to the limit of Lovejoy's critical realist method, where it demands a fulfillment in terms of psychoanalytical, genetic inquiry. For what if the historical or philosophical relativist cannot be budged from his determination to maintain that his "history" or "philosophy" is neither true nor false, but a private perspective? The historian is involved in what Lovejoy calls a "presenticentric predicament"; Dewey's philosopher would similarly be involved in a "culturocentric" predicament. None of these "predicaments," egocentric, presenticentric, culturocentric, are predicaments for the plain man who takes it for granted that he has the power to make statements whose reference is to objective facts external to himself, his present, and his culture.[17] Our bounded psychosocial existence as individuals becomes a "predicament" only for those affected by some neurotic anxiety. If Lovejoy repeats to no avail the assumptions we all make, if the relativist still represses somehow the distinction between historical fact and fiction, and blurs the difference between his psychological wishes and objective realities, if the consistent relativist is prepared to jettison the moorings of evidence, then we are clearly confronted by the workings of a revolt against the "reality-principle," by an irrational, emotional fixation. We must then proceed to a psychogenetic analysis of the kind of disruption of reality-sense which makes for cultural relativism; we must enquire into what disturbance of the time-sense makes for historical relativism. To Lovejoy's "temporalistic realism," the reality of the past is "pure postulate"; in that case, the relativist may claim the privilege of an alternative postulate. From the viewpoint, however, of a more completed psychoanalytical realism, no philosophical doctrine can provide an alternative postulate if adherence to it can be explained and predicted as the outcome of a desire to repress the very elements in our framework which make such an explanation possible. From this standpoint, the reality of the past is not a pure postulate, but a necessity of our experience which is in our sane lives always spontaneously acknowledged.

(2) Lovejoy's critique of behaviorism culminates likewise in the simple issue of the philosophic sincerity of the behaviorist. In other words, he is finally compelled to make a genetic analysis of the behaviorist, and to conclude that the latter does not really believe what he says. Lovejoy's analysis, "the paradox of the thinking behaviorist," published at the height of the behaviorist vogue in 1922, followed his customary lines of the confrontation of the paradoxical philosopher with the plain man. The behaviorist, he noted, denies that he finds any images or ideas in his experience. How then shall one resolve this disputed question of fact? Lovejoy then tries to show that the consistent behaviorist will then be in the position of not being able to know anything about anything outside himself. For no matter how minutely the behaviorist describes his muscular movements, Lovejoy reminds the behaviorist, "you will have said no word suggesting that the organism is also apprehending objects external to itself." "There is nothing in the shifting of a set of molecules which in the least resembles what we mean—and what the behaviorist manifestly means—by a knowledge of the existence of bodies not identical with these molecules."[18] The fact of occurrence of a muscular contraction, says Lovejoy, is not equivalent to an awareness of that occurrence.

But what, we might ask Lovejoy, shall we say if the behaviorist is resolute in his denial, and proposes to define "awareness" as a tensional direction in his muscles, accompanied by some cortical response isomorphic with the external object in question? Lovejoy's argument, an "obvious" one, he recognizes, may be of avail with behaviorists who have only a slight impairment of their sense of reality. When we are dealing, however, with a behaviorist who has very deep resistances against the recognition of consciousness and mental events, resistances which actually thwart the avowal of his perceptions, we are faced with a situation where the mere emphasis of the plain man's standpoint will resolve nothing. How then shall we show, in Lovejoy's words, that "the behaviorist manifestly means" what we mean by awareness, images, and ideas? Clearly we shall have to proceed to the deeper levels of the behaviorist's consciousness and unconscious. We shall have to show that on these levels he does acknowledge the existence of mental events; to do so, we shall have to bring into his consciousness the character of his aggressive and repressive motive against consciousness. Why does he wish to negate the existence of his own and others' mental lives? What variety perhaps of sadistic motivation is involved? Our explanatory framework will take for granted the existence of people's mental lives, with all their drives, desires, anxieties, conflicts; the behavioristic repression of consciousness is explained by the laws of the very mental life which it would deny. Genetic analysis, in this sense, not only provides a scientific account of repressional errors, and "explains them away"; it constitutes also at the same time the most powerful method for restoring the sense of reality to the behaviorist by bringing him self-knowledge, and removing the fixation which thwarts his perceptual reports. In short, the rudimentary genetic judgment which Lovejoy does make concerning the behaviorist must be pursued; only when critical realism merges with psychoanalytical realism can we resolve such paradoxes as that of the "thinking behaviorist" on a common basis of scientific self-understanding which even the latter can accept.

A decade after the behaviorist vogue, the philosophy of logical positivism was engaged in a similar revolt against consciousness. Lovejoy, at a memorable discussion in Baltimore with Rudolf Carnap before the American Philosophical Association on December 31, 1935, cogently probed the positivist theses which, in effect, he found absurd. It is probably evidence of the irrational forces and the herd spirit which move young philosophers into given "schools" at certain times that Lovejoys criticisms fell on ears virtually insensitized with resistances (at that time) to the notion of the reality of the physical world and others' consciousness. Twenty years later, the same young positivists, having gone through several emotional stages, had rediscovered these same objections, and positivism, as a distinctive philosophical mood or school, seemed again an extinct force. The ultimate basis of Lovejoy's criticism of positivism once more tended to merge his method with psychogenetic analysis.

Psychoanalytical factors were very much in evidence in the examples, arguments, counterexamples, and counterarguments, of Carnap and Lovejoy.[19] There was Carnap's "worm argument," for instance, against the dualistic realists. The realist, said Carnap, claims that when he says, "This animal—say a worm—has consciousness," he is expressing "something more than the fact that the animal, under a given stimulus, is exhibiting certain observable reactions; for the proposition has an influence upon my conduct." Thus the antipositivist will say: "If I know that the worm feels pain, I avoid treading on it." To such tender-hearted epistemologists, Carnap replied that an emphatic feeling of pain "is not knowledge; it contains no theoretic content, nothing that can be expressed. It is an act, not a cognition, which puts a feeling in the place of a cognition, . . ." Lovejoy replied by speaking not of worms but human beings. According to Carnap's criterion, he observed, a mother who is moved by her crying child to say the child is suffering is talking nonsense: "A very great part of human speech, therefore, is by this theory, expressly declared to consist of meaningless propositions. The most obvious reply to this is, of course, that the emotional efficacy of these propositions should, and would, disappear when and in so far that they were recognized to be meaningless." We would not respond to "appeals for funds to relieve the hungry" if they were taken to be "insensible automata." In Lovejoy's view, Carnap did not really believe what he was saying: "Anyone who can count up to two can think of a perceiving or a thinking not his own."

To show that the positivist inwardly rejects his own philosophy, that it is a matter of formal utterances peripheral to his inner convictions, Lovejoy would have had to enter upon a psychoanalysis of the positivist unconscious. For on the level of superficial overt consciousness, no argument from social consequences need necessarily budge the positivist. If he is consequential in his positivist aggression, he will say that he never responds to appeals for the hungry, and finds an utter zero in himself when he tries to think of another's feelings. If he does respond to appeals from the needy, he may regard it either as an inconsistency on his part or as an act born of emotion or as the outcome of severe social conditioning. Lovejoy appeals beyond the notion of the merely "empirically verifiable" to that of the "socially verifiable"; rather than be bound by the doctrine that "sentences referring to that which is by me unexperienceable are meaningless," we must frankly recognize what we do as a matter of fact take for granted as human beings living on this world with each other. Are we then confronted, we must ask Professor Lovejoy, simply with an arbitrary choice between the "empirically verifiable" and the "socially verifiable"? Is our epistemology then a matter of taste?

Clearly, the notion of the "socially verifiable" houses a multitude of psychoanalytical judgments; it is that criterion for meanings which is exercised by personalities who are not warped or maimed by neurotic compulsions, by narcissist, or sado-masochist anxieties. That the positivist in his unconscious rejects his own theory of meaning would again be shown by the familiar modes of psychoanalytical evidence, slips of the tongue, jokes about his philosophy, uncensored remarks, dreams, free associations. In Lovejoy's words: "This queer affectation, a hypertrophy of the logic of scientific empiricism, is manifestly belied at every moment by the behavior and speech of the philosophers who assume it; it denies the meaningfulness of a belief which every creature of our kind seems inevitably to hold and from which ... all the tang and poignancy of his social experience derive."[20] The character of the aggressive drive which would deny meaning to statements about other people's consciousness is illumined by the example which comes to Carnap's mind, that of the worm which will be trod upon, an example which, from the psychoanalytical standpoint, is redolent with sado-masochist overtones. The positivist theory became an ideology of intellectuals precisely during that period in Central Europe especially when they showed a tragic impotence before the rise of Nazi power. In some curious way, their sado-masochist theory of meaning substituted a pseudointellectual potency, an act of philosophic destruction, for their own lack of physical potency. Behind what Lovejoy called "the neo-positivist's taste in terminology," his "eccentricities of private terminology" lie all the imperious anxieties which make his philosophy an aggressive instrument against segments of existence. This is not a matter of alternative, eccentric labels.

(3) Now Lovejoy has not only been America's most outstanding critical philosopher, but he has also been its most creative historian of ideas. And as such, he has frequently written of the history of philosophy as a history founded on irrational motives, with presumably periodic efforts to attain rationality. "The entire history of philosophy is there to remind us," he wrote fifty years ago, "what part has again and again been played by unconscious emotional bias, by inexplicit yet controlling presuppositions ..."[21] Lovejoy has studied the "undulancy

of philosophical opinion," and pondered the "transient epidemics of intellectual blindness" which affected his teachers, contemporaries, and students. He has written down some of the pithiest generalizations in the sociology of philosophic ideas: "Every philosophy thus far discovered can be shown to have been generated by a more or less complicated interworking of diverse dialectical motives and of temperamental inclinations." He has been aware of "the psychology of philosophical and literary fashions," and suggested that behind the vogue of philosophical fashions is a desire to partake of a kind of mystic cult with its sacred words and rituals: "the craving to be mystified is a perennial human craving, which it has, in the more highly civilized ages, been one of the historic functions of philosophy to gratify. What the public wants most from its philosophers is an experience of *initiation;* what it is initiated into is often a matter of secondary importance. Men delight in being ushered past the guarded portal, in finding themselves in dim and awful precincts of thought unknown to the natural man, in experiencing the hushed moment of revelation, and in gazing upon strange symbols, of which none can tell just what they symbolize. The need for a new sort of philosophic *Eleusinia* is recurrent among the cultivated classes every generation or two; it is a phenomenon almost as periodic as commercial crises."[22] Perhaps what one misses most in Lovejoy's sociology of philosophic ideas is an account of why one philosophy rather than another will be more adapted at a given time to perform this Eleusinian function, why, for instance, pragmatic futurism subserved certain ideological-emotional needs. The very absence in Lovejoy's critical method of such ideological components is the chief reason why its influence has been small. The language of philosophy is logical, but its content is ideological.

Few things render a doctrine more attractive to many minds, Lovejoy writes, than "an air of ineffability." He has described the experience of discipleship in the case of Bergson in terms which are applicable to other variants of this experience among other schools, masters, and disciples: ". . . they have felt the fascination of this quality of mystical unutterableness in his doctrine. The adepts of this philosophy have often the air of going about with monitory fingers on lips and an expression of wondering rapture."[23] In his own philosophical evolution, it was thus "unpleasant and disillusioning," writes Lovejoy, to perceive that his many teachers and contemporaries who held to the principle of idealism were simply "the victims of a specious pseudo-axiom" in their conviction that "to be real is to be sentient, or psychical." During the last half-century, mathematical logic, Wittgenstein's aphorisms, the grammar of existentialism, identity, either-or, subject and object, Thou and I, and the parsings of ordinary language have all served respective Eleusinian functions; Lovejoy, three years ago, hearing of the latest vogue reflected that "philosophy is on its last legs."

What Lovejoy's philosophical method requires to fulfill its promise is precisely a fruitful union with his insights as a psychological historian of ideas. His final basis, we have seen, for rejecting the philosophies of the paradoxians is psychogenetic; the latter assail those "beliefs which underlie the whole of human life as it is actually lived by us; genuinely to disbelieve them appears to be impossible to man, certainly to those whom we are accustomed to classify as sane." The lovers of philosophic paradox who profess such disbelief, he writes, "seem to me merely to display an unconvincing affectation.[24] These are quite clearly psychogenetic considerations.

To reject a view because only the nonsane, or the neurotic, would persist in it is equivalent, in psychoanalytical terms, to judging that the person's sense of reality, his adherence to the "reality-principle," is disrupted. To judge a view as an "affectation" is to make the psychoanalytical diagnosis that the person does not "really" believe what he is saying, that there is sufficient evidence from his unguarded comments, informal remarks, dreams, and actual behavior that in a deeper level of consciousness, or in his unconscious, he rejects the doctrine he consciously affects. The convincing evidence that certain beliefs of man's "working epistemology" are truly universal and indispensable can come only with psychoanalytical facts concerning their operative status in the unconscious, even among those who would deny beliefs, combined with the added evidence that the motivation for their denial proceeds from some deep-seated emotional hostility against those components of the real world. He who would deny the reality of the past, or the other person's consciousness, or his own consciousness, or the physical world, would be shown by psychoanalytical evidence to be motivated by aggressions arising from or heightened by traumatic experience with those very segments of existence. Unless, in short, critical realism were itself simply to declare itself the relativistic perspective of the "sane" and "unaffected," it must go on to a psychoanalytical realism which locates the operation of repressional and overdeterminative epistemological neuroses, and reinstates from the unconscious to the consciousness the person's underlying repressed "working epistemology." Given a psychoanalytical realism, we can hope to go on to understand the phenomena of the alternation of philosophies, the cravings for initiation, and the mystical power which certain tautologies, trivialities, phonemes, and "principles" acquire at different historical periods.

(4) There are obscurities in Lovejoy's philosophical method which would be clarified by psychoanalytical study. Lovejoy appeals, as we have seen, to the substratum of our indispensable instinctive beliefs. But it might be argued, the solipsistic and behavioristic theories of meaning have their instinctive roots as well. For if there are constructive, preservative, reality-emphasizing instincts, there may also be destructive, if not self-destructive ones. It is possible that the solipsistic positivist theory has its roots, for instance, in an instinctive death-wish, which denies reality to others and then to one's self. It was indeed Jacob Loewenberg who queried the critical realists acutely many years ago: "And could not an instinctive or innate tendency be discovered for a genuine doubt in the existence of an external world?"[25]

When there are conflicting instinctive sources for philosophies, which instinctive source shall we credit? Here the crucial significance of the psychogenetic method is not merely that it traces a view to its psychological origin. Rather there is the fact that the whole framework of explanation, which enables us to make the genetic account, involves reference to the physical world, our consciousness, others' consciousness, interactions, causal sequences. The operations of a destructive instinct, as an element in the explanatory framework, would presuppose the existence of the self, others, and the physical environment. We might explain, for instance, how the redirections of aggressive instinct from self to others caused a corresponding oscillation in positivism between solipsism and behaviorism. In any case, the explanatory framework we use is consistent with the truth in the standpoint which arises from the life-maintaining, reality-instincts; it is inconsistent with the philosophy which stems from the self-aggressive drive. The genetic account of the rechannelings and fluctuations of the self-destructive impulse is precisely what leads us not to credit the world which it posits. The world which the "reality-principle" describes is identical with the world which provides our framework for genetic explanation; the worlds which the "pleasure-principle" conjures up are inconsistent with the world which is the framework of genetic analysis. How is a philosophy "explained away" by psychoanalytical realism? When adherence to a given philosophical doctrine is shown to be an instance of a sociopsychological law, whose variables or relations are denied by that philosophical doctrine, when, in other words, the truth of that philosophical doctrine is incompatible with the truth of the explanatory sociopsychological law, then the former doctrine is "explained away." Psychoanalytical realism thus vindicates the "working epistemology of the human race."

Without a psychoanalytical foundation, Lovejoy's own notion of the "socially verifiable" as the criterion of meaningfulness would lead into an extreme cultural relativism. The concept of God, for instance, may be "culturally significant" in some cultures, but devoid of significance in others. Clearly, Lovejoy does not mean "socially verifiable" to be taken as equivalent to "culturally significant." By "socially verifiable," he means to denote those meanings which are universal and indispensable in all societies. These universalistic meanings, we might say, are conditions for the biological and social existence of man. Genetic analysis, however, is required to determine whether the meaningfulness of propositions concerning God is "socially verifiable" in this universal sense or limited to the "culturally significant." If all the "meaning" that a person attaches to "God" turns out to have been culturally imposed, through anxiety-mechanisms and the desire to please adults, rather than spontaneously acknowledged, and if no ingredient of its "meaning" is the outcome of his autonomous experience, then the concept of God would not be meaningful by the criterion of the "socially verifiable." That will remain "socially verifiable" which cannot be genetically "explained away," but genetic analysis is required precisely to delimit the "culturally significant" from the "socially verified."

Lastly, the question arises as to the theory of meaning which Lovejoy derives from our "natural and spontaneous epistemological creed."[26] Whether it is the eternalistic idealism of his teachers, Royce and Howison, or the diffused multiple perspectivism of Russell, or Samuel Alexander's monistic realism, Lovejoy's basis for rejection is the fact that he finds such philosophies "absurd," "inconceivable," "meaningless," "self-contradictory," or "unimaginable." He finds that the notion of a conjunction of eternity with temporal process in a single human person or a single Absolute Mind simply cannot be rendered intelligible: "For the eternal to enter into relations with aught that becomes or changes is *ipso facto* to lose its eternity."[27] This eternalistic conception, "when taken in its entirety," says Lovejoy, "seemed impossible."[28] Does Lovejoy's method of "descriptive epistemology" really provide him with a criterion for reaching such far-reaching metaphysical conclusions? Would we on Lovejoyan grounds be ready to affirm that Niebuhr's theory of the conjunction of eternity with time in human history is plainly meaningless? What if the historical eternalist maintains that the limitations of human imagination should not be taken as legislation on the modes of existence? Let us grant, he might say, that our "natural and spontaneous epistemological creed," assisted by genetic analysis, provides abundant grounds for rejecting the solipsist and behaviorist epistemologies. Does the "working epistemology of the race," however, have anything really definite to say about the truth or falsehood, meaningfulness or meaninglessness, of eternalistic metaphysics?

There is indeed a certain ambiguity in Lovejoy as to his basis for the rejection of eternalistic metaphysics. Usually he regards it outright as self-contradictory or absurd. The notion of an absolute eternal experience, made up of temporal parts, he writes, "even in my student days, seemed to me no better than a flat contradiction—and so it seems to me still." At other times, however, he is prepared to entertain this "preposterous" paradox, and to elaborate its consequences in a kind of imaginary experiment in speculative psychology. A Cosmic Consciousness, he argues, could not really include the unique quality of a human being's experience: "If the Absolute knows my error, but at the same time sees beyond it to the truth, it is not my experience that he is having; and if he is aware of my grief, but at the same time knows that it is transitory and that some joy is to follow, he is not sharing my grief, since of that the very essence is its seeming finality." At such times, Lovejoy is prepared to accept the Absolute as a meaningful philosophical speculation, although one personally uncongenial to him: "That there may be, apart from the temporal world, some reality corresponding to that Boethian definition of eternity ... I will not deny, though, as a temporal creature, I find the conception, to say the least, difficult, irrelevant, and unappealing."[29] Lovejoy's own philosophical method becomes indecisive in this zone, wavering between a decision of eternalistic metaphysics as nonsense, or unimaginable, or humanly irrelevant.

We reach a point indeed where our "natural and spontaneous epistemological creed" becomes indeterminate in its standpoint. We can deal genetically with those, such as the solipsist, who would repress different portions of existence. We are less assured, however, in venturing judgment on speculations about what lies beyond our experience. Metaphysical theories, of course, have their psychogenetic roots too; Lovejoy remarks, for instance, (perhaps with an autobiographical nuance) that eternalism is a "distemper of adolescent metaphysics."[30] We would be able indeed to cast doubt on the presumed meaningfulness of metaphysical ideas by finding genetically that in our unconscious we reject their meaningfulness, and that what we have done is to submit to coercive parental and social forces, accepting consciously as meaningful that which we unconsciously reject. Metaphysical "meanings" are, from the genetic standpoint, largely (what we may call) "coercive meanings" rather than "spontaneous" ones. For although spontaneous metaphysical cravings do seem to arise in children, the effort to articulate them involves at least self-coercive meanings. The child who is resolved to believe in an Eternal God does so even though it involves violating its usual reality and meaning criteria.

The peculiarity of many metaphysical assertions is, however, that they propose additions to the inventory of existence, rather than subtractions as do behavioristic or solipsistic epistemologies. For this reason, though arguments for metaphysical views can usually be shown, logically and genetically, to be projective superimpositions on fact, or rationalizations founded on a neurotic conviction in what Freud calls the "omnipotence of thought," still the truth or falsehood of these metaphysical views seems to stand wholly outside the purview of all analysis; they seem to be wholly indeterminate. The "unformulated working epistemology of our race" seems to contain an ingredient of agnosticism with respect to the meta-experiential.

Notes

1. Arthur O. Lovejoy, "Dualisms Good and Bad (I)," *The Journal of Philosophy,* Vol. XXIX, (1932), p. 337.
2. *Collected Papers of Charles Sanders Peirce,* Vol. V, *Pragmatism and Pragmaticism,* ed. C. Hartshorne and P. Weiss, Cambridge, 1934, pp. 354–355.
3. Arthur O. Lovejoy, "The Meanings of Emergence and its Modes," *Proceedings of the sixth International Congress of Philosophy,* New York, 1927, pp. 20–33.
4. Arthur O. Lovejoy, *The Revolt Against Dualism,* New York, 1930, pp. xi, 15, 24. Also, Arthur O. Lovejoy, "The Discontinuities of Evolution," *University of California Publications in Philosophy,* Vol. 5, 1924, pp. 208–209.

5. Arthur O. Lovejoy, "A Temporalistic Realism," in *Contemporary American Philosophy*, eds. G. P. Adams and W. P. Montague, New York, 1930, Vol. II, p. 98.
6. John Dewey et al., *Creative Intelligence*, New York, 1917, pp. 13–14.
7. Arthur O. Lovejoy, "Pragmatism Versus the Pragmatist," *Essays in Critical Realism*, London, 1920, p. 65.
8. Letter of Sidney Hook to the author, August 13, 1956.
9. John Dewey, *Philosophy and Civilization*, New York, 1931, pp. 24–25.
10. John Dewey, "Realism without Monism or Dualism—I," *The Journal of Philosophy*, Vol. XIX, (1922) pp. 310–31.
11. "It is of the nature of really genetic theories not to explain," says Lovejoy, though the reason he assigns is the emergence of unpredictable novelties. Arthur O. Lovejoy, *Reflections on Human Nature*, Baltimore, 1961, p. 85.
12. Arthur O. Lovejoy, "Time, Meaning and Transcendence—I," *The Journal of Philosophy*, Vol. XIX, (1922), p. 515.
13. John Dewey, *Philosophy and Civilization*, p. 21.
14. John Dewey, *Philosophy and Civilization*, pp. 32–34.
15. John Dewey, *Philosophy and Civilization*, pp. 4–5.
16. Arthur O. Lovejoy, "Present Standpoints and Past History", *The Journal of Philosophy*, Vol. XXXVI, (1939), pp. 486–487.
17. Arthur O. Lovejoy, "Dualism and the Paradox of Reference," *The Journal of Philosophy*, Vol. XXX, (1933), pp. 530–533.
18. Arthur O. Lovejoy, "The Paradox of the Thinking Behaviorist," *The Philosophical Review*, Vol. XXVI, (1922), pp. 139–142. Also cf. Arthur O. Lovejoy, "The Anomaly of Knowledge," in *University of California Publications in Philosophy*, Vol. 4, (1923), pp. 24–26.
19. Arthur O. Lovejoy, "On the Criteria and Limits of Meaning," *Philosophical Essays in Honor of Edgar Arthur Singer, Jr.*, ed. F. P. Clarke and M. C. Nahm, Philadelphia, 1942, pp. 8, 20–22.
20. Arthur O. Lovejoy, *The Revolt Against Dualism*, p. 14.
21. Arthur O. Lovejoy, "On Some Conditions of Progress in Philosophical Inquiry," *The Philosophical Review*, Vol. XXVI, (1917), pp. 151, 138–139. Also, Arthur O. Lovejoy, "William James as Philosopher," *The International Journal of Ethics*, Vol. XXI, (1911), p. 133.
22. Arthur O. Lovejoy, "The Practical Tendencies of Bergsonism," *The International Journal of Ethics*, Vol. XXIII, (1913), pp. 254–258.
23. Arthur O. Lovejoy, *The Reason, the Understanding and Time*, Baltimore, 1961, p. 40.
24. Arthur O. Lovejoy, "A Temporalistic Realism," op. cit., p. 96.
25. Jacob Loewenberg, "The Metaphysics of Critical Realism," *University of California Publications in Philosophy*, Vol. 4, (1923), p. 173. Lovejoy himself has suspected "that there is something secretly suicidal for science in an unrestrained craving for the elimination of intrinsic diversity from nature." Arthur O. Lovejoy, "The Discontinuities of Evolution," op. cit., p. 197.
26. Arthur O. Lovejoy, *The Revolt Against Dualism*, p. 15.
27. Arthur O. Lovejoy, "The Obsolescence of the Eternal," *The Philosophical Review*, Vol. XVIII, (1909), pp. 490, 495.
 The Revolt Against Dualism, pp. 75, 219–220, 238, 256, "A Temporalistic Realism," op. cit., pp. 90, 101.
28. A. O. Lovejoy, "A Temporalistic Realism," op. cit., p. 93.
29. Arthur O. Lovejoy, "A Temporalistic Realism," *op. cit.*, pp. 90, 95.
30. Arthur O. Lovejoy, "The Obsolescence of the *Eternal*," *op. cit.*, p. 501.

10

Lawless Sensations and Categorial Defenses: The Unconscious Sources of Kant's Philosophy

Kant has long seemed to be, on the basis of Heine's (1835) celebrated description, the archetype of the pedantic, academic professor, pursuing on a strict schedule his passionless studies. That is only, however, because we have left aside the Kant who has told us of his inner strife (1746b, 1797b, 1798). It is from such documents that we are enabled in part to reconstruct the emotional sources of Kant's philosophy.

Now the basic tenets of a philosopher generally reflect assumptions which are held because they are the ones that best express his emotional standpoint toward reality. Every philosopher has what George Boas (1948) calls a "protophilosophy," the set of his underlying axioms which are not themselves called into question. The protophilosophy issues from the philosopher's decision-base as the projection in propositional form of the philosopher's sense of reality, with all its emphases, conflicts, repressions, exaggerations. And Kant's protophilosophy is what we seek to subject to psychoanalytic study, and thereby to ascertain the deepest emotional, nonlogical determinants of his philosophy. The *Critique of Pure Reason* (Kant, 1781) rests on two important propositions: (1) that we do not know things as they are in themselves, but only as they appear to us, as phenomena; and (2) that our mental functioning imposes on our sensations certain necessary and universal forms and structures (Royce, 1919).

Kant's Sensory Life: His Estrangement from Sensation

That we do not know things as they are but only as phenomena is Kant's basic notion. We may describe this notion as the "estrangement" assumption; it asserts that our sensations do not convey information concerning the external world, that we are from the beginning estranged from reality, that sensations are to be mistrusted, that only through naive common sense do we assume that they tell us something about the external world. There is thus a denigration of the status of sensation, which becomes a phenomenal wall between ourselves and genuine reality. Schopenhauer (1818), indeed, said that Kant's greatest service was to demonstrate the "dreamlike creation of the entire world" (p. 430). The first impression which the *Critique* made on its readers was, as a matter of fact, its destruction of naïve realism. That was why Moses Mendelssohn called Kant the total pulverizer, "Den alles Zermalmenden" (Paulsen, 1902). As against common-sense materialism, Kant (1781) was prepared to entertain the transcendental hypothesis that "our present mode of knowledge hovers before us, and like a dream has no objective reality, . . . that this life is an appearance only" (p. 619).

What type of person is inclined, on emotional grounds, to denigrate the role of sensation? It is the kind of person who, in various ways, is at odds with his own sensuality or sexuality, who seeks to repress or contain his sexuality with various defense mechanisms. It is, above all, the kind of person in whom sensory reception evokes an experience of guilt. It is characteristic of persons engaged in such intense repression of sensation that the world of things in themselves recedes for them as something which has been shorn of its reality. Thus Kant developed an intense dislike for all varieties of sensations proportionate to the degree that they evoked some sort of sexual association.

To begin with, the sense of smell awoke Kant's resistances. It was too obtrusive and could not be shut out easily; it infringed on freedom more than the other senses. In his *Anthropology,* a work invaluable for his self-portrayal, Kant (1798) tells us: "Smelling is, as it were, a tasting in the distance, and forces others to partake, whether they will or not. . . . Which organic sense is the most ungrateful and also the least useful? That of smelling. It does not pay to cultivate, or perhaps even to refine it; for there are more objects of nausea—especially in populous places—than of enjoyment, which the sense can procure us, and our enjoyment through this sense can at the best be only fleeting and temporary, if it is to give us pleasure" (11:311).

A hostility to the sense of smell, Freud (1930) has noted, is an especial trait of those who are offended by the notion of sexuality. The first "depreciation of his [man's] sense of smell," Freud believed, took place in the course of human evolution when man adopted an erect posture; thereby "the whole of his sexuality" was threatened with an organic repression, "so that since this, the sexual function has been accompanied by a repugnance which cannot be accounted for . . . All neurotics and many others besides, take exception to the fact that 'inter urinas et faeces nascimur' [we are born between urine and feces]. The genitals, too, give rise to strong sensations of smell which many people cannot tolerate and which spoil sexual intercourse for them" (p. 106).

What is remarkable is that Kant in his speculative anthropology was reaching for a similar explanation of the origin of his own sensory neuroses. Kant, like Freud, believed that the transition to an erect posture was part of a process in which reason became so dominant that it "meshed" man into "discomforts and diseases." He linked the dominance of reason in its physiological consequences to man's assumption of an erect stance. The Italian thinker, Moscati, had proved, wrote Kant in 1771, that "the upright walk of man is forced and unnatural: that . . . if he makes it a necessity and constant habit, he must look forward to discomforts and diseases, which show beyond dispute that he has been misled by reason and imitation to diverge from the original animal arrangement . . ." (Wallace, 1882, p. 112; Lovejoy, 1911; Kant, 1786a, pp. 31–32). Reason, said Kant, had led man to assume the biped posture; thereby, he rose above the animals, but in return, he was "obliged to endure certain disorders that afflict him in consequence of his having raised his head so proudly above his comrades." Reason, in other words, was the agency for the repression of natural man, and inevitably brought illness in its wake.

The sense of vision too could torment Kant with its intrusive sexual suggestion. He complained that when he lectured, his thoughts were "interrupted continually because a button was wanting on the coat of one of his hearers. . . . Peculiarities in the appearance of students were apt to disturb him, such as a bare neck, an exposed breast, or long hair carelessly over the neck and brow, which were regarded by some youths as evidences of genius" (Stuckenberg, 1882, p. 81). A missing button on the coat seemed especially to upset his equanimity, for he referred to it explicitly in his *Anthropology* (1798) as making for difficulties in the way of "abstraction" or, as we would call it today, "repression":

"Many men are unhappy because they cannot abstract. The wooer might contract a good marriage if he could only overlook a wart in the face of his sweetheart, or a missing tooth in her mouth. But it is a particularly naughty feature of our power of attention to fasten itself, even involuntarily, upon the very defects of others, to direct one's eye upon a missing button on the coat right opposite to one's eye..." (Kant, 1798; 9:21). The missing button with its import of undress would disturb Kant in the course of his lecture with its portent of lawless, uncontrollable sensation: the missing button threatened the stability of the categorial defensive system.

The sense of sight was to Kant's mind, "the noblest" of the senses, because it was the least "sensory" of the senses. It dephysicalized the physical world, and (we might add) desexualized reality. "It comes nearest to a pure contemplation of the immediate representation of the given object, without any mixture of perceptible sensation," wrote Kant (1798; 10:322). Vision too, the noblest of the senses, could however become ignoble and lowly, and then would have to be repressed for what it brought to mind. And Kant in the course of his life twice sustained a blindness which was evidently neurotic in its origin. "In earlier life,' wrote his friend and biographer, Wasianski, "he had two remarkable affections of the eyes: once, on returning from a walk, he saw objects double for a long space of time; and twice he became stone-blind" (De Quincey, 1863, p. 147). Now, as Freud (1910) has illumined, what underlies psychogenic visual disturbances are the retaliatory measures which the ego imposes on "sexual pleasure in looking [scopophilia]...so that the ideas in which its desires are expressed succumb to repression and are prevented from becoming conscious..." (p. 216). The unconscious self, the transcendental ego, as it were, does not allow the conscious empirical self to see objects of sexual interest. "The ego refuses to see anything at all any more, now that the sexual interest in seeing has made itself so prominent" (p. 216). It is "as though a punishing voice was speaking from within the subject and saying: 'Because you sought to misuse your organ of sight for evil sensual pleasures, it is fitting that you should not see anything at all any more'" (pp. 216–217). We might say that a categorical imperative underlies Kant's epistemology: "Act so that your senses can sense no object which provokes your sexual instinct."

Kant first began to have visual disorders in his fortieth year, the same year in which his interest in mental illness became so great that he wrote his essay (Kant, 1764b), *Versuch über ie Krantheiten des Kopfes* ("Essay on the Diseases of the Mind"). The visual disorder would threaten him as he lectured. This was the time when Kant was going through a difficult intellectual and emotional transformation. He was still outwardly the sceptical, Voltairean satirist who, in 1766, published *The Dreams of a Spirit-Seer,* but he was also undergoing those changes which were to lead him to revolt against the psychological method, and embark on the *Critique.* The visual disorders may well have been the first overt reproaches of his conscience against the empirical. psychological method.

Kant (1797b) wrote in later years: "Among the morbid accidents of the eyes (not sicknesses, properly speaking), I experienced one which affected me for the first time in my fortieth year, afterwards, at intervals of a year, and which now happens to me a few times in the same year. Here is what its symptom was: when I was reading on a page, suddenly all the letters became confused, and to that was added a certain feebleness of sight which rendered them entirely unreadable to me" (pp. 430–431, author's translation).

Later in his life, Kant was surprised to discover that he had become blind in one eye. One cannot help observing that a favorite term of ridicule for Kant (1762) was to refer to somebody as "Cyclopean," one-eyed (p. 36). He used this term especially in referring to pedantic scholars: "mere poly-history is a cyclopean learning which lacks one eye—the eye of philosophy" (Kant, 1803, p. 170).[1] But Kant too found himself at last a Cyclopean. Was he, as it were, semi-castrated in self-punishment, rendered like the pedants who had lost contact with reality? In truly antinomian fashion, he was only semi-incapacitated; half of himself kept an access to physical reality which the other half denied, like thesis and antithesis.

A sensory demand was for Kant a threat to the rational will. He extended this attitude to inordinate lengths. Thirst, with its suggestion of maternity and birth, was particularly threatening. It could be overcome, Kant (1797b) maintained, by the resolution not to yield to it. The imbibing of fluids was associated with shame. Thus, for more than half a century, he refused to have anyone present when he drank tea. Once, when a friend did intrude, Kant asked the friend to sit down in a place where his own tea-drinking would be unobservable.

Sound was equally perturbing to Kant. He resented music on the curious ground that it was "like an odor, which spreads in every direction and must be breathed even when not wanted." He also thought that listening to music tended to make one effeminate (Stuckenberg, 1882, pp. 142–143). This dual anxiety over sensory experience for its sensual suggestion, on the one hand, and the fear for his masculinity, on the other hand, was a prototype for Kant's antitheses. His sensory life was divided against itself. A man of boundless scientific curiosity, he longed to know the richness of the world's varied phenomena, yet he avoided any first-hand contact with its sensory fullness. He could amaze listeners with his meticulous knowledge of the structure of Westminster Bridge or St. Peter's in Rome; he could grow rapturous upon the joys of travel: "To sit upon a piece of the wall of an old Roman theatre (in Verona or Nismes), to have under one's hands the house-furniture of that people, discovered after so many years in Herculaneum... to be able to exhibit a coin of the Macedonian kings... all this arouses the senses of a connoisseur to profound attention" (Stuckenberg, 1882, pp. 109, 146–147). But then, it turned out (Kant, 1798) that it was all derived from travel books which were his favorite reading for relaxation. He read every important book of this kind which he could find (11:316). "The first academic teacher of physical geography," remarked Paulsen (1902), "never saw a mountain with his own eyes" (p. 45). He could never bring himself to travel more than a few miles from his native city.

Indeed, the ocean remained for Kant a fearful philosophical symbol. "A wide and stormy ocean," he wrote in the *Critique of Pure Reason* (1781), "surrounded the island of truth." He called it, "the native home of illusion, where many a fog bank and many a swiftly melting iceberg give the deceptive appearance of farther shores, deluding the adventurous seafarer ever anew with empty hopes, and engaging him in enterprises which he can never abandon and yet is unable to carry to completion" (p. 257). The ocean's water carried all the valences of an overwhelming feminine enticement; its maternal water brought bewilderment. He would be no adventurous seafarer, and he assured himself that only illusion lay beyond. He was convinced (1798) that simply seeing the sea was enough in his case to bring on seasickness: "Seasickness—of which I myself have had an experience in a voyage from Pillau to Koenigsberg, if, indeed, anyone chooses to call it a sea voyage—what with the tendency to vomit, that arose in my case, as I believe to have observed, solely through the eyes. For when the ship began to roll, and I looked out of the cabin, my eyes caught now the lowness and in the next moment the highness of the shore"; so that through his "power of imagination" there was stirred up "an antiperistaltic movement on the part of the intestines" (11:358).

Kant preferred to substitute a world of reports for the world of sensations, maintaining a safe distance from what might impinge on his senses. To ascertain the source of this protective self-withdrawal from the world of sense, we must seek more deeply into Kant's unconscious. "Kant," recorded Hippel, the burgomaster of Koenigsberg, "often says that if a man were to write and say all he thinks, there would be nothing more horrible on God's earth than man" (Stuckenberg, 1882, p. 133). What horrible things would Kant have written had he dared to write all he thought?

The Categorical Defenses: The Psychology of "Transcendental Deduction"

The second of Kant's protophilosophical postulates is that our minds impose on our experience certain necessary forms and categories (Lovejoy, 1907). We may call this the assumption of the "law enforcement" agency of the human mind. Here we are confronted by the imposing argument of the so-called "transcendental deduction of the categories." The so-called "deduction" was, in effect, as we shall see, a concealed, highly personalized psychological statement. Kant was compelled to denigrate the world of sensations, to shut out the sexual suggestions of the senses, to render for himself a desexualized, shadowy version of reality. He had to struggle to hold on to his sanity against the lawless data, the sensations which threatened to emerge with any upheaval. And as a neurotic, intent on preserving order, Kant imposed on the lawless data his *a priori* categories of the understanding. The "transcendental deduction" projected his own introspective experience of controlling what otherwise would be a subversive world. It expressed his own struggle to hold on to rationality in an experiential world which would otherwise break apart. The "transcendental deduction," for all its rejection of the psychological method, is an example of crypto-psychologism.

All his life Kant struggled to master his melancholy and gloom. There is a stereotyped myth of Kant as a purely rational man, always in equilibrium, undisturbed by emotion. To dispel this myth, the noted Kantian scholar, Hans Vaihinger (1898), was once moved to write an essay entitled "Kant als Melancholiker." Kant, indeed, drew abundantly in his writings from his own experiences with neurosis. The opening sentence of his essay on mental therapy (Kant, 1797b) is a frank avowal of self-analysis: "As I cannot illustrate this proposition by examples drawn from the experience of others, I must necessarily consult my own; and when I have made known the result, I may then put the question to others—Whether or not they have made similar observations?" (Colquhoun translation, p. 246).

Kant had gone through a youth without joy, a tormented youth, which had led him to long to quit life. He had, he wrote (1797b), "always had a natural disposition towards hypochondriasis; which, in my earlier years, rendered me even disgusted with life." He tried to attribute this suicidal feeling to his "flat and narrow chest, which leaves little room for the motion of the heart and lungs." But clearly, his youthful suicidal despair had more than a physiological basis, as his own spontaneous comments on the ordeal of adolescence indicate: "Many people," wrote Kant, "think that their youth was the happiest and the most agreeable time of their whole life; but this is certainly not so. It is the hardest period, because one is under discipline, and can seldom have a true friend and less rarely freedom" (1803, p. 196).[2] How strange is this! For youth to almost all people is preeminently the age of friendship, comradeship, and mutual confidence, and if there is a discipline imposed from above, a generational comradeship of protest and evasion usually unites the young. But in Kant, the discipline of curbing himself, coping with his awakened sexual desires, was evidently of such a character that he could confide to no friend. The bed became associated in Kant's mind with the source of morbid illness. Descartes had loved the comfort of his couch, and had enjoyed taking his ease there with books and meditations. Not so Kant (1797b), to whom bed was "the nest of numberless diseases" (p 249), and for youth especially it was the place for sexual self-abuse: "The impulses to this habit can be escaped by continuous occupation, which keeps one from spending more time in bed and in sleeping than is necessary. Thoughts about it can be banished from the mind by these occupations; for, so long as the subject is even in the imagination, it gnaws at one's vital powers" (Kant, 1803, p. 219).

The *Critique of Pure Reason* (Kant, 1781) abounds with metaphors which depict reason as imposing law on lawless data. Reason is like the "police," preventing the violence of which citizens live in fear; reason is like "an appointed judge who compels the witnesses to answer questions" and secures general order. Without the control of the *a priori* categories, there is the threat of "barbarism" (p. 8). Against this menace of "anarchy" (p. 9), we must "institute a tribunal [to] assure to reason its lawful claims" (pp. 20–21). The lawless sceptics, "a species of nomads, despising all settled modes of life, broke up from time to time all civilized society" (p. 27). The aim of the transcendental deduction is to secure a "clear, legal title" for the categories. Kant likened the senses to the vulgar lawless mob; their domination over the understanding constituted insanity; they were akin to the Freudian id. Kant (1798) warned: "The understanding should rule without weakening sensuousness—which in itself has a mob-characteristic, since it does not reflect . . ." (9:407). The senses had a potential for mad uprising against their superiors: "The senses prefer no claim upon them (the judgments of the understanding), but resemble the common people, who, if they are not a mob (ignobile vulgus), submit readily to their superior, the understanding, though they certainly also want to be heard in the matter. Hence if certain judgments and insights are regarded as proceeding immediately (and not through the mediation of the understanding) from the internal sensuousness, and if the latter is, consequently, presumed to wield a rule of itself, this is mere extravagance of fancy closely allied to insanity" (9:409).[3]

In Kant's own life, the function of the mind was to achieve mastery over the lawless propensities of the body. He felt his will had triumphed in a number of cases. Afflicted by a respiratory cough, he "resolved" to conquer it by closing his mouth completely, and breathing only through his nose. It was hard at first, but the nose triumphed, and he fell asleep. Kant concluded (1797b) that to conquer the cough all that had been required was "an immediate action of the mind . . . for which a firm resolution is highly necessary. . . ." So well did he train himself that if he opened his mouth while asleep, he would awake. He believed likewise that he had overcome nightly thirst by training himself to breathe deeply "with full chest and thus, so to speak, drink in air through the nose" (pp. 424–427, author's translation). The categories of understanding were similarly the means by which the mind imposed order on the disorderly sensations. Where the categories for any reason broke down in a more than temporary way, the person was insane. Underlying Kant's epistemological concern for preserving "the unity of self-consciousness" was a psychological concern, an anxiety lest the lawless data and emotions rend that stability which he maintained with such difficulty. The manifest argument was epistemological; the latent significance was the control of a psychological anxiety.[4]

Evidence of Kant's anxiety concerning the unity and order of his mental processes is to be found in the details of his social and personal habits. Even during his moments of relaxation at social parties, Kant was concerned with the maintenance of the unity of the mind against centrifugal distractions. A successful social party in his view (1798) had to be controlled by the categorial framework: "such entertainments must not skip abruptly from one thing to another . . . , for, in that case, the social party disperses in a condition of distraction of mind . . . unity of conversation lacking altogether, and the mind thus finding itself utterly confused . . ." (16:48).

All of Kant's personal eccentricities derived from an "overdetermined" preoccupation to exclude "bestial" elements from his experience, that is, whatever was suggestive of sexuality. An object that had a bodily or sexual association tormented him. He went to great lengths never to perspire, wearing thin, silk clothes in the summer. Then, if he did chance to perspire, "he had a singular remedy in reserve. Retiring to some shady place, he stood still and motionless—with the air and attitude of a person listening, or in suspense—until his usual *aridity* was restored. Even in the most sultry night, if the slightest trace of perspiration had sullied his nightdress, he spoke of it with emphasis, as of an accident that perfectly shocked him" (De Quincey, 1863, p. 119). Perspiration was a defiling bodily secretion, too suggestive of sexual secretions. He had to restore his "aridity," that is, to desexualize his bodily life. This was the physical counterpart of categorial control. Kant apparently suffered from an unconscious fear of impotence. For example, he had to see the church tower from his window in order to think. Once, when his neighbor's poplars shut the tower from his view, he found himself unable to think. Measures had to be taken to trim the poplars. The erect church tower sustained him.

Kant went to the most extraordinary lengths to avoid wearing garters, so much so that one might infer that the enclosed leg had a tremendous emotional valence for him—his manhood enveloped in female sexuality. True, he argued that garters obstructed the circulation, but his substitute for them suggested an inordinate reaction. "In a little pocket . . . there was placed a small box, something like a watchcase, but smaller; into this box was introduced a watch-spring in a wheel, round about which wheel was wound an elastic cord, for regulating the force of which there was a separate contrivance. To the two ends of this cord were attached hooks, which hooks were carried through a small aperture in the pockets, and, so, passing down the inner and the outer side of the thigh, caught hold of two loops which were fixed on the off side and the near side of each stocking. As might be expected, so complex an apparatus was liable, like the Ptolemaic system of the heavens, to occasional derangements . . ." (De Quincey, 1863, p. 119). Here was a virtual mechanical analogue of the table of categories and analogies of experience designed at a tremendous expenditure of effort and energy to avoid the garter around the leg.

Kant also took a strong stand in the controversy over vaccination; he opposed it, questioning Jenner's evidence in its favor (De Quincey, 1863). He regarded vaccination as an "inoculation of bestiality," and in response to two queries in 1800, when the subject was being warmly debated, indicated that it was, in his view, morally unjustifiable (Wallace, 1882). Curiously, at this time, opponents of vaccination were arguing that "cowpox inoculation was comparable to incest, introducing into the human body a disease of bestial origin similar to syphilis" (Stern, 1927, p. 22). It was evidently under Kant's influence that his close Jewish friend and former pupil, Dr. Marcus Herz, published in 1801 an open letter entitled, *Über die Brutalimpfung* (On the Inoculation of Bestiality). The "bestial" was denied admission by Kant into his body even as its threatening incursion was held in check by the controls of perception and the categories.

The struggle with sexuality was moreover a theme which provoked Kant to his deepest reflections on history. The basic model of conflict in man, the exemplar of his inner strife, was, according to Kant (1786a), that between his sexuality and the demands of the surrounding culture, "All evils which express human life, and all vices which dishonor it, spring from this unresolved conflict." Like Freud, he traced many human ills to the postponement of marriage which was the outcome of social requirements. "But as society increases in complexity nature does not alter the age of sexual maturity. She stubbornly perseveres in her law. . . . Hence manners and morals, and the aim of nature inevitably come to interfere with each other. . . . The civilized state comes into inevitable conflict with that disposition. . . . But the space of time during which there is still a conflict is as a rule filled with vices and their consequences—the various kinds of human misery." The basic cause of human misery, according to Kant, was in the "inevitable conflict between culture and the human species" (pp. 60–61). He defined the "perfect civil constitution" as the one which would terminate this conflict. In the mechanisms of repression attendant to this conflict we look for the source of Kant's protophilosophical postulates.

With the repression of sexuality, Kant's sense of the reality of the external world diminished. The world which was sexually unknowable became truly unknowable. Indeed, the unknowable world of things in themselves, the abyss in which categories and science foundered, had the aspect for Kant of the "abyss" of sexuality. The mystery of sexuality was a principal example of the unknowable reality from which he drew back in literal fear. When the poet Schiller, in 1795, sent Kant a magazine with an article on sex differences in organic nature, Kant replied that "the natural arrangement that all impregnation in both of the organic kingdoms requires two sexes . . . opens up an abyss for the human reason." If the origin of sex were not simply the playfulness of Providence, but if this arrangement were as reason indicated "the only possible one," then Kant said, "an infinite prospect lies before us, of which we can make simply nothing, about as little as Milton's angel tells Adam of the Creation: 'Male light of distant suns mingles with female for ends unknown.' "[5] Sexuality was, as Kant (1786b) indicated in the *Prolegomena*, the exemplar of the incomprehensible: "It therefore only sounds paradoxical and is really not strange to say that in Nature there is much that is incomprehensible [for instance, the faculty of procreation]" (p. 98).[6]

Kant's Analysis of "Insanity"

According to Kant, insanity[7] is a categorial disorder, the rebellious data of sense having overcome the categories which made a unified experience possible. Kant postulated three specific modes of insanity corresponding to the principles of each of the major divisions of the *Critique of Pure Reason*.

Breakdowns in the operation of the transcendental aesthetic, transcendental analytic, and transcendental dialectic each define a corresponding type of insanity. Kant thus distinguishes between three types of insanity—tumultuous, methodic, and systematic.

The tumultuous insanity, as Kant (1798) explains it, includes that which most clearly exhibits the breakdown of the most basic connections necessary for experience: "Craziness [amentia] is the incapacity to put our representations even into that connection which is necessary for the mere possibility of experience. In the insane asylums the female sex is, by reason of its talkativeness, especially subject to this disease; that is, to intersperse with their narration so many productions of their lively imaginations that nobody can understand what they really wish to say. This first class of insanity is *tumultuous*." Then there are those suffering from a fragmentary methodical insanity in which the power of judgment is disordered. The imagination of such persons "causes a play of the connection of dissimilar things"; they can be "very jolly, rave absurdly, and please themselves in the enjoyment of so extensive a relation of conceptions which, in their opinion, rhyme together." Less deep-seated a disorder, according to Kant, is *methodic* insanity (dementia) which includes, for instance, those persons afflicted by persecution complexes, "who imagine that they have everywhere enemies." In such cases, everything which the afflicted person says is "conformable to the formal laws of thinking necessary for the possibility of experience." Last, there is *systematic* insanity in which the patient "flies beyond the whole ladder of experience ... and believes that thus he comprehends the incomprehensible"; he can square the circle, and knows the mystery of the trinity (16:398–399). The last is evidently the metaphysical mode of insanity.

This whole discussion of insanity in the *Anthropology* has a remarkable philosophical consequence which Kant overlooks. For if the causation of insanity, the disruption of the categories of the understanding, is a question of psychological anthropology, if a psychological method would explain the modes of insanity, then likewise one would expect the psychological method to explain the origin and development of the categories of the understanding. We should have a genetic account of the categories rather than a transcendental deduction of them.

Insanity indeed was only the extreme case of categorial disorganization for Kant. The categorial system, the precondition of sanity, evolves from experiences in infancy and childhood. It is threatened by drunkenness, drugs, and dreams. The precategorial stage of childhood is not really experience, according to Kant (1798), because it consists "of scattered perceptions that have not yet been united in the conception of the object" (9:18). Drunkenness disrupts the categorial framework because it makes it impossible "to regulate our sensuous representations in accordance with the laws of experience" (11:353–354). Also, the imbibing of such drugs as the chica of the Peruvians or the ava of the South Sea Islands brings on "mis-states of the senses" (11:356). Then too there is the involuntary play of our imagination which we call dreams, "unloosening" (11:358) in sleep, in which we spend "unconsciously . . . probably one-third part of our life" (11:360). Kant's categorial table, his defense mechanism against sensation, was like a beleaguered fortress, holding in check the lawless, nomadic, barbarian invasions.

The Lawless Element in Kant's Psyche

A chance association of Kant's affords us an initial clue to his sensoriphobia and to the meaning of the categories. All his life, Kant suffered from insomnia. It grew especially bad in later years when it was accompanied by spasmodic attacks. He should have called a doctor, Kant (1797b) stated: "But then, impatient at not being able to sleep, I soon had recourse to my Stoic remedy of occupying my mind intensely with whatever object I chose at will (I concentrated for example, on the numerous associations which the name of Cicero brought to mind), thereby turning my attention from this sensation and making it ineffective" (author's translation). In his insomnia, the name of Cicero brought a curious subsidence. "Cicero," we may reasonably infer, had a most significant valence for Kant's psyche. Its very sound would have suggested directly his mother's breasts which the anxiety-ridden old man still cherished in his unconscious. The word, Cicero, as pronounced in German, is similar to the word *Zitze* (teat). His associations to the first two syllables would have carried Kant to the memory of his mother's softness, and brought him sleep, as his mother did when he was an infant. Thus an oedipal fantasy was disguised as an irreproachable classical allusion which had the unusual property of liberating Kant from anxiety and sleeplessness.

Still another free association of Kant's pointed toward his emotional tie with his mother as the source of his neurosis. Kant felt that his youth was so miserable that he wished to die. He attributed this youthful death-wish to his flat chest. But in his *Pedagogy,* he states that flat chests are caused by leading-strings, in other words, that they are the physiological consequence of over-involvement with one's mother. Kant resented leading-strings strongly; they had bound him, as with umbilical cords, to his mother, and brought about his physiological and psychological illness: "Leading-strings are very injurious. A certain author once complained of being narrow-chested, which he attributed entirely to leading-strings. . . . Children do not learn to walk with the same steadiness by the use of such means as when they learn it by their own efforts" (Kant, 1803, p. 143). In other words, Kant's own unsteadiness, his inability to remain erect, were tied to his over-involvement with his mother. An oedipal longing, an impaired sexuality, and a death-wish are all suggested by this passage. Curiously, "leading-strings" (the maternal umbilical cord) remained for Kant the symbol of an emasculating influence on reason. In the Preface to the second edition of the *Critique of Pure Reason,* he (1781) stated that reason "must not allow itself to be kept, as it were, in nature's leading strings," but must constrain nature "to give answer to questions of reason's own determining" (p. 20).[8] There is a suggestion here that his emotional fixation on his mother, who had bound him with leading-strings, had inhibited Kant's intellectual development.

Kant was indeed sexually fixated on his mother. He exhibited a distaste for physical sexuality, which seemed to him out of keeping with a civilized, refined person. There is a section of Kant's essay, *Observations on the Feeling of the Beautiful and Sublime,* which might be subtitled: "Why I Never Married." Kant (1764a) analyzes the cause of bachelorhood, how there "arises the postponement and finally the full abandonment of the marital bond." He confesses that he does not know how noble souls can manage to combine refinement of taste with a natural simplicity of sexual response: "if only I saw how this were possible to achieve. . . . A very refined taste serves to take away the wildness of an impetuous inclination," makes it "modest and decorous," and limits it to "few objects"; thus refined, "such an inclination usually misses the great goal of nature." A very refined taste, says Kant, "demands or expects more than nature usually offers"; therefore, "it seldom takes care to make the person of such delicate feeling happy." It becomes "oversubtle because actually it is attracted to none"; its sexuality atrophies (pp. 90–91)

Kant seemed to have some awareness that his attachment to his mother was at the basis of his so-called "refinement of taste." In a remarkable discussion of the grounds of men's sexual preferences among women, he tended to agree with Buffon that the mother's traits determine the son's later sexual preferences. "The figure that makes the first impression," wrote Kant (1798), "at the same time when this impulse is still new and is beginning to develop, remains the pattern all feminine figures in the future must more or less follow so as to be able to stir the fanciful ardor, whereby a rather coarse inclination is compelled to choose among the different objects of a sex." Therefore, a woman with "tender feeling and a benevolent heart" was superior to one with "merriment and wit in laughing eyes."[9] As a coy bachelor, he added: "I do not want to engage in too detailed an analysis of this sort, for in doing so the author appears to depict his own inclination" (16:49). That, of course, was precisely what Kant was doing.

Above all, Kant's emotional fixation on his mother gave rise to a passionate resistance to the evolutionary hypothesis. This curious aberration in his thinking deserves a special discussion.

Kant's Emotional Antagonism to the Evolutionary Hypothesis

In studying the psychoanalytic basis of a philosophy, a most important indicator is found in the points of inconsistency in the given philosophy, i.e., a place where the philosopher's underlying emotion is constrained by his own doctrine. The formal philosophy has then failed to express adequately the informal emotions—the underlying basis of the philosopher's decision. Thereupon, the emotion breaks through and imposes its will on the philosophy with a crude directness, demanding a certain doctrine without any further ado. The point of inconsistency is a place where a quantum of the philosopher's raw emotion has not yet been rationalized into philosophical form. For this reason, it helps us especially to perceive the character of the philosopher's emotion and unconscious. The hypothesis of organic evolution awakened in Kant a resistance which was quite out of keeping with his formal philosophy, and which was evidently based on intense, unresolved oedipal feelings.

Lovejoy (1911) showed most clearly how Kant disapproved of "any attempt to inquire into the origin, the laws of genesis, of organisms in general, or of the original stock 'from' which any species is descended"

(p. 47). And Lovejoy observed the significance of certain traits of Kant's character in relation to his concept of nature: ". . . it was because of certain temperamental peculiarities of his mind—a mind with a deep scholastic strain of its own, one that could not quite endure the notion of a nature all fluent and promiscuous and confused, in which a series of organisms are to an indefinite degree capable of losing one set of characters and assuming another set." Why, however, did Kant have such a strong emotional resistance to the notion of a nature "all fluent and promiscuous"? Lovejoy's metaphor indeed grasps at those very elements of the sensory world with which Kant's conscious self was at war—a world lawless, indeed, in a sexual way, which he endeavored to contain with his categories.

The concept of a promiscuous Nature awoke in Kant an anxiety which he called "a not unmanly terror." A promiscuous "Mother Nature" seemed to open her womb to all adventures. She threatened all the defenses which Kant had laboriously constructed against his own sexuality and oedipal longings. The concept of evolution implied a mother whose promiscuity would bring forth what could only be "monstrosities." Kant depicted such a mother as "the earth in travail, giving birth to animals and plants from her pregnant womb, fertilized by the sea-slime." To believe, he wrote, "that one species should originate from another and all from one original species, or that all should spring from the teeming womb of a universal Mother—this would lead to ideas so monstrous that the reason shrinks before them with a shudder." Again, in his review of Herder's *Philosophy of the History of Mankind,* Kant spoke of such hypotheses as "so monstrous that reason recoils from them." He abhorred the idea of a "single primordial womb" (Lovejoy, 1911, pp. 46–47).

There was nothing in Kant's theory of knowledge which forbade a hypothesis of organic evolution.[10] The categorial structure, Kant thus perceived, was not strict enough to exclude all lawless tendencies. Nevertheless, Kant felt compelled to ostracize evolutionary theories as metaphysics and sought to close any opening in the philosophical-defensive system: "A hypothesis of this kind," he wrote (1790) in his *Critique of Teleological Judgment,* is a "daring adventure," and there were "few investigators of Nature, even of the most acute minds, to whom the hypothesis has not at times presented itself. For absurd it is not. . . . *A priori,* in the judgment of reason alone, there is nothing self-contradictory in this. Only, experience shows no example of such a thing" (pp. 79–80).

But Kant's revulsion against the evolutionary hypothesis did not stem from the fact that he could point to no experienced cases of it. A hypothesis of organic evolution was "metaphysics" not because it violated the categories' limits but because it aroused anxiety. "These ideas [Kant wrote inconsistently and defensively] will not, indeed, cause the investigator of nature to shrink back from before them with a shudder, as from before a monstrosity (for there are many who have played with them for a time, though only to give them up as unprofitable). But the investigator *will* be frightened away from them upon a serious scrutiny, by a fear lest he be lured by them from the fertile fields of natural science to wander in the wilderness of metaphysics. And for my part I confess to a not unmanly terror in the presence of anything which sets the reason loose from its first and fundamental principles and permits it to rove in the boundless realms of imagination" (Lovejoy, 1911, p. 47).

The evolutionary hypothesis, with its evolving Mother Nature, thus filled Kant with disgust. Evolutionary fertility, in Kant's mind, was like that of a mother's which ends with the menopause. Kant conceded that such a theory might well suggest itself to the paleontologist, the "archeologist of nature": "He can suppose the womb of Mother Nature to have given birth at first to creatures of less purposive form," then to better adapted ones, "until finally, Nature's womb, grown torpid and ossified" (Lovejoy, 1911, p. 48), would produce no new species; her "potency" would be at an end. Kant saw Nature as his mother writ large, and he projected the anxiety which she aroused in him on Nature as a whole. From the scientific standpoint, there was no reason why evolution should terminate in a cosmic menopause, why the emergence of novel species should be brought to an end. It was only when Nature was personified as Mother Nature that it seemed natural to say her womb would grow "torpid and ossified." Evolution, genesis, awoke in Kant's mind all the repressed anxieties of his own relationship to his mother. The study of biological origins trespassed on the domain of Kant's repressions, and he preferred to see the problem put aside.

The underlying function of "reason" was thus more than epistemological; it was the censor of lawless emotions. The role of reason indeed was to repress all lawless sexual associations. "The fig leaf," Kant wrote (1786a) "was a far greater manifestation of reason than that shown in the earlier stage of development." It reflected the "consciousness of a certain degree of mastery of reason over impulse" (p. 53). The emergence of the primacy of reason in Kant was thus linked to its origin in the repression of sexuality. Reason represses lawless sexual data: this is the first *a priori* categorial imposition on the sensuous manifold. The categories were based on the repression of the id by the superego.

Why, we may inquire, did the thought of his mother exercise such a tremendous hold on Kant's unconscious? The answer might well lie in the traumatic impact which his mother's death had on the onset of his adolescence. "The thirteenth or fourteenth year," wrote Kant (1803) "is usually the time when the sexual instinct is developed in a boy" (p. 219). Kant's mother, it should be remembered, died in 1737, when Kant was in his thirteenth year. Toward her, he felt a warmth which he did not feel toward his father who lived on until 1746. What would be the effect of the death of his mother on a Pietist boy of thirteen? His first sexual stirrings were concomitant with the death of his mother. Thus, we may surmise, a deep guilt was placed upon Kant's sexuality. In his unconscious, he would associate his own sexual lust with the death it brought his beloved mother. Herein, we may surmise, was the source of that hypochondria which afflicted Kant in his youth so that he hated life itself. And here as well was the basis of his attitudes toward marriage and women. He atoned for his guilt by remaining faithful to his mother's memory, and ridiculed the learned and polite women whom he met. They were compared unfavorably with his own ignorant mother. Kant never, therefore, became a full Voltairean child of the Enlightenment. Instead, he wrote (1803) a diatribe against bluestockings: "A woman who has her head full of Greek, like Mme. Dacier, or who carries on profound discussions in mechanics, like the Marquise de Chastelet, may just as well have a beard beside; for a beard would perhaps express still more unmistakably the air of profoundness which she is trying to acquire.... So far as learned women are concerned: they use their books in something like the way they use their watch—namely, carry it, in order to let it be seen that they have one" (p. 86). The pietistic mother whose breasts brought him sleep ruled in Kant's unconscious, and finally prevailed against all the new fashions of the salons.

Kant's Dreams

Persistant oedipal longings and guilt were evident in Kant's dreams, and became a torture for him. During his later years, said his Boswell, Wasianski, "his dreams became continually more appalling: single scenes, or passages in these dreams, were sufficient to compose the whole course of mighty tragedies, the impression from which was so profound as to stretch far into his waking hours. Amongst other phantasmata more shocking and indescribable, his dreams constantly represented to him the forms of murderers advancing to his bedside; and so agitated was he by the awful trains of phantoms that swept past him nightly, that in the first confusion of awaking he generally mistook his servant, who was hurrying to his assistance, for a murderer. In the daytime we often conversed upon the shadowy illusions; and Kant, with his usual sort of stoical contempt for nervous weakness of every sort, laughed at them; and, to fortify his own resolution to contend against them, he wrote down in his memorandum book, 'No surrender now to panics of darkness' " (De Quincey, 1863, p. 142).[11]

Kant became terrified of the dark. He began to burn a light in his room to ward against its powers, "an expression of the great revolution accomplished by this terrific agency of his dreams." Previously, he had barricaded the windows of his room, night and day, to keep out lawless data, like his categorial structure. "If he saw but a moonbeam penetrating a crevice of the shutters, it made him unhappy." Now with darkness a terror, and silence an oppression, he placed a repeater in his room as well as a lamp (De Quincey, 1862, p. 142).

The content of Kant's dreams, we are told, would have provided the scenes of mighty tragedies. He did not wish to be held responsible for the motivations indicated by his dreams. "That was a cruel saying," wrote Kant, "and utterly opposed to experience, which is attributed to the Greek emperor who condemned a man to death that had been reported as having had a dream wherein he murdered the emperor: 'Well, he would not have dreamed it, if he had not thought about it while awake' " (De Quincey, 1862, p. 142). Evidently Kant too was murdering his own emperor, his father, in his dreams (Freud, 1916); and the father as monster would come to Kant every night to wreak his punishment. The themes of Kant's dreams suggest why every night in his life was filled with the reproaches of his conscience. Kant (1798) recorded them in a remarkable passage of the *Anthropology:*

Lawless Sensations and Categorial Defenses 231

"Thus, I well remember, have I, being a boy, tired out by play, laid me down to sleep, and in the moment of dropping off to sleep was quickly awakened by a dream, as if I had fallen into the water, and near drowning, was being turned around in a circle; but all in order to fall soon asleep again, and more quietly—probably because the activity of the chest muscles in breathing, which depends altogether upon the will, relaxes, and must therefore (the movement of the heart being checked by the stoppage of the breath) be revived by the imagination of the dream. To this we may also count the beneficial effect of dreams in the so-called nightmares (incubus). For without this terrible imagination of a monster that oppresses us, and the exertion of all our muscular power to change our position, the stoppage of the blood would soon put an end to our life. This seems to be the reason why nature has so arranged matters that most of our dreams involve difficulties and dangerous circumstances, since such pictures excite the forces of our soul more than dreams wherein everything happens according to our desire. We often dream that we cannot lift ourselves on our feet, or that we have lost ourselves, or stopped in the middle of a sermon, or through forgetfulness, put on a nightcap instead of a wig on entering a large assembly, or that we can fly in the air like a bird, or burst out in joyful laughter without knowing why. But it will probably remain a mystery forever, how it happens that in our dreams we are often transported back to long vanished times, and speak with people long since dead; and that, although we are tempted to look upon the whole occurrence as a dream, we nevertheless feel ourselves compelled to consider the dream an actuality. But we may probably accept it as certain that there can be no sleep without dreaming, and that a person who thinks he has not dreamed, has only forgotten his dream" (14:162).

From the vantage point of our knowledge of Kant's philosophical writings and biography, what can the themes of Kant's dreams tell us? Without exception, they all involve what we might call "categorial collapses,"—a breakdown in the causal sequence or time-order. We often dream, says Kant, (1) "that we cannot lift ourselves on our feet," (2) "that we have lost ourselves or stopped in the middle of a sermon," (3) "that through forgetfulness we put on a nightcap instead of a wig on entering a large assembly," (4) "that we can fly in the air like a bird," (5) that we "burst out in joyful laughter without knowing why,"

(6) that "we are . . . transported back to long vanished times," (7) that we "speak with people long since dead."

A certain continuity runs through Kant's dream themes. He can't lift himself on his feet, that is, he has a fear for his capacity for being erect, his masculinity. He is stopped mid-sermon; something disturbs him (we have already learned) as the clothes in disarray, the missing button in his lecture. Some lawless sensation is obtruding. What is the character of that sensation? It is indicated in the next theme; he would inwardly prefer to put on a nightcap rather than the wig. "The bed is the nest of numberless diseases," Kant (1797b) warns. "To sleep much at a time, or at intervals is a method of avoiding those cares to which we are exposed, when awake" (p. 249). The bed is the source of sin, or moral weakness. But in his dream, his unconscious speaks and he prefers the nightcap to the wig. Next, he flies like a bird—the classical symbol for sexual intercourse. He has overcome his anxieties and achieved orgasm. Then he bursts out in joyful laughter without knowing why. The preceding theme tells us why: He has enjoyed the liberation of sexual experience. Then he is back in vanished times, speaking with persons long since dead—not only his mother, but his father, we may infer, in view of the numerous murderer-robber dreams which Kant had. Indeed, Kant's dreams brought the reassurance that his father was dead— though guilt feelings always remained. Kant's father had died after a palsy-stroke incapacitated him for 18 months (Wallace, 1882). Kant, then not quite 20 years old, was too poor to pay burial dues, and there was no service at the grave. A man who has nursed his father during his last illness, and felt his death keenly, will often dream he is conversing with him as though he were alive. The dreamer, says Freud (1900) may have actually desired that death terminate his father's miseries. But when the father dies, the son bears the unconscious self-reproach that his wish contributed to shorten the dead man's life. The dream then consoles the son that now that his father is dead he is actually free, liberated from discipline.

The flying bird became for the old philosopher, in his last years, a pathetic symbol for the freedom he had never had. He had a "childlike love for birds in general," tells Wasianski, and "he took pains to encourage the sparrows to build above the windows of his study." He would then watch them with delight and tenderness. But one trait in Kant especially impressed his friend: "Of all the changes that spring carries with it, there was one only that now interested Kant; and he longed for it with an eagerness and intensity of expectation, that it became almost painful to witness: this was the return of a little bird (sparrow was it, or robin red-breast?) that sang in his garden, and before his window." The bird, or one like it, had sung there for years, "and Kant grew uneasy when the cold weather, lasting longer than usual, retarded its return" (De Quincey, 1863, p. 138).

Kant's dream as a boy seems to have revolved around his oedipal affection for his mother, his guilt before his father, and resultant potency fears; there was an alternation of fear and joy, like thesis and antithesis. Thus: Kant as a boy dreams of falling into water, nearly drowning, and being turned around in a circle, all in order to fall asleep again more contentedly. Falling into water, of course, is a familiar symbol for birth. In addition, however, the dream curiously identifies Kant with his mother. As Freud (1915–1916) observed, "If one rescues somebody from the water in a dream, one is making oneself into his mother" (p. 161), and Kant himself is being rescued from drowning in his dream. But here the surcease of quiet sleep comes in the aftermath of being turned around in a circle; he has had, as it were, intercourse with his mother, turning around in her watery circle. The dream brings contentment to Kant. He proposes a fanciful explanation for the dream's satisfaction: Its very traumatic aspects, like nightmares generally, serve to excite us and awaken our will to breathe and survive.

Dreams, according to Kant, keep us alive. Without their traumatic stimuli, we would relax into death. Kant's strange theory of dreams is actually a remarkable evidence of his own weariness with life, his longing for death. Unless our will, he says, were revived by the traumatic content of dreams, our chest muscles would relax and we would stop breathing. Just as the categorial understanding polices the sensations, so the will is supposed to stand guard over the chest muscles. An act of will is thus required to keep us breathing and alive. This curious theory that every breath requires an act of will is indicative of a strong underlying death-wish; he breathed as a matter of duty. The tremendous guilt associated with his oedipal fantasies made him long to die, yet he rationalized his dream-anxieties by saying they stirred in him the will to live.

Kant's Attack on the Psychological Method

The distinctive method of Kant is the so-called transcendental method. As always, the "method" that a philosopher adopts constitutes the most important of his protophilosophical postulates. And in Kant's case, what is important about the transcendental method is what it is not. What A. C. Ewing (1938) has called the "anti-psychological side" of Kant's thought is nowhere more evident than in his "transcendental deduction" of the categories. This, Kant insists, has nothing to do with what he calls an "empirical deduction," namely, a genetic account of the categories. The striking fact, however, is that though Kant devotes a huge labor to explaining what he means by a "transcendental deduction," it turns out to be a covert application of the psychological method. Every transcendental demonstration of a category consists of an attempt to show that "experience" would be impossible without it. This appeal is precisely to an experiment in imaginative introspection. What, asks Kant, keeps your experience together and differentiates it from irrational, uncategorized presentations? Uncategorized presentations are, according to Kant, what we have when we are insane, drunk, dreaming, or infants; the categories are the conditions of sanity, the conditions for the possibility of experience. Their "deduction" is an argument that we would be like insane people if we didn't accept them. This is an exploration into the descriptive psychology of sanity.

Experience itself, Kant (1781) acknowledges, could conceivably defy being subsumed under the categories: "Everything might be in such confusion that, for instance, in the series of appearances nothing presented itself which might yield a rule of synthesis and so answer to the concept of cause and effect" (p. 124). Such is presumably the experience of an insane man, or possibly the infant's experience—"the blooming, buzzing confusion" (James, 1890). One might refuse to call it "experience" in Kant's terms, and reserve "experience" only for the categorized contents of consciousness. The insane man's or child's experience would then require another word.

The actual items in Kant's "transcendental deduction" are much the same as those in genetic investigations (e.g., of Jean Piaget) into how the child gradually differentiates between objective and subjective reality. Kant (1781) argues that the workings of causality enable us to distinguish between objective and subjective successions of appearances: "Were it not so, were I to posit the antecedent and the event were not to follow necessarily there-upon, I should have to regard the succession as a merely subjective play of my fancy; and if I still represented it to myself as something objective, I should have to call it a mere dream" (p. 227).[12]

Yet Kant (1781) repeatedly denied that his transcendental undertaking had anything to do with psychological, genetic analysis: "We are indebted to the celebrated Locke for opening out this new line of [psychological] enquiry. But a *deduction* of the pure *a priori* concepts can never be obtained in this manner; it is not to be looked for in any such direction." We must show for the categories, he continues, "a certificate of birth quite other than that of descent from experiences." A "physiological" (psychological) derivation, he stated, would be an empirical enterprise, not a deduction. "Plainly the only deduction that can be given of this knowledge is one that is transcendental, not empirical." And with respect to pure *a priori* concepts, the effort of genetic psychology "is an utterly useless enterprise which can be engaged in only by those who have failed to grasp the quite peculiar nature of these modes of knowledge" (p. 122).

Kant, however, was in a quandary as to just what status to assign the categories with respect to genetic analysis. *A priori* concepts, he argued, cannot have an empirical origin, but, he acknowledged, they did have some sort of origin. Searching for an origin which was not empirical in character, an alternative which would preclude genetic analysis, he seized on an embryological concept; he postulated an "epigenesis of pure reason." The categories were thus exempt from causal law. Their origin was, so to speak, immaculate.

The use of the term, "epigenesis" shows to what desperate straits Kant had been driven. He was trying, on the one hand, to fend off genetic analysis, and, on the other hand, to preserve his "transcendental deduction" from the charge that he was engaged in reading off the decrees of the noumenal ego, an enterprise of rational psychology which was illicit from the standpoint of the *Critique*. Wolff, a young embryologist, had published a booklet, *Theoria generationis,* which introduced the concept of epigenesis. He showed that the organs of plants develop by differentiation from undifferentiated tissue at the tip of the growing shoot or root. The growth of the bud, Wolff wrote, was not an "unfolding" of something already preformed; rather, it was an epigenesis, a novel emergence. Something appeared which was not there before (Singer, 1931). This theory was a landmark in its time.

One must remember that the preformationist school held such notions as that Eve's ovary had within it the forms of all subsequent men and women (27 million of them), and that spermatozoa were packages of complete homunculi which could be observed under the microscope. Nevertheless, an epigenetic origin for the categories of the understanding signified a break within the causal order which violated the categories themselves. For an epigenetic origin was a minor creation of something out of nothing, and would have violated Kant's (1781) principles that "in all changes of appearances substance is permanent" (p. 212) and that "all alterations take place in conformity with the law of the connection of cause and effect" (p. 218).

Actually, in every case of transcendental deduction which Kant adduced, we find that what he did was to try to discover a psychological causal law of rational experience. Then, he attempted to repress the fact that he had been engaged in genetic psychology, and laid claim to a "pure deduction." In a minimal sense, every transcendental deduction is empirical in the sense of being a "deduction" by reference to the possibility of human experience, but Kant denied that this had any relevance to particular psychological causal laws (Ewing, 1938). Then, we may ask: what is this "transcendental deduction" which is empirical without being empirical, which discusses the conditions which make rational human experience possible but which has nothing to do with the causal laws of psychology?

Essentially, what Kant has done is to present in transcendental language the outcome of a long series of experiments commenced in infancy. The infant has to discover the distinction between material and psychic reality, between the objective and subjective. He discovers substance, the enduring, for instance, as he reaches in anxiety for his mother's breast and milk, and finds, at recurrent intervals, the same persisting breast and the same joy of milk.

Kant's antipsychological bent became so strong that in later years he denied that psychology could be a science (Kant, 1786b, p. 141). Yet he himself had studied anthropology, the customs of people, the sexual preferences of men, in the spirit of a psychologist looking for causal explanations. Why then this hostility to psychology? It was part of Kant's fear of the consequences of genetic analysis. Recognize a science of psychology, and the whole pseudopsychological apparatus of the "transcendental deduction," the whole law-enforcement agency of the categories, is endangered. As Ewing (1924) emphasizes, with the denial of a science of psychology, Kant's conception of a creative synthesis was entrapped "in a hopeless contradiction" (p. 66).[13] The creative synthesis can have one of two statuses: on the one hand, it can be taken as an event within the phenomenal world, an event of the psychical order, or on the other hand, the synthetic unifying activity may be regarded as noumenal, although operating from the noumenal realm on the sensations. If the first alternative is taken, "it should be the study of psychology not of epistemology"; if the second alternative is taken, "if it is noumenal, it is on Kant's own principles unknowable" (Ewing, 1924, p. 66). Actually, Kant treats synthesis as both phenomenal and noumenal; it is phenomenal when he regards it as a knowable psychical process; it is noumenal when he takes it as a condition for the existence of the phenomenal world. This was what happened when genetic facts were dephysicalized for a transcendental derivation. As William James wrote (1890): "I call this view mythological, because I am conscious of no such Kantian machine-shop in my mind, and feel no call to disparage the powers of poor sensation in this merciless way" (2:275).[14]

We cannot, however, rest with James' statement that he is not "conscious" of a Kantian machine-shop in his mind, for we can imagine Kant replying that James is not conscious of a machine-shop of his categories because they derive from an unconscious source. He would say that James talks about the contents of consciousness in the empirical self, while he was talking about the *a priori* legislation of the transcendental self.

Thus we must shift the gravamen of our analysis to the question of who is misreporting or repressing the contents of his unconscious? Is James repressing from consciousness the evidence for a transcendental mind which legislates *a priori* categories? Or did Kant's own unconscious motives lead to a tremendous reaction against the psychological method and the natural self so that he projected a transcendental self, which like all such projections coexisted inconsistently and precariously with knowledge?

The Antipsychological Revolution in Kant's Life

Kant, in his middle years, came to reject the psychological method of which he had been previously such an enthusiast. In his revulsion lay the basis for the separation of the transcendental from the empirical self. The transcendental self never fitted well into Kant's system. In fact, it was one of those "points of inconsistency" which cast a light on the philosopher's underlying conflicts. It was solely a formal presupposition; you knew the transcendental "I" existed, but you knew nothing more about it. To say that such a noumenal ego existed, without knowing some identifying characteristics about it, contravened Kant's own restriction of knowledge to the phenomenal realm. Yet Kant was convinced that this unknowable transcendental self existed. His transcendental self was a formal equivalent for the unconscious, repressed self. It was the unconscious subject which, as Kant insisted, could never be made object. That is, in psychological language, it could never be brought from the unconscious into consciousness. The theory of the transcendental self, in other words, reflected the tremendous self-repression on Kant's part. The estrangement of the transcendental from the empirical self corresponded to the intensity of the repression of his unconscious. Like the transcendental self, it was reduced in conscious experience to the vague sense of its presence somewhere.

Let us try to trace briefly the circumstances of the antipsychological revolution in Kant's life. The period in Kant's life when the British psychological method most appealed to him was that when, in Wenley's (1910) words, he was "the brilliant young *Docent,* a familiar in the *beau monde,* a frequenter of my lady's *salon,* a witty and courted conversationalist, a great reader of poetry and travels, a fastidious dresser even" (p. 80). Kant at this time, moreover, enjoyed the close friendship of the beautiful and witty Countess Keyserling. She was a young woman when Kant, a few years older, became tutor to her sons. The Countess evidently felt warmly towards her friend; she painted a highly sentimental portrait of Kant. He, on his part, treasured to his old age the wit with which she mocked at stuffed shirts. He called her "the ornament of her sex," and recorded in his *Anthropologie* a joke she had told him about a pompous Grandmaster of the Knights of Malta. Despite their differences in social origin, Kant occupied at table the place of honor next to the young Countess (Fromm, 1898, pp. 144–155).[15]

Kant evidently began to read the British moralists around 1756 (Schlipp, 1939). Under their influence he wrote *Observations of the Feeling of the Beautiful and Sublime* (1764a), and *Dreams of a Spirit-Seer* (1766) which gave him the reputation of a stylist, as a German *La*

Bruyère. Kant was in so much of a "social whirl" that his friend, the gifted Hamann, worried that social distractions would interfere with his philosophical labors.[16] He rejected his Pietist upbringing, the church of his father; he refused to go to church even on the university's ceremonial occasions.

At this time, Kant was emphatically a follower of Shaftesbury (1711) according to whom a key philosophical method was ridicule, raillery, and humor: "Truth, 'tis supposed, may bear all Lights: and one of those principal Lights or natural Mediums, by which Things are to be view'd, in order to a thorough recognition, is Ridicule itself, or that Manner of Proof by which we discern whatever is liable to just Raillery in any subject' (p. 61).[17] In a Shaftesburyan vein, Kant stated that he would not blame the reader if he were to dispatch spirit-seers as "candidates for the hospital." With excretory sarcasm, Kant declared metaphysics was the outcome of disordered digestions. He thought that perhaps purgatives might be the therapy for metaphysics. Perhaps as "the keen Hudibras" surmised, "visions and holy inspirations are simply caused by a disordered stomach." One passage of Kant's analysis, "the outspokenness of which is hardly bearable in English," according to the translator, was rendered in a censored version. The original (1776) states: "if a hypochondriacal wind rages in the bowels, then it depends on which direction it takes, if it goes downwards, then comes thence a ____, if it rises however upwards, then it is either a vision or a holy inspiration" (p. 84). The reference to the *Hudibras* of Samuel Butler (1612–1680) is noteworthy, for *Hudibras,* "the most memorable burlesque poem in the English language" (according to Sola Pinto), mocks at the militant Puritans, their metaphysics and fanaticism. The peculiarity of insanity, to which metaphysics was akin, said Kant, was "that the confused individual places objects of his imagination outside himself, and considers them to be real and present objects." This confusion of the subjective and objective, of the psychical with the physical, was, in Kant's diagnosis, the defining characteristic of the metaphysical disease.

Metaphysics was regarded by Kant, during his psychological period, as belonging to the pathology of the human mind.[18] "No metaphysician," Kant declared, "could do as much good in the world as Erasmus of Rotterdam and the celebrated Montaigne of France had accomplished" (Stuckenberg, 1882, p. 120). A sense of humor was the surest antidote to metaphysics. Because to Kant, metaphysics was a mental aberration with physiological causes, he was interested, throughout his life, in the psychology of crackpots and clairvoyants, from Swedenborg to Cagliostro.[19] "The number of crackpots increases from year to year," Kant (1798) observed. His first essay on this subject, "Versuch über die Krankheiten des Kopfes" (1764b) was called forth by the appearance in Königsberg of a wild prophet from the woods, in nomadic accouter, and bearing a Bible. His name was Jan Pawlikowicz Zdcmonyrskich Komarnicki, and he was accompanied by a herd of cattle and a boy of eight who charmed Kant as a Rousseauan savage who had "none of the bashful awkwardness caused by bondage or compulsory lessons of attention" (Wallace, 1882, p. 111). Kant was much interested in studying the crackpot whom the people called the goat-prophet. Noting that an insane man could still be very intelligent, he classified mental illness into three categories—hallucination, delirium, and mania—and noted that by a common blind spot men do not see what exists but what conforms to their inclinations. A woman looks at the moon through a telescope and thinks she sees the shadow of two lovers, but her pastor sees the two towers of a cathedral. But melancholia, said Kant, was found perhaps in everyone. It was, according to him, accompanied by what we would now call anxiety, and it led people to fear that they were laughable, or to laugh without seeming cause. These very themes (as we have seen) appeared in Kant's dreams. "All sorts of obscure representations excite in him a violent desire to do something evil, by the idea of which he is much tormented, although it never comes to the fact" (Kant, 1764b, p. 310). This was Kant commenting on his own melancholy and distinguishing it from the visionary, exalted fanaticism of those like Mohammed and John of Leyden, whose cult of divine inspiration was truly dangerous.

As a mature man of the Enlightenment, Kant insisted at this time that mental disorders had physiological causes. Indeed, he housed their etiology, as we have seen, in the digestive system: "I cannot persuade myself," he wrote (1764b), "that the disorders of the mind must result, as is commonly believed, from pride, love, too hard study, and other misuses of the spiritual resources.... This confuses the cause and effect." Medicine, with some help from philosophy, could cure these disorders with proper dietetic rules for the spirit. Some intellectual laxative or cathartic could be devised: "For if, according to Swift's observations, a bad poem is simply a purging of the brain ... why shouldn't the writing of a pitiable chimera have the same virtue? But it would then be prudent to indicate to nature an alternative way of cleansing, so that the sickness could be basically and quietly cured, without everybody else being disturbed by it" (p. 315).[20]

The raillery of Kant in *Dreams of a Spirit-Seer* was part of his own effort at psychological therapy to rid himself of metaphysical tendencies. As he confessed to Moses Mendelssohn, he had to struggle with tendencies in himself much like those of Swedenborgian mysticism: "the attitude of my own mind is inconsistent and, so far as these stories are concerned, I cannot help cherishing an opinion that there is some validity in these experiences in spite of all the absurdities involved in the stories about them. . . ." (Kant, 1766, p. 162).[21] His work was thus in large part, as Kant was aware, one of self-criticism.

His own struggle for liberation from metaphysics had been a hard one: "I have purified my soul from prejudices. . . . [Now] I observe my judgments, together with their most secret causes, from the point of view of others" (pp. 85–86). It was a psychological inquiry into "the most secret causes" of his beliefs which had liberated him from them—accomplished the transition from dogmatism to scepticism. Within a few years, however, Kant's therapy of raillery was confronted with new, far deeper resistances than before (Wallace, 1882).

Kant's lectures on ethics at this time still presented him forthrightly as one working in the method and fashion of the English psychological school. The announcement for his lectures for the winter semester of 1765–1766 stated: "The attempts of Shaftesbury, Hutcheson, and Hume, which although unfinished and deficient, have nonetheless progressed farthest in the search for the first principles of all morality, will receive that precision and supplementation which they now lack. And since in ethics I always consider historically and philosophically what happens before I point out what ought to happen, I shall make clear the method by which one must study man. . . . This method of ethical investigation is a pretty discovery of our times and, if one considers it in its entire plan, was altogether unknown to the ancients" (Schilpp, 1939, p. 8). Thus, at this time, Kant, like his master, Shaftesbury, used the concept of taste to write in a naturalistic fashion concerning such questions as "the relation of the sexes." "The European alone," wrote Kant (1764a), "has found the secret of decorating with so many flowers the sensual charm of a mighty inclination," whereas the Oriental "is of a very false taste," and "thrives on all sorts of amorous grotesqueries" (pp. 112–113). Underlying the principle of justice for Kant at this time was not an emotion-free categorical imperative but the more Humean-Shaftesburyan "universal affection toward the human species" (p. 58).

When Kant wrote that Hume had awakened him from his "dogmatic slumber," what he probably had in mind was that the genetic-psychological method had first put to rout that of the Leibnizian metaphysics.[22] Genetic analysis had undermined the metaphysical tendency. Kant's favorite metaphor for metaphysics became one which characterized it as a desexing activity. In his *Inaugural Dissertation* (1770), he wrote that "it commonly happens that one of the disputants appears as it were to be milking a he-goat and the other to be holding a sieve," and again in the *Critique of Pure Reason* (1781) he averred that the propounding and answering of absurd questions not only brings "shame" on the former but presents one, "as the ancients said, the ludicrous spectacle of one man milking a he-goat and the other holding a sieve underneath" (p. 97). A metaphysical questioner thus brings a "shame" upon himself akin to that of a man who blunders sexually, who mistakes male for female. Rationalistic metaphysics in other words operated to feminize one, to shame one. This was the "dogmatic slumber," i.e., the extinction of one's manhood.

Kant, struggling to rid himself of melancholy and guilt, was assisted by the psychological method of Hume. Here was a method which was bold, clear, masculine, without chimeras. Kant was led to revel in the English novels, in Fielding's *Tom Jones* and *Jonathan Wild*, in Richardson's *Clarissa*, in Laurence Sterne's *Tristram Shandy*, in Jonathan Swift's satires and Butler's *Hudibras*. In the spirit of an anthropological relativist, Kant pondered the cultural variations of the Malays, Peruvians, Jews, American Indians, Otaheite tribe, Connecticut Puritans, Scottish Highlanders, Swedes, Tunguses, Russians, Siberian Shamans, and Tahitians (Kant, 1798, 2:20–24, 32; 9:411; 14:159–165; 15:66).[23]

But in the period around 1769, there took place what Paulsen (1902) refers to as "the revolution of 1769" (p. 311). It was not so much a revolution as a counter-revolution against the psychological method. "The year 1769," wrote Kant, "brought me a great light" (Ward, 1922, p. 8).[24] A year before, he had still believed that space was something external to the mind, and he had derived a theory of absolute space from the orientations of the left and right hands; the following year, he believed it was solely a form of intuition (Stuckenberg, 1882; Ward,

1922). In his *Inaugural Dissertation,* Kant (1770) announced his definitive break with the English psychological school. Together with Epicurus, Shaftesbury and his school were to be "quite rightly condemned," for trying to reduce the criteria of morals to the feelings of pleasure or unpleasantness (pp. 60–61). As Paulsen (1902) stated: "Pure reason alone is to be considered. As contrasted with it all empirical principles are 'impure.' " From this time, we may surmise, dates Kant's transformation into a person "in whose heart nature has placed little sympathy, who is naturally cold and indifferent to the sufferings of others; perhaps, being endowed with great patience and endurance, he makes little of his own pains, and presupposes or even demands that every person should do the same" (p. 335).

We are at a loss as to what events during Kant's middle years produced this moral revolution. That he was returning from the Voltairean-Shaftesburyan ethics to the Pietist standpoint is clear when we note that it was precisely the latter which held as Kant did (1803, p. 269; 1794, p. 43), that moral change had to be revolutionary. As Pinson (1934) wrote: "The most characteristic experience and central point in the life of a Pietist was the *Wiedergeburt* or the regeneration. The *Wiedergeburt* became the dominant motif . . ." (p. 115).[25] Furthermore Kant's change of attitude toward lying was indicative of the reinstatement, in his mind, of paternal authority. The Kant of the psychological method was reflected in the tolerant statement: "Many children have a disposition to lie, which has no other cause than a vivacious imagination." The Pietist Kant now added (1803): "It is the father's affair to see to it that they break off this habit . . . " (pp. 194–195). He wrote that lying was "the really corrupt spot in human nature," though the psychologist in Kant recognized that the inclination to "guileless lying" is always found in children and "now and then in adults, otherwise good" who when telling "alleged adventures" cannot prevent their imagination from "growing like an avalanche . . . with no intention whatsoever than merely to be interesting" (p. 194).[26] Again, we are struck by Kant's reference to the father as the one who has to condition the child, a natural liar, to become a truth-teller. The tremendous importance Kant came to assign to truth-telling, the horror which he came to attach to lying, suggest the reinstatement of his father's authority over that of the psychological culture. However, the source of the crisis of paternal reinstatement, of the restoration of the superego, eludes us. Freud (1913), in studying children's lies, said that they can be designed to conceal reminders of oedipal feelings or hidden incestuous love. Later, he stated that "Kant's Categorical Imperative is thus a direct heir of the Oedipus complex" (Freud, 1924, p. 167).

Kant's expulsion of all feeling from the categorical imperative was the basic characteristic of his moral doctrine; its repressive rigorism was part of the same struggle to contain lawless sensations which was the aim of the controls of the *a priori* categories. Kant's Voltairean, hedonistic tolerance was replaced with the principle that: "To act morally is to do what one does not want to do" (Paulsen, 1902, p. 332). Good acts, as Kant wrote, could be carried out without any feeling: "pardon without sympathy, conjugal faithfulness without love" (Schilpp, p. 115). A world of ethical automata, emasculated from feeling but obedient to principle, became the highest moral projection. One wonders with Freud if an ethic of pure "formal" principle is not one dominated by an underlying castration complex, in which emotion has been extirpated.

Kant himself recognized that he had gone through a series of psychological-intellectual stages, and he believed that his psychological evolution recapitulated the psychological history of mankind. Comte and Freud both proposed laws of psychological evolution, and Kant was their precursor. "The first step in matters of pure reason, marking its infancy, is *dogmatic*. The second is *sceptical*: and indicates that experience has rendered our judgment wiser and more circumspect. But a third step, such as can be taken only by fully matured judgment, is now necessary.... This is not censorship but the criticism of reason..." (Kant, 1781, p. 607). This law of evolution, from dogmatic to sceptical to critical, Kant believed, was inherent in the nature of the human mind. He wrote in his notes: "Can a history of philosophy be written mathematically (this must mean dogmatically, or from concepts)? Can we show how dogmatism must have arisen; and from it scepticism, and that this necessarily leads to criticism? Yes, if the idea of a metaphysic inevitably presses on human reason, and the latter feels a necessity to develop it; but this science lies entirely in the mind although only outlined there in embryonic form" (Paulsen, 1902, p. 290). Kant's law of intellectual evolution and its three sequential stages was patterned on stages of his emotional development, which he subsequently transposed into logic. We have not, however, been able to unravel the source of the transitional drive, the emotional revolution in Kant's later life.

Perhaps Kant developed resistances to the British psychological method to the degree that it tended to uncover so much of what he found "loathsome" in his own nature. Why, for instance, was his own nature so cold? Why was he so lacking in the benevolent affection that Shaftesbury regarded as natural in human beings (Fowler, 1882)? Shaftesbury had traced the origin of the moral sense to feelings of fellowship and family affections. By analogy, Kant would have had to inquire into the origins and consequences of his own disturbed family life, e.g., his coldness to his sisters and his condescension toward his father. It would have obliged him to look deep within himself to investigate his attitudes toward mother and father, something which he preferred not to do. He must have been aware of the latent threat in Hume's "experimental method" to the very basis of his moral life. Hume (1738) had, for instance, blandly found that a child's murder of its parent was as natural as a sapling destroying its parent tree: "I ask, if, in this instance, there be wanting any relation which is discoverable in parricide or ingratitude? Is not the one tree the cause of the destruction of the former, in the same manner as when a child murders his parent?" And Hume had gone on even further to query the moral turpitude of incest, asking "why incest in the human species is criminal, and why the very same action, and the same relations in animals, have not the smallest moral turpitude and deformity?" (pp. 175–176). These explorations must have seemed as "monstrous" to Kant as did the evolutionary speculations concerning a promiscuous Mother Nature, and would have led to his finally drawing back in horror from the psychological method.

Rousseau, the revolutionary primitivist-populist, symbolized in Kant's mind his emotional revulsion against the psychological method. During the latter sixties, what de Vleeschauer (1962) calls a "change in tonality" in Kant's thinking was taking place, and Rousseau came to represent the reinstatement of his father's moral supremacy (p. 41). Kant described this change in himself in an oft-cited passage: "There was a time when I despised the masses who know nothing. Rousseau has shown me my error. This dazzling advantage vanishes, and I should regard myself as of much less use than the common laborers if I did not believe that this speculation (that of the Socratic-critical philosophy) can give a value to all others, and to restore the rights of humanity" (Paulsen, 1902, p. 39; Schilpp, 1939, p. 40).

Kant shared none of Rousseau's optimism, sentimentalism, and worship of feeling. But in Rousseau, Kant found the reverence for the "common laborer," i.e., his own father. The son's pilgrimage was over. His early revolt against his Pietist father had been social as well as philosophical. Kant had found his way into the social life of the Königsberg aristocracy. He had been tutor to their children and a guest at their parties. He had put aside his associations with his humble family. Now, however, he recalled the deep moral character of his father: "I still remember how once disputes arose between the harness-making and saddler trades regarding their privileges, during which my father suffered much. But, nevertheless, this quarrel was treated by my parents, even in family conversation, with such forbearance and love towards their opponents, and with so much trust in Providence, that the memory of it, although I was then a boy, has never left me" (Paulsen, 1902, p. 39). The Voltairean-Shaftesburyan standpoint was essentially that of the enlightened middle classes and aristocracy. Kant partially renounced their genetic-psychological method. The critical philosophy would come to terms with the Pietist-Rousseauan values, with the outlook of the "common laborer." Such was the psychological drama which underlay Kant's revolution of 1769.

In a sense, this essay is written in the tradition of the psychological method which Kant followed in his middle years. Only rarely has any scholar proposed that the psychoanalytic method be applied to Kant's philosophy. Hall (1912), a lonely pioneer, thought that unconscious feelings and motives determined Kant's postulates (p. 414), and Pfister (1923, p. 78) spoke of the *Zwangsneurose des Schematisieren* (the compulsion to schematize).[27] But most philosophers have shared the view of Weldon (1958): "There is no reason to suppose that his [Kant's] experiences, except in the strictly intellectual sphere, were of the slightest interest or importance; and if they were, we shall certainly never know it, since his biographers could discover nothing but the most meagre trivialities to record of him" (p. 1). Actually "trivialities" can be psychologically momentous, and Kant's own psychological writings belie the notion that he was a creature of pure intellect. We have tried to unravel the nonlogical, emotional, and unconscious determinants of Kant's philosophy.

Notes

1. On the symbolic use of "Cyclopean" as indicative of physical and intellectual one-sideness, see Freud (1900, p. 443).
2. "Moral friendship," wrote Kant, "is the complete confidence of two persons in revealing their secret thoughts and feelings to each other, in so far as such disclosures are consistent with mutual respect for each other." This "need to reveal himself to others is strong," but "hemmed in and cautioned by fear of the misuse others may make of the disclosure of his thoughts, he finds himself constrained to lock up in himself a good part of his opinions . . ." (Kant, 1797a, p. 143). Kant warned: "Even to our best friend we must not reveal ourselves, in our natural state as we know it ourselves. To do so would be loathsome." Friendship, said Kant, was "man's refuge in this world from the distrust of his fellows," but his loathsomeness in his own eyes, always unexplained, made the communion thus sought impossible (Kant, posth., 1930, pp. 206–207). Cf. also Paton (1956), pp. 54–55.
3. It is noteworthy that the one American philosopher who struggled all his life to retain mastery over his lawless, wayward impulses also had a tremendous emotional fixation on Kant's categorial system. Peirce wrote: "I was a passionate devotee of Kant. . . . I believed more implicitly in the two tables of the Functions of Judgment and the Categories than if they had been brought down from Sinai" (Murphey, 1961, p. 33).
4. Tillich has interpreted the categories subjectively as the union of anxiety and courage. This view is not unlike the analysis of the Kantian categories which I am presenting. See Thomas (1963, p. 116).
5. The article, unsigned, was by Wilhelm von Humboldt, and entitled "Über end Geschlechsunterschied und dessen Einflusz auf die organische Natur."
6. Poet-advocates of sexual revolution have intuitively seized on the Kantian terminology to express their aim: "the barrier of noumenon-phenomenon/transcended/ the circle momentarily complete/" Kandel (1966, p. 5).
7. Kant (1798) used the German word, *Verrückung*.
8. Again, in 1794, Kant referred to "the leading-string of holy tradition" which "becomes bit by bit dispensable, yea, finally, when man enters upon adolescence, it becomes a fetter" (p. 112).
9. Kant always mocked at women for talking too much, for their *übermundigkeit* (verbosity).
10. The youthful Kant had united an evolutionary cosmology with an optimistic conception of the rule of a First Cause effectuating "harmonies and beauties." In his student days, when Kant was in bold revolt, Lucretius, the Roman materialistic evolutionist, was his favorite poet, and he memorized long passages of the *De Rerum Natura*, and delighted in the grandeur of its imagery. Kant (1755, pp. ix, 22, 24, 26, 148–150, 154).
11. According to Freud (1900), in every case, "the robbers stood for the sleeper's father. . . . Robbers, burglars, and ghosts, of whom some people feel frightened before going to bed, and who sometimes pursue their victims after they are asleep, all originate from one and the same class of infantile reminiscence" (pp. 403–404). Freud had noted, too, that the moon in the child's consciousness was used to signify the place of its birth. According to Stekel (1935), "The pale moon symbolizes the white rump out of which in the infantile theory of sex, the baby is born" (p. 115). Thus, the moonbeam would have awakened Kant's oedipal feelings and anxieties.

12. Also, cf. Ewing (1924). pp. 156–164.
13. See also Ward (1922), pp. 48, 58, 59, 154, 155, 166.
14. See also James, *The Principls of Psychology,* Vol. 1, p. 363.
15. Also, cf. Wallace (1882), p. 21; Paulsen (1902), p. 34; Cassirer (1923), 3:152–153. The exact circumstances, place, and date of Kant's tutorship to the Countess's sons are not clearly known.
16. It was Hamann who wished to be an Alcibiades to Kant's Socrates (Smith, 1960, p. 238). Kant confessed himself too much a "poor son of the earth" to fathom his friend's "divine language of the intuitive reason" (Smith, 1960, pp. 46–47). On Kant's refusal to attend church, see Wallace (1882, p. 46) and Stuckenberg (1882, p. 354).
17. Kant wrote: "According to Shaftesbury, it is a valuable touchstone for any truth if it can stand mockery" (Rabel, 1963, p. 291). Kant also greatly admired Lichtenberg who was celebrated for such observations as: "In woman the seat of the *point d'honneur* coincides with the center of gravity. . . . Everyone should study at least enough philosophy and *belles lettres* to make his sexual experience more delectable" (*The Lichtenberg Reader,* pp. 49, 50; Stuckenberg, 1882, p. 119).
18. Kant, according to Hamann, regarded Mendelssohn's lectures as a system of illusion; "they are to him similar to Mendelssohn's description of a lunatic" (Stuckenberg, 1882, p. 404).
19. Cagliostro inspired the essay in 1790, "Über Schwârmerei und die Mittel dagegen." The Fifth Monarchy men provided an example of crackpots in his *Versuch.* Kant referred to the "wonderful sayings" of Holberg that "the number of crackpots grows from day to day," and that they might take it into their heads "to found a Fifth Monarchy." Mohammed and John of Leyden were prototypes of crackpots (Cassirer, 1923, 2:307–311).
20. Author's translation.
21. Broad (1949–1952, p. 86) has noted the evidence in Kant's correspondence of his unresolved attraction toward Swedenborgian doctrines of the supernatural.
22. Kant at different times gave different explanations for what had awakened him from his "dogmatic slumber." To Professor Christian Garvé, he wrote in 1798 that it was thinking about the antinomies "that awakened me from my dogmatic slumber and drove me to a critique of Pure Reason, in order to remove the scandal of an apparent contradiction of Reason with itself" (Rabel, 1963, p. 356). Kant's account of the influence of Hume is set forth in his *Prolegomena to any Future Metaphysics* (Kant, 1784, p. 7).
23. See also Kant (1803), pp. 132, 139, 162.
24. Cf. Kant (post.), p. 4 and Stuckenberg (1882).
25. Pietist nurture of children inculcated an intimate sense of God's Presence. The biographies of Kant's older contemporary, Zinzendorf, give a vivid picture of what Pietism must have meant in Kant's upbringing. See Meyer (1928), pp. 132, 202; Weinlick (1956), p. 28.
26. Cf. Jean Piaget (1932), pp. 160–165.
27. See also Loewenberg (1953).

References

Boas, G. (1948), The Role of Protophilosophies in Intellectual History. *J. Philos.,* 45:673–684.
———. (1957), *Dominant Themes of Modern Philosophy.* New York: Ronald Press.

Broad, C. D. (1949–1952), Immanuel Kant and Psychical Research. *Proc. Soc. Psychical Res.*, 49:79–104.
Cassirer, E., ed. (1918), *Briefe von und an Kant, 1749–1789*. Berlin: Bruno Cassirer.
———, ed. (1921–1923), *Immanuel Kants Werke*. Berlin: Bruno Cassirer.
Clark, R. T., Jr. (1955), *Herder: His Life and Thought*. Berkeley: University of California Press.
De Quincey, T. (1862), The Last Days of Immanuel Kant. In: *De Quincey's Works*, Vol. 3. Edinburgh: Adam & Charles Black.
Ewing, A. C. (1924), *Kant's Treatment of Causality*. London: K. Paul, Trench, Trubner.
———. (1938), *A Short Commentary on Kant's Critique of Pure Reason*. London: Methuen.
Fowler, T. (1882), *Shaftesbury and Hutcheson*. London: S. Low, Marston, Searle and Rivington.
Freud, S. (1900), The Interpretation of Dreams. *Standard Edition*, 4 & 5. London: Hogarth Press, 1953.
———. (1910), The Psycho-Analytical View of Psychogenic Disturbance of Vision. *Standard Edition*, 11: 209–218. London: Hogarth Press, 1957
———. (1913), Infantile Mental Life: Two Lies Told by Children. *Standard Edition*, 12:305–309. London: Hogarth Press, 1950.
———. (1915–1916), Introductory Lectures on Psycho-Analysis. *Standard Edition*, 15 & 16. London: Hogarth Press, 1963.
———. (1924), The Economic Problem in Masochism. *Standard Edition*, 19: 159–172. London: Hogarth Press, 1950.
———. (1930), Civilization and Its Discontents. *Standard Edition*, 21:64145. London: Hogarth Press, 1961.
Fromm, E. (1898). Das Kantbildnis der Gräfin Karoline Charlotte Amalia von Keyserling. In: *Kantstudien,* 2. Cologne: Kölner Universitätsverlag.
Hall, G. S. (1912), Why Kant Is Passing. *Amer. Journal Psychol.*, 23:370–426.
Heine, H. (1835), *Religion and Philosophy in Germany*. London: Trübner, 1882.
Herz, M. (1801), In: *Jewish Encyclopedia*, 6:368. New York: Funk & Wagnalls, 1901–1906.
Hume, D. (1738), *A Treatise of Human Nature*. New York: Dutton, 1930.
James, W. (1890), *The Principles of Psychology,* 1 & 2. New York: Holt, 1910.
Kandel, L. (1966), *The Love Book*. San Francisco.
Kant, I. (1755), *Kant's Cosmogony*. Glasgow: J. Maclehose, 1900.
———. (1762), *Kant's Introduction to Logic*. Philadelphia: 1886.
———. (1764a), *Observations on the Feeling of the Beautiful and Sublime*. Berkeley: University of California Press, 1960.
———. (1764b), Versuch über die Krankheiten des Kopfes. In: *Immanuel Kants Werke*. 2:301–311. Berlin: Bruno Cassirer, 1922.
———. (1766), *Dreams of a Spirit-Seer, Illustrated by Dreams of Metaphysics*. London: New-Church Press, 1900.
———. (1770), *Kant's Inaugural Dissertation and Early Writings on Space*. Chicago: Open Court, 1929.
———. (1781), *Immanuel Kant's Critique of Pure Reason* (Second Edition). London: Macmillan, 1963.
———. (1783), *Prolegomena to Any Future Metaphysics*. Chicago: Open Court, 1929.
———. (1786a), Conjectural Beginning of Human History. In: *Kant on History*. Indianapolis: Bobbs-Merrill, 1963, pp. 53–60.
———. (1786b), *Prolegomena and Metaphysical Foundations of Natural Science*. London: G. Bell, 1909.

———. (1790), *Kant's Critique of Teleological Judgment*. Oxford: Clarendon Press, 1928.
———. (1794), *Religion Within the Limits of Reason Alone*. New York: Harper, 1960.
———. (1797a), *The Doctrine of Virtue*. New York: Harper, 1964.
———. (1797b), Von der Macht des Gemüths durch den blossen Vorsatz seiner krankhaften Gefühle Meister zu Sein. In: *Immanuel Kants Werke,* 7:411–431. Berlin: Bruno Cassirer, 1916. On the Power of the Mind in Overcoming Unpleasant Sensations by Mere Resolution. In: J. Sinclair, trans., *The Code of Health and Longevity,* 3:245–259. Edinburgh: 1807.
———. (1798), Anthropology of Immanuel Kant. *J. Speculative Philos.,* 9 (1875):16–27, 239–245, 406–416; 10 (1876):319–323; 11 (1877): 310–317, 353–363; 14 (1880):154–169; 15 (1881):62–66; 16 (1882):47–52, 395–413.
———. (1803), *The Educational Theory of Immanuel Kant*. Philadelphia: Lippincott, 1904.
———. (posth.), *Lectures on Ethics*. London: Methuen, 1930.
———. (posth.), *Reflexionen Kants zur kritischen Philosophie,* 2. Leipzig: Fues's Verlag, 1882–1884.
The Lichtenberg Reader. Boston: Beacon Press, 1959.
Loewenberg, R. D. (1953), From Immanuel Kant's Self-Analysis. *Amer. Imago,* 10: 307–322.
Lovejoy, A. O. (1907), Kant's Classification of the Forms of Judgment. *Philos. Rev.,* 16:588–591.
———. (1911), Kant and Evolution. In: *Forerunners of Darwin: 1745–1859*. Baltimore: Johns Hopkins Press, 1959, pp. 173–207.
"Marcus Herz." *Jewish Encyclopedia,* 6:368. New York: Funk & Wagnalls, 1901–1906.
Meyer, H. W. (1928), *Child Nature and Nurture*. New York: Abingdon Press.
Murphey, M. G. (1961), *The Development of Peirce's Philosophy*. Cambridge, Mass.: Harvard University Press.
Paton, H. J. (1956), Kant on Friendship. *Proceedings of the British Academy*. London: Oxford University Press, pp. 45–66.
Paulsen, F. (1902), *Immanuel Kant: His Life and Doctrine*. New York: Scribner.
Peirce, C. S. (1933), *Collected Papers of Charles Sanders Peirce,* 4. Cambridge, Mass.: Harvard University Press.
Pfister, O. (1923), *Zur Psychologie des philosophischen Denkens*. Bern: Bircher.
Piaget, J. (1927), *The Child's Conception of Physical Causality*. New York: Humanities Press, 1930.
———. (1932), *The Moral Judgment of the Child*. Glencoe, Ill.: Free Press, 1948.
Pinson, K. S. (1934), *Pietism as a Factor in the Rise of German Nationalism*. New York: Columbia University Press.
Rabel, G. (1963), *Kant*. Oxford: Clarendon Press.
Royce, J. (1919), *Lectures on Modern Idealism*. New Haven: Yale University Press.
Schilpp, P. A. (1939), *Kant's Pre-Critical Ethics*. Evanston: Northwestern University Press, 1960.
Schopenhauer, A. (1818), *The World as Will and Idea*. New York: Doubleday, 1961.
Shaftesbury, A. (1711), *Characteristicks* (5th ed.). Birmingham, 1773.
Singer, C. (1931), *A Short History of Medicine*. Oxford: Clarendon Press.
Smith, R. G. (1960), *J. G. Hamann; 1730–1788: A Study in Christian Existence*. New York: Harper.
Stekel, W. (1935), *The Interpretation of Dreams*. New York: Grosset, 1962.

Stern, B. J. (1927), *Should We Be Vaccinated? A Survey of the Controversy in its Historical and Scientific Aspects.* New York: Harper.
Stuckenberg, J. H. W. (1882), *The Life of Immanuel Kant.* London: Macmillan.
Thomas, J. H. (1963), *Paul Tillich: An Appraisal.* Philadelphia: Westminster Press.
Vaihinger, H. (1898), Kant als Melancholiker. *Kantstudien,* 2:139–141.
Vleeschauwer, H. J. de (1962), *The Development of Kantian Thought.* London: T. Nelson.
Wallace, W. (1882), *Kant.* Edinburgh: Blackwood.
Ward, J. (1922), *A Study of Kant.* Cambridge: Cambridge University Press.
Weinlick, J. R. (1956), *Count Zinzendorf.* New York: Abingdon Press.
Weldon, T. D. (1945), *Kant's Critique of Pure Reason.* Oxford: Clarendon Press, 1958.
Wenley, R. M. (1910), *Kant and His Philosophical Revolution.* Edinburgh: T. T. Clark.

11

The Dream of Benedict de Spinoza

A metaphysical system is a projection which issues from the deepest anxieties of men. Perhaps the most truthful way of writing a philosophy would be in the form of an autobiography. We would understand why a man saw reality as he did, what traits of reality he wished to repress, what he was motivated to exaggerate. But philosophers are great rationalizers. Each of them prefers to see his philosophy as the rational, scientific system. Each of them erects an edifice of argument, but tends to be most sparing as to personal details concerning himself. We are permitted to see only the finished play, never the actor as he really is, adjusting his make-up, rehearsing his lines. But here and there, in uncensored asides, in letters to their friends, in dreams, we get an invaluable clue which leads into the labyrinth of the philosopher's unconscious life. We even get a glimpse of components in the philosopher's unconscious which rebel against his own conscious metaphysics, and condemn it as false. And such is the one dream which Spinoza recorded. It was in a letter to his friend Balling in 1664:

"I can confirm and at the same time explain what I have just said by something which happened to me at Rhynsburg last winter. When I awoke one morning, when the sky was already growing light, from a very heavy dream, the images which had come to me in my dream remained as vividly before my eyes as if they had been real things, especially the image of a certain black and scabby Brazilian whom I had never seen before. This image for the most part disappeared when, in order to divert myself with something else, I fixed my eyes on a book or some other object: but as soon as I again turned my eyes away from such an object and fixed them on something inattentively, the image of the same Ethiopian again appeared with the same vividness, and that again and again until gradually it disappeared from my presence." (16, p. 139–140)

This dream made a tremendous impression on Spinoza. A half-year after its occurrence, he could still recall vividly its dominant threatening image. What were the circumstances of this dream? Shortly after his excommunication from the Amsterdam Jewish synagogue in 1656, Spinoza went to live among the Collegiant-Mennonite community at Rhynsburg. The Mennonites were a pacifist mystical group, with a vaguely communistic heritage derived from their Anabaptist forerunners. The community at Rhynsburg in their theology was much akin to the English Quakers, though they expressed their mysticism in the language of Descartes. Pieter Balling was a noble representative of this group. In 1662 he wrote a tract on his spiritual mysticism and self-discovery, *The Light on the Candlestick,* which later found its place in English Quaker literature. (14, p. 713–723) He was devoted to Spinoza, and prepared the Dutch translation of his first published work, *The Principles of Descartes' Philosophy.* At this time, furthermore, Balling was the intermediary between Spinoza and a society of young men who met at Amsterdam to discuss Spinoza's ideas. Balling would bring them portions of Spinoza's writings when he went to Amsterdam on his business journeys, and the society would send back its questions to Spinoza. They were a bold, questioning group of minds, and they looked to Spinoza for intellectual leadership in their struggle with Calvinist orthodoxy. As one of them, Simon de Vries, wrote to Spinoza: ". . . under your leadership we may be able to defend the truth against those who are superstitiously religious or even Christian, and to stand firm against the onslaught of the whole world." (16, p. 102)

Now in 1664, Balling's child had sickened and died. The father turned, grief-stricken, to Spinoza, and wrote how he had heard omens of his child's death even when the child was well,—"Sobs like those it uttered when it was ill and just before it died." Spinoza wrote with warmth to comfort his friend: "It has caused me no little sadness and anxiety, although this has certainly diminished now that I dwell on your good sense. . . . But my anxiety daily grows more; and therefore I beseech and adjure you by our friendship not to mind writing to me fully." Then Spinoza undertook to discuss the significance of Balling's premonitions. Are dreams omens of some future event? He compares Balling's experience with his own dream, and says, "Your case was an omen, but mine was not." Spinoza's theory of dreams, as we analyze it, seems to have been largely a resistance device to prevent the analysis of his own dream. His repression-mechanism, as we shall see, is inconsistent with his own philosophy, and indicates an anxiety as to the underlying significance of the symbol of the "black and scabby Brazilian."

Spinoza distinguishes between two kinds of dreams—those which have a bodily cause and those which have a mental cause. Of the first kind, he writes: "We find that fevers and other physical changes are the causes of delirium, and that those who have thick blood imagine nothing but quarrels, troubles, murders, and the like." Dreams of this kind, Spinoza affirms, cannot (in our terms) be psychoanalyzed, for they are not the outcome of psychical anxieties. But the second kind of dreams are those which are largely determined by "the constitution of the soul". In this variety, the imagination "follows the traces of the intellect and concatenates its images and words in a certain order. . . . so much so that there is almost nothing that we can understand of which the imagination does not form some image from the trace thereof." These dreams do indicate basic anxieties, and in that sense, can be regarded as ominous of some future event. Spinoza states the distinction clearly, "all the effects of the imagination that proceed from physical causes can never be omens of future events, because their causes involve no future thing. But the effects of imagination, or images that derive their origin from the constitution of the mind can well be omens of some future event; since the Mind can confusedly have a presentiment of something yet to come."

Now Spinoza held that his own dream was not an omen because it was the consequence of purely physical causes. Yet, from the standpoint of Spinoza's philosophy, it is an error to say that a dream can have either a mental or physical cause. According to Spinoza's metaphysics, the order and connection of extended modes is the same as that of ideas. Every object, Spinoza held, was both a mental and physical reality. Nature could be regarded from the standpoints of either the attributes of thought or extension, and every constituent within it was at once both a mental and physical mode; every event is psychophysical. Within the confines of his system, Spinoza is not justified in saying that the imaginations of delirium issue from physical causes. Rather he should have said that what takes place is a weakening of mental inhibitions corresponding to the weakening of certain physical structures.

And the images which emerge in delirium would be just as much "omens" as those of ordinary dreams. For they would reveal anxieties hitherto repressed in the unconscious. Spinoza's assertion that non-ominous dreams are purely physically caused can therefore be regarded as a symptom of his own resistance to an analysis of his unconscious. For his argument that his dream was non-ominous rests on a blatant departure from the psychophysical parallelism which is so central to his system of the universe. And when a philosopher, so loyal to the geometrical method as Spinoza was, is led to such an obvious inconsistency in applying his doctrine, we may infer that the phenomenon in question has awakened strong resistances, strong enough to bend his metaphysics into contradiction rather than yield.

We are therefore led to enquire: what was the significance of the traumatic "black and scabby Brazilian" in Spinoza's dream?

The Holland of Spinoza's time was beginning to awaken to the problems of race. Dutch involvement in the slave trade was great, and on rare occasions, a Negro might even be seen in Amsterdam. People were beginning to speculate as to why Negroes were colored. The great founder of microscopy, Leeuwenhoek, citizen of Amsterdam in 1684, recorded: "there are many people in our country (as I have often expressed) who are firmly convinced that the Moors become black merely by rubbing their bodies with a certain oil, for they say the children are red when they come into the world, just like our children." (7, Vol. IV, p. 249) The problem had begun to interest scientists as well. The secretary of the English Royal Society asked Leeuwenhoek in 1677 to "closely examine the skin of Moors, also called negroes." (7, Vol. II, p. 239) The Dutch scientist promised to do so when an occasion offered. Seven years later, in 1684, Leeuwenhoek finally reported to the Royal Society that he had made observations on "several parts of the arms of a black Moorish girl, about 13 years old, . . ." She was the descendant of West African Negroes who had been enslaved in Brazil. "But a certain old lady of this town to whom the grandmother of this Moorish girl had been a slave in Brazil, told me that this Moorish woman was the child of parents from Angola, . . ." Leeuwenhoek dutifully reported that the popular theory of color differences was mistaken: "it is impossible to dye the scales forming the epidermis, so as to keep it black, . . ." And he found that the Negro skin "consisted of no other parts than little scales, . . . in the manner I have described of my own skin," but less transparent. (7, Vol. IV, p. 245–247)

Spinoza was a microscopist even as Leeuwenhoek, and they both had one common friend, the good Lutheran pastor, Dr. Cordes, whose sermons Spinoza much admired. (4, p. 176–177) There is no indication, however, that the Negro was among Spinoza's scientific interests, and Negroes, as we have seen, were exceedingly rare on the streets of Amsterdam. The "black and scabby Brazilian" was, as Spinoza said, one "whom I had never seen before." We must look for the genesis of the traumatic symbol elsewhere. And we find it, as far as available evidence can suggest, in the tragic circumstances which surrounded Spinoza's excommunication.

The figure of a black Brazilian had loomed in one terrifying episode in the history of the Dutch Jewish community.

The Amsterdam Jews had a large share in the slave trade. They had made immense profits, but they also experienced the horror of a slave uprising. The history of the Dutch Jewish community in Brazil has only recently been studied, but its salient facts are now established. It was in 1630 that the Dutch West India Company conquered Pernambuco. Jewish holdings in this company were extensive; although only four per cent of the shareholders were Jewish, they evidently owned a much more considerable proportion of the stock. The Company's guarantee of religious liberty induced the migration of several hundred Jewish families, many of whom came from Holland. (19, p. 1–2)

The slave trade, in which the Company held a monopoly, was extremely lucrative. The Company made a net profit of as much as 240% on each slave. Among the largest plantation owners, slave-holders, and slave traders in the valley of Pernambuco were Jews. Jorge Homen Pinto, for instance, owned 370 slaves and 1000 oxen. (2, p. 49, 63, 76) The Jewish community itself was supported partially by a tax on the slave trade. A rule of the community in 5409 (1648–1649 A.D.) provided that "Negroes bought from the Company shall be taxable at five soldos a piece." (19, p. 74) The conversion of Negroes to Judaism had to be discouraged because of the Biblical laws against the enslavement of fellow Israelites. The Jewish community therefore enacted a law prohibiting the circumcision of Gentiles without the prior consent of the Mahamad, under penalty of fine and excommunication. The ordinance was especially designed for application to slaves: "And if that person be a slave, he shall not be circumcised without first having been freed by his master, so that the master shall not be able to sell him from the moment the slave will have bound himself (to Judaism)." (19, p. 69, 22) To circumcise a slave, one would have to free him. The slave holder would hesitate at this step.

The Brazilian Jewish community was not destined to survive. The Portuguese fought to regain their colony. Their cruelty towards Jewish prisoners was frightful even for that age so that the Dutch asked: " Why are Jewish prisoners of war martyred unto death in so beastly a manner? Are they worse people than we?" When Isaac de Castro, of Amsterdam, fell into the hands of the Portuguese in Brazil, he was sent to Lisbon for trial before the Inquisition. Upon his refusal to renounce his faith, De Castro was burnt at the stake in 1647 in the presence of three fellow Jewish captives. (2, p. 94, 96)

Finally, in 1654, came catastrophe. The Portuguese captured Pernambuco. They were aided by a slave uprising. Negro troops under a Negro commander fought beside the Portuguese in what was for them a war of liberation. Henrique Diaz, the Negro leader, became the colony's governor, and was further rewarded by the Portuguese commander in 1656 with a deed to the lot of the Jewish cemetery. (19, p. 56) The Synagogues on the Street of the Jews were given as spoils to the man who, in 1645, had led a rebellion of the Portuguese and Brazilians against the Dutch.

The " black and scabby Brazilian " of whom Spinoza dreamt, was, I shall try to show, probably none other than Henrique Diaz, Negro commander of the slave insurrectionists. In personal appearance, Diaz's blackness marked him out from other fairer Negroes. (11, p. 176) His warlike qualities were a byword among the Dutch. He gloried in the fierceness of his slave followers. This New World Spartacus described his Negro legion of gladiators in 1647: "This regiment is composed of four nations: Minas, Ardas, Angolas, and Creoles. These last are so malevolent that they know neither fear nor duty; the Minas so fierce that what they cannot come at with their brawn they come at with their name; the Ardas so fierce that they would slash everything at a single stroke; and the Angolas so robust that no labor tires them." (5, p. 301) Such was the rebel leader and his warriors. They had a cause to inspire them; slavery under the Calvinist Dutch was more cruel than under the Catholic Portuguese. (18, p. 65) Captured at one time by the Dutch, Diaz was released because the Dutch thought "because he was a Negro slave he wasn't worth the cost of feeding." He returned to battle over the years, was wounded, and finally triumphant, became a member of the Portuguese nobility and holder of the Ordem de Christo. (11, p. 91, 176)

Among the Dutch Jewish subjects in Brazil, none had been braver than the Rabbi Isaac de Fonseca Aboab. Now he, like most of the Brazilian Jews, returned sadly to Amsterdam where his co-religionists at once welcomed him to a rabbinate in their synagogue. For Isaac, prior to his departure for Brazil, had served the Amsterdam community for many years as Rabbi and teacher. He was an admired Talmudist, and an eloquent preacher. The community preferred him to the more restless Menasseh ben Israel, who was indeed briefly excommunicated in 1640 for offending the community's wealthy magnates. Aboab probably presided at that ceremony. (13, p. 49–50, 56) It was the same Rabbi Isaac de Fonseca Aboab who on July 25, 1656, read to the Jewish community the decree excommunicating the young Baruch de Spinoza from its midst. (3, p. 391)

Ten years before, Rabbi Aboab had stood with the besieged Jews in the siege of Pernambuco. A Portuguese expedition, inspired by priestly words, had in 1646 launched an attack with the hope of exterminating the Jews. Rabbi Aboab ordered his countrymen to fast, he exhorted them to stand their ground, he prayed to God. The Jews' hour of trial was great. Many of them died in battle, many starved. Famine had almost forced them to surrender, when at the last moment, on the ninth of Tammuz, a Dutch fleet of rescue appeared on the horizon. The Jews were saved by the deliverer as their ancestors at the Red Sea. They sang: "Who is like Thee among the gods, O Lord! " All these events Aboab had commemorated in a poem of history and thanksgiving. (6, p. 128)

Rabbi Aboab was a man of culture. His library included Latin, Greek, Spanish, and Portuguese works; Homer, Virgil, Aristophanes, Plutarch, Cicero, the church fathers, Montaigne, Hobbes and Machiavelli had a place on his shelves. (9, p. 210–11) He was not the sort of man to deny to Spinoza the right to study Latin and Descartes. But to Rabbi Aboab, the Jews were God's Chosen People; the Portuguese and Brazilians, he had written, were " an abomination of Amalek." Rabbi Aboab had known the exiled wanderings of his people. Born in Portugal in 1605, he had been taken to France as a child, then studied at Amsterdam. He had seen his people driven and unwelcome in the New World and the Old. And now before him and his fellow-judges on the Beth Din, he heard witnesses testify that Spinoza scoffed at the Jews as " 'superstitious people born and bred in ignorance, who do not know what God is, and who nevertheless have the audacity to speak of themselves as His people, to the disparagement of other nations,' . . ." (20, p. 48) This young man of twenty-four was proud and bold. He was something new in the history of Jewry. He was the first of modern Jews to be taking his stand in the ranks of a radical political party. He had learned his Latin from Van dan Ende, a sceptical former Jesuit, who railed at religions, was called "lucianist", and engaged in revolutionary political plots. In 1674, Spinoza's master was executed for trying to help start a revolution in France which aimed to found a democratic republic with liberal welfare legislation. Spinoza's personal friends were not Jews but apparently Collegiants or Mennonites, strange mystics, who were pacifist in politics, without an organized ministry, vaguely communistic in their economic outlook. This young man of twenty-four was also an admirer of the Republican leader, John de Witt, who had set himself to reduce the power of the house of Orange, to which the Jews felt themselves indebted. He spoke heresy, political, economic, religious, and when Aboab's colleague, Morteira, remonstrated with him, Spinoza replied sardonically "that he knew the gravity of his threats and that in return for the trouble which he (Morteira) had taken to teach him the Hebrew language, he was quite willing to show him how to excommunicate'!" (20, p. 50) This mocking young man believed in the equality of all peoples, but he had no respect for his own.

So Rabbi Aboab hurled his curses at Spinoza with the same fury of invective which he had unleashed against the "children of Amalek", the Portuguese and their black Brazilians and their Negro slave commander. Many times he had described those horrors to the Amsterdam Jews, but Spinoza evidently felt himself not as one of his people. May he be cursed then with all the terrors that the Jews of Brazil had known. May the black Brazilian slave commander visit vengeance upon him. Rabbi Aboab read the decree: . . .

"The chiefs of the council do you to wit, that having long known the evil opinions and works of Baruch de Spinoza, . . . that the said Espinoza should be excommunicated and cut off from the nation of Israel; and now he is hereby excommunicated with the following anathema:

"With the judgment of the angels and of the saints we excommunicate, cut off, curse and anathematize Baruch de Spinoza, . . . : by the 613 precepts which are written therein, with the anathema wherewith Joshua cursed Jericho, . . . and with all the curses which are written in the law. Cursed be he by day and cursed be he by night. Cursed be he in sleeping and cursed be he in waking, . . . The Lord shall not pardon him, the wrath and fury of the Lord shall henceforth be kindled against this man, . . . The Lord shall destroy his name under the sun, and cut him off for his undoing from all the tribes of Israel, . . .

"And we warn you, that none may speak with him by word of mouth, nor by writing, nor show any favor to him, nor be under one roof with him, nor come within four cubits of him, nor read any paper composed or written by him." (12, p. 17–18)

The experience of excommunication is perhaps the most traumatic that a religious person can undergo. He is suddenly alone, utterly so, and in this region where no one knows certainty, told that his opinions are evil and damned. All the forces of the community, all the symbols which he has been taught to respect from childhood years, are arrayed against him. It is the act by which the full power of the community's super-ego is turned against the individual's solitary, rational self. The haunting threat of excommunication by the Catholic Church was enough to paralyze the great energies of Lord Acton, and to make him the author of the greatest historical work that was never written. And in recent years, George La Piana has written movingly on his life spent as an excommunicate, seeking sources of value within that very experience of intellectual rejection and isolation. Spinoza himself denied that he was affected emotionally by his excommunication, saying: "All the better; they do not force me to do anything that I would not have done of my own accord if I did not dread scandal;" . . . (20, p. 51) None the less, he had contested his accusers before the Beth Din, "feeling that his conscience had nothing to reproach him, he went cheerfully to the Synagogue", and he had denied before his judges the charge of "the most awful of all crimes, namely, contempt for the Law." (20, p. 48)

Spinoza knew how excommunication could arouse all the hatreds of a community against a man, and destroy him. As a child of eight, he had seen how the Amsterdam Jews had broken the spirit of the proud Uriel Acosta. That troubled man had left Portugal and his juristic career to identify himself openly with the Jewish people at Amsterdam. But he was disillusioned with the Jews as he found them in actuality, mean, petty, money-seeking, intolerant. He rebelled against the community's leaders and was excommunicated. In the streets of Amsterdam, Acosta narrated, "many of them spit upon me as they passed by me in the Streets, and encouraged their Children to do the same, . . . set their Children upon me in the Streets, who insulted me in a body, as I walked along, abusing and railing at me, crying out, 'There goes a Heretick, there goes an Apostate.'" The child Spinoza witnessed the horror of Acosta 's recantation: . . . "I stripped myself naked down to the Waste, tied a napkin about my Head, pulled off my Shoes, and holding up my Arms above my Head, clasped a sort of Pillar to my Hands, to which the Door-keeper tied them with a Band. Having thus Prepared myself for my Punishment, the Verger came to me, and with a Scourge of leather Thongs gave me nine and thirty Stripes, . . . During the Time of my whipping they sang a Psalm . . ." Then, "I prostrated myself, the Door-keeper holding up my Head, whilst all both old and young passed over me, stepping with one Foot on the Lower Part of my Legs, and behaving with ridiculous and foolish Gesture, more like Monkeys than human Creatures." (8, p. 21–30). Soon afterwards, Acosta killed himself.

Spinoza probably remembered all this. His own ideas on ethics and natural law were indeed much like those of Acosta. Spinoza would never recant. He fortified himself against the communal pressures for what is now called brainwashing. The excommunication stirred within him the deepest anger. He could withstand it with the help of his Mennonite friends. He came indeed to hate the Jews for what they had done to him, and could never write of them without invective touching his pen. This hatred, in a man like Spinoza, indicates how deeply the trauma of excommunication went. "Now the Hebrew nation", he wrote, "has lost all its grace and beauty (as one would expect after the defeats and persecutions it has gone through) . . ." He referred to the revered Maimonides' writings as "harmful, useless, and absurd", "mere nonsense", an effort "to extort from Scripture confirmations of Aristotelian quibbles." He repudiated the ancient rabbis, writing that it was "grievous to think that the settling of the sacred canon lay in the hands of such men." As for the later Rabbis, they "let their fancy run wild, . . . dream, invent, and as a last resort, play fast and loose with the language." The Kabbalists among them, of whom his teacher Morteira was one, indulged in "childish lucubrations", "whose insanity provokes my unceasing astonishment." (15, p. 108, 118, 17, 147,

56, 139–140) The Jews were simply "Pharisees" to Spinoza, arrogant, intolerant men, and the "foundations of their religion". he wrote, might well have "emasculated" the minds of the Jews. (16, p. 353; 15, p. 56) And he, Spinoza, would not allow himself to be emasculated. The Jewish community and its religion were summed up finally by Spinoza in this metaphor of "emasculation", the threat of intellectual and spiritual sterility conveyed under the symbol of castration. His reason had met the threat of the communal super-ego, "though I was imbued with the ordinary opinions about Scriptures, I have been unable to withstand the force of what I have urged."

There were violent radical political emotions in Spinoza. Usually, he kept them under close control. An excommunicate Jew, suspected of political heresy, he had to guard his words well. His feelings once expressed themselves in a strange way, in a drawing, where one can see more clearly the emotion behind the geometrical façade. His early biographer, Colerus, tells the story:

". . . he apply'd himself to Drawing, which he learn'd of himself, and he cou'd draw a Head very well with Ink or with a Coal. I have in my Hands a whole book of such Draughts, amongst which there are some Heads of several considerable Persons, who were known to him, or who had occasion to visit him. Amongst those Draughts I find in the 4th Street a Fisherman having only his Shirt on, with a Net on his Right shoulder, whose Attitude is very much like that of Massanello the famous head of the Rebels of Naples, as it appears by History, and by his Cuts. Which gives me occasion to add, that Mr. Vander Spyck, at whose House Spinoza lodged when he died, has assured me, that the Draught of that Fisherman did perfectly resemble Spinoza, and that he had certainly drawn himself." (3, p. 392)

In fantasy, Spinoza projected his own face in the costume of Massaniello. If a philosopher today were to take joy in drawing his face beneath Lenin's famous cap, or in the austere Asian garb of Mao Tse-tung, we should draw a simple conclusion concerning his underlying political sympathies. And likewise with Spinoza's self-projection into the person of Massaniello. Tomas Aniello, for that was his full name, was a young illiterate fisherman of Naples. In 1647, Tomas led an insurrection against the Spanish viceroy. During the few days that he held power, he revoked the hated fruit tax which bore on the poor, and promulgated a decree for constitutional reform. Though he was soon murdered by his enemies, his movement continued to gain, and when the Spanish navy attacked Naples, it was met with the cry "Long live the Republic!" The rebels defeated the Spanish forces, but their Republic did not long survive. Europe was amazed at these events. Massaniello for six days shook the world with the spectacle of the illiterate fisherman who had humbled Europe's proudest monarchy. (10, p. 19–26)

The radical republican Spinoza was a Jew, disfranchised from practical politics. His deepest feelings, which he had to repress in the Holland around him, where Calvinist clerics waited to denounce him, he placed into a drawing. Repressed aggressions moved within him, and in an amusement of a strange kind, he gave expression to them: ". . . he look'd for some Spinders, and made 'em fight together, or he threw some Flies into the Cobweb, and was so well pleased with that Battel, that he wou'd sometimes break into Laughter." (3, p. 395)

His philosophy grew into a curious network of ambivalences. He detested the masses for their irrationality, but he defended them against the advocates of absolutism. He conceived a vision of God which was the apotheosis of the sublimation of love, and which he felt would bring him a mystic's release,—"And only in this union, . . . does our blessedness consist . . . For even the knowledge that we have of the body is not such that we know it just as it is, or perfectly; and yet, what a union! what a love!" (17, p. 133) And yet this love, which was supposed to bring freedom, was suffused with internalized self-aggression. The frustrated revolutionist turned his aggression inwards in a masochist passion,—"it follows therefore that we are truly servants, aye, slaves of God, and that it is our greatest perfection to be such necessarily." We are like a "Hatchet" says Spinoza, in the hands of a Carpenter, a tool to render good service. "For the sole perfection and the final end of a slave and of a tool is this, that they duly fulfill the task imposed on them". This "true service of God" is "our own eternal happiness and bliss." "But if God should (so to say) will that man should serve him no more, that would be equivalent to depriving him of his well-being and annihilating him; . . ." (17, p. 115–117) We are like "hatchets",—in this symbol once more of destruction taken on willingly, Spinoza naturally expresses his philosophy. God is the mathematical despot who geometrizes with sadistic indifference to human beings, while we must delight in our status as theorems. Recently, when a writer on philosophy died, his eulogy said that he had conformed to Spinoza's saying: "He who loves God cannot strive that

God should love him in return." Was a metaphysics of self-immolation, of self-aggression, ever more simply stated?

As an excommunicate Jew, Spinoza would never quite fit into any ordinary social group. He conceived himself finally, despite his social and political aspirations, and numerous friendships, as essentially solitary. "And he would say sometimes to the people of the House, that he was like the Serpent, who forms a Circle with his Tail in his Mouth; to denote that he had nothing left at the years end." (12, p. 393–394) His self-chosen symbols were ultimately less those of the love of God, than of the auto-erotic recluse. His seal was an oval ring, containing a rose with his initials, and the word *caute* (*beware*) to his fellowmen (16, p. 441) He longed in later years to speak the Spanish language of his childhood; it was the tongue of his earliest, most intimate feelings, but excommunicate, he spoke Dutch and wrote Latin, never at home in these languages of the alien world. To a correspondent in 1665, he confessed: "I do indeed wish that I might write the language in which I was brought up. I might possibly express my thoughts better." (16, p. 151) And then there was the haunting doubt which besets every man who has staked his life on an idea. What if others are right and he is wrong? What if his understanding had misled him? In one such moment, Spinoza acknowledged this possibility, and said: "Even if I was once to find untrue the fruits of my natural understanding, they would make me happy since I enjoy them, and I endeavor to pass my life not in sorrow and sighing but in peace, joy and cheerfulness," ... (16, p. 173) What a confession for a votary of reason! An ultimate pragmatism, that his philosophy brings him happiness, and that this rather than truthfulness is its final appeal! It is the mission of truth, however, which leads a man to defy excommunication. The pragmatic path is one of doubt, not the direction to martyrdom.

But we must return to the dream of the "black and scabby Brazilian." The mystic in Spinoza struggled with the frustrated political radical. No revolutionist ever frees himself from his earliest cultural values. Pietro Spina, the revolutionary protagonist in Silone's great novels, found that the Catholic idealist in him was not expunged by his conscious Marxist dedication. The residuum of an ineradicable, intractable conservative remains in the soul of the most radical reformer. The commandments we heard in childhood reverberate through our most adult years. Liberation is always fractional. Karl Marx disliked secularism, and advised his wife and daughter to seek religion in the Jewish prophets, if they felt a metaphysical need; he would yield them that opiate. (1, p. 74) And in the hours of sleep, the rational discipline of adequate ideas failed Spinoza, nor was the intellectual love of God of any avail.

The figure of the Negro terrorist, the spectre which Rabbi Aboab had described to the Amsterdam Jews, came to menace Spinoza in his dream. He was the symbol of all the hostile forces that await a Jew in the external world, all the forces of hatred, and Spinoza, excommunicate, would have to deal with them alone. The Negro Terrorist was the embodiment of all the curses of the world's powers which Rabbi Aboab had summoned up against him.

Clear, independent thinking is a kind of parricide. It is an act which is revolutionary in essence, for by free thinking, the individual sunders the bonds of tradition, and hurls himself against the cultural super-ego. He dares to rise with only the power of his rational ego against the hydra-like tentacles with which he is enveloped by the unconscious superego.

Then arises the neurosis of the revolutionist. Guilt returns to him in the hours when he seeks sleep. The superego comes back to menace him in his weariness, when his philosophy fails him, when even the intellectual love of God is far too peripheral, confined to the consciousness. Anxieties emerge from his unconscious, which reproaches him for disloyalty. The subversive is subverted by the sense of guilt with which the communal super-ego lies in ambush. And all the resources of his philosophy are, for a moment, helpless against forces from the irrational deep. Such, we may surmise, was the meaning of Spinoza's dream in 1663. The trauma of the excommunicate never healed.

References

1. Beer, Max, *Fifty Years of International Socialism,* London, George Allen & Unwin, 1935.
2. Bloom, Herbert I., "A Study of Brazilian Jewish History 1623–1654, based chiefly upon the Findings of the late Samuel Oppenheim," *Publications of the American Jewish Historical Society,* Vol. 33, 1934.
3. Colerus, John, *The Life of Benedict de Spinoza,* translated from the French, London, 1740, reprinted in Pollock, *Spinoza: His Life and Philosophy,* pp. 385–418.
4. Dobell, Clifford, *Antony van Leeuwenhoek and his "Little Animals",* New York, Harcourt, Brace and Company, 1932.
5. Freyre, Gilberto, *The Masters and the Slaves,* transl. by Samuel Putnam, New York, Alfred A. Knopf, 1946.

6. Kayserling, M., "Isaac Aboab: the First Jewish Author in America", *Publications of the American Jewish Historical Society,* 1897, No. 5.
7. Leeuwenhoek, Antoni van, *Collected Letters,* edited by a committee of Dutch scientists, Amsterdam, Swets & Zeitlinger, Vol. II, 1941; Vol. IV, 1952.
8. Limborch, Philip, *The Remarkable Life of Uriel Acosta,* London, 1740.
9. Marx, Alexander, *Studies in Jewish History and Booklore,* New York, Jewish Theological Seminary, 1944.
10. Merriman, Roger, *Six Contemporaneous Revolutions,* Oxford, Clarendon Press, 1938.
11. Pierson, Donald, *Negroes in Brazil,* Chicago, University of Chicago Press, 1940.
12. Pollock, Sir Frederick, *Spinoza: His Life and Philosophy,* Second Ed., New York, The MacMillan Co., 1899.
13. Roth, Cecil, *A Life of Menasseh ben Israel,* Philadelphia, Jewish Publication Society, 1945.
14. Sewel, William, *The History of the Rise, Increase, and Progress of the Quakers,* transl. by himself from Low-Dutch into English, Lond., 1722.
15. Spinoza, Benedict de, *Chief Works,* Vol. I. transl. by R.H.M. Elwes, London, George Bell and Sons, 1883.
16. Spinoza, Benedict de, *Correspondence,* transl. by A. Wolf, London, George Allen & Unwin, 1928.
17. Spinoza, Benedict de, *Short Treatise on God, Man, and His Well-being,* transl. by A. Wolf, London, Adam and Charles Black, 1910.
18. Tannenbaum, Frank, *The Slave and Citizen,* New York, Alfred A. Knopf, 1947.
19. Wiznitzer, Arnold, *The Records of the Earliest Jewish Community in the New World,* New York, American Jewish Historical Society, 1954.
20. Wolf, A., editor and translator, *The Oldest Biography of Spinoza,* London, George Allen & Unwin, 1927.

12

The Dreams of Descartes

On the night of November 10, 1619, the young René Descartes, soldier in the army of the Holy Roman Emperor, "had three consecutive dreams which he believed could have come only from above." During that day, in a stove-heated room in the little town of Neuberg on the Danube River, the fundamentals of science, Descartes believed, had been revealed to him; the dreams conveyed a mystical experience of God's blessing on his new insights as well as on his vocation as a scientist. "He believed he saw through their shadows the indications of the road which God had traced for him to follow in his choice of life, and in the search for that truth which was the subject of his anxieties."[1] Modern philosophy, we might say, was born during the night of Descartes' three dreams.[2] They have drawn the attention of several philosophers and psychoanalysts, including Freud himself, but the connective lines between Descartes' philosophic ideas and the underlying anxieties which his dreams expressed still remain obscure. To this effort we shall address ourselves.

The night of November 10, 1619 was St. Martin's Eve. People in France used to spend it in debauchery. Descartes, however, had passed the whole day with sobriety; indeed, he hadn't had any alcoholic drink for three months. That it was St. Martin's Eve, however is a circumstance to be noted for this was a holiday customarily given to liberating repressed sexual emotion. Descartes' thought was liberated on this ritualistic evening of sexual freedom. Descartes wrote down his dreams in a manuscript now lost called OLYMPICA. His faithful biographer Baillet used this manuscript in giving his detailed account of them. We shall narrate the three dreams (with a few paraphases) in Descartes' own words as transmitted by Baillet. Then we shall undertake to explicate their significance.[3]

Narrative of Descartes' Dreams

The First Dream:

Several phantoms presented themselves and frightened Descartes. Walking through the streets, he had to turn to the left side to advance to where he wanted to go, because he felt a great feebleness on the right side, on which he couldn't support himself. Ashamed of walking this way, he tried to straighten himself, but an impetuous wind carried him in a sort of whirlwind ("tourbillon") and whirled him about three or four times on his left foot. He thought he would fall at each step, until having seen an open college on the road, he entered to find a retreat and a remedy for his sickness. He wished to reach it with the thought first of prayer, but having passed a man he knew without greeting him, he tried to retrace his steps to do so, but was pushed violently by the wind against the Church. In the middle of the college court, he saw another person who called him courteously by name, and told him that if he was looking for Monsieur N., he had something to give him. Descartes imagined it was a melon which had been brought from a foreign country ("un melon qu'on avait apporté de quelque pays étranger"). What surprised him more was to see that those who were gathered around this person stood firm on their feet, while he always remained bent and unsteady ("courbe et chancelant"). The wind meanwhile subsided. He then awoke, and felt a real pain, which made him fear that some evil genius had wished to seduce him ("quelque mauvais génie qui l'auvait voulu séduire"). He had been sleeping on his left side, and turned to the right, praying God to save him from misfortune and punishment of his sins.

The Second Dream:

He had fallen asleep after meditating for two hours on the goods and evils of this world. Then he dreams that he hears a sharp and piercing noise which he took for a peal of thunder ("un bruit aigu et éclatant, qu'il prit pour un coup de tonnerre"). The fright awoke him, and he perceived many sparks of fire scattered in the room ("beaucoup d'étincelles de feu répandues par la chambre"). This had often happened to him before. He would awake in the middle of the night his eyes quite glittering ("les yeux assez étincellans"). But this time he desired philosophical reasons, and he drew conclusions favorable to his mind ("favorables pour son esprit").

The Third Dream:

This dream was not accompanied by feelings of anxiety. Descartes finds a book on the table opens it, discovers it's a Dictionary, and is delighted with it. Then, that very instant, he finds another book under his hand, and doesn't know where it had come from. He ascertains it is an anthology of poems of different authors entitled CORPUS POETARUM. He was curious to read something, and at the beginning of the book, happened upon the verse: *Quod vitae sectabor iter?* At the same time he saw a man he didn't know but who presented him with a verse beginning *Est et Non,* which he praised. Descartes told him he knew what it was, and that it was one of the Idylls of Ausonius found in the big collection of poets on the table. He wished to show it himself to this man, and began to leaf through the book whose order and arrangement he had boasted he knew perfectly. While he was looking for the place the man asked him where he had obtained the book. and Descartes answered that he was unable to tell him how he had it; but that a moment before he had held still another which had disappeared without his knowing who had brought it or taken it back from him. He hadn't finished when he saw the book appear once more at the other end of the table. But he discovered that this Dictionary was not complete as it had been when he saw it the first time. However, in perusing the Anthology of Poets, he came upon the poems of Ausonius; but not finding the piece which began with *Est et Non,* he told the man he knew another one by the same poet which was still more beautiful than the former, and which began with *Quod vitae sectabor iter?* The person asked him to show it to him, and Descartes set about looking for it, whereupon he found several small portraits engraved in copperplate; this made him say that the book was very beautiful, but that it was not of the same printing as the one he had known. There he was when the books and man disappeared, and faded from his imagination, without nevertheless re-awakening him.

What was singular in the circumstances of the third dream, continues the account, was that "Descartes, wondering whether what he had just seen was a dream or vision, not only decided that it was a dream, but he rendered its interpretation even before he awoke. He judged that the Dictionary represented all the sciences gathered together; and that the Anthology of Poems, entitled CORPUS POETARUM, indicated particularly, and in a distinct fashion, Philosophy and Wisdom joined together. For he believed one shouldn't be much surprised to see that the poets, even those who fool their time away, have many maxims which are deeper, more sensible and better expressed than those which are found in the writings of philosophers. He attributed this wonder to the divinity of Enthusiasm and the force of Imagination, which emitted the seeds of wisdom (which are found in the minds of all men, like the sparks of fire in flint-stones) with much more ease and brilliance even than Reason can among the philosophers. Descartes, continuing to interpret the dream in his sleep, thought that the verse on the uncertainty of the way of life which one must choose, which begins with *Quod vitae sectabor iter,* marked the sound advice of a wise person, or even of Moral Theology.

"Thereupon, uncertain whether he was seeing or imagining, he awoke without emotion, and his eyes open, continued to interpret his dream along the same line. By the poets gathered in the anthology he understood Revelation and Enthusiasm, which he hoped would still favor him. By the piece of verse *Est and Non,* which is the Yes and No of Pythagoras, he understood Truth and Falsehood in human knowledge and the secular sciences. Seeing that his work in all these fields was succeeding so well, he was bold enough to persuade himself that it was the Spirit of Truth which had wished to open for him (by this dream) the treasures of all the sciences. There was left for him only to explain the little portraits in copper plate, which he had found in the second book. He didn't try to explain this any more after an Italian painter visited him the next day.

"This last dream, which was altogether very pleasant and agreeable, indicated to him the future and what must happen to him for the rest of his life. But he took the two preceding ones as threatening warnings concerning his past life, which may not have been as innocent before God as before men. And he believed that it was the reason for the terror and fright with which these two dreams had been accompanied. The melon, which was offered him in the first dream, signified, he said, the charms of solitude, but available through purely human attractions. The wind which pushed him towards the Church of the college, when he felt bad on the right side, was nothing but the evil Genius which tried to throw him by force into a place he had voluntarily planned to go."

Analysis of Descartes' Dreams

Let us now try to analyze these three dreams of Descartes item by item.

Analysis of the First Dream:

He is walking through the streets. Descartes, always seeking solitude, avoided walking through the streets of Paris out of fear that he might be seen by friends. For two years, from 1614 to 1616, he lived, according to Baillet, almost without leaving his house. The streets were a threat to his "mask" (as he called it), to the defenses which he had constructed for himself against his inner anxieties. *He has to turn to his left side.* He had to deviate from the approved ways of church, family, and society. *He feels feeble on the right side.* He is not sustained by any traditional authority, neither church nor father, as he undertakes his work of doubt and destruction, his lonely revolt against the super-ego. His feebleness, moreover, his inability to stand erect, suggests a lack of confidence in his own manhood, his own potency. He has been bearing a guilt which has disabled him sexually. *A kind of whirlwind turns him about three or four times on his left foot.* The whirlwind is the traditional Biblical symbol for God's speaking to man. Thus: "Then the Lord answered Job out of the whirlwind, and said, Who is this that darkeneth counsel by words without knowledge?" (Job, 38:1; 40:6).[3a] And God's power overwhelms and bewilders Descartes in rebuke for his choice of an independent role, for his intellectual parricide, for standing on his own. Perhaps he challenges too Descartes' manhood; the fear that he does not exist would be the philosophic counterpart of Descartes' emasculation, his enfeeblement, even as God rebuked Job: "Gird up thy loins as a man; for I will demand of thee, and answer thou me. Where wast thou when I laid the foundations of the earth?" *He sees an open college and enters to find a retreat, and remedy for his sickness.* The memories of the Jesuit college of La Flèche where he knew a kind of retreat come back to him.

The "open college" is a maternal vaginal symbol where he hopes to achieve the security which he lost as a child when his mother died. God the Father assails him in the whirlwind, and he longs for the comfort of a mother's love to sustain him. At La Flèche, indeed he had been the favored child. His best friends were among the school-fellows. His "perpetual director", Father Charlet, and his prefect, Father Dinet, had allowed him freedom to pursue independently his mathematical studies. Father Charlet had granted him furthermore the privilege of remaining late in bed. Descartes acquired the habit of sleeping long and late, ten to twelve hours a night, and of staying in bed till noon. His philosophy was born in the maternal comfort of a bed's warmth, as if he were seeking the substitute physical solaces for his anxieties that a child seeks. If Socrates' philosophy was conceived in the market-place, Nietzsche's on mountain walks, and Marx's in a journalist's office, Descartes' philosophy was nurtured in a bed. *He wishes to reach it, but snubs a man on the way, then wishes to go back and greet him, but is pushed by the wind violently against the Church.* Descartes, as we have mentioned, was always contriving ways to safeguard his solitude. This led many of his friends to think he was snubbing them. Baillet tells a typical episode which occurred, during Descartes' two years' retreat at the Faubourg Saint-Germain in Paris from 1614 to 1616. The young Descartes, obliged to leave his house for his needs, had finally, despite all his detours and precautions, been found by one of his friends whom he could not shake off, till he gave him his address ("il fut recontré par un de ses amis, qui ne voulut pas le quitter qu'il ne lui eut decouvert sa demeure").[4] The snubbed man we may take as the generic symbol for Descartes' father, who had desired his son to prepare himself for a councillor's bureaucratic career. Descartes had rejected his father's way of life and profession. He is now stricken with doubts as to the rightness of his decision, and even begins to reverse it. God's anger meanwhile drives him violently to the Church, to the traditional institutions. But then the irony of a subversive enclave within an established institution asserts itself. *In the middle of the college court, he is asked courteously if he is looking for Monsieur N. If so, he is to be given a melon from a foreign country.* In this setting, as if re-united with the maternal source, Descartes is asked if he wishes to see Monsieur N. Who is this personage? The name that suggests itself is his closest personal friend, Father Mersenne, who though eight years older than Descartes, was already friendly with him at La Flèche, and was known later as "le doyen de ses amis et de ses sectateurs" (the chief of his friends and followers). The two friends renewed their association when Descartes came to Paris at the age of seventeen. Mersenne's influence on Descartes was deep; he liberated him from a predilection for gambling and a wastrel's life. If we accept Freud's equation of gambling with masturbational activity, this would signify that Mersenne helped his friend to surmount auto-eroticism, the sexuality of the solitary, and to find a greater delight and sublimation in communal scientific activity. "They began to taste the sweetness of their innocent habits, and to ease each other in the search for truth, "writes Baillet ("Ils commencaient à gouter les douceurs de leurs innocentes habitudes, est à s'entresoulager dans la recherche de la vérité"). Mersenne, however, had joined the Order of the Minims, and was ordained as a priest six months after Descartes' arrival in Paris. Their close association was broken up in a traumatic way for Descartes when Mersenne in 1614 was assigned by his order to teach at Nevers. "This separation affected Descartes quite deeply," says Baillet ("Cette séparation toucha M. Descartes assez vivement").[5] He didn't return to a life of dissipation, but instead withdrew even more into himself. "Monsieur N.", we might infer, is his friend at Nevers; "M" of Mersenne and the "N" of Nevers are phonetically and associationally interchangeable.

The "melon from a foreign country" which is offered Descartes as a friend of Monsieur N. is again a classical religious symbol. The "melon" is the fruit from the Tree of Knowledge in the Garden of Eden; as forbidden fruit, it represents that eroticizing of the study of mathematics and of the sciences which Mersenne stimulated in Descartes. It is also as fruit the maternal symbol of the female breast, linked to knowledge.[6]

Descartes indeed felt the bond between mother and child was the most intimate intellectually and emotionally that there was. The psychological union of mother and child was so close, Descartes held, that feelings and imaginations in a mother's mind transmitted themselves to the child in its womb. ("La même disposition qui était dans le cerveau de la mère et causait son ennui, se trouve aussi dans le sien"). The sense of his own deprivation of his mother's supporting love remained so keen that his memory later distorted the date of her death, accentuating her loss. His mother, said Descartes, died a few days after his birth ("peu de jours après ma naissance") of a lung ailment caused by certain misfortunes ("causé par quelques déplaisirs"). From her he had inherited, he said, a dry cough and a pale color which had remained with him till he was more than twenty years old ("que j'ai gardée jusques à l'âge de plus de vingt ans") so that all the doctors who saw him before that time had condemned him to an early death.[7] His own natural optimism and self-sufficiency, said Descartes, had kept him alive. It is a remarkable circumstance that Descartes' age at the time of his dreams, twenty-three, and that of his recovery from his physical ailments, "more than twenty years old", coincided. The dreams had brought liberation from a force which had previously destroyed the mother with whom Descartes identified himself. And that hostile, destructive force had been his father whose grief over his wife's death had been false. For there was a distinct personal note in Descartes' discussion of the "secret joy" of a bereaved husband:

"when a husband laments his dead wife whom (as sometimes happens) he would be sorry to see brought to life again, . . . it may be that some remnants of love or pity which present themselves to his imagination draw sincere tears from his eyes, notwithstanding that he yet feels a secret joy in the inmost parts of his heart, the emotion of which possesses so much power that the sadness and the tears which accompany it can do nothing to diminish its force."[8]

Descartes, with the ever-present pain of his early loss of his mother, remembered her death as having occurred almost at his birth. As a matter of fact, she had died more than a year afterwards, but it was as if he had never had a mother.[9] Love was, after joy, said Descartes, the first passion experienced by the infant, and it was directed to that being which gave it nourishment. ("l'âme, se joignant de volonté a cette nouvelle matière, a eu pour elle de l'amour"). The infant's first sadness, wrote Descartes, came with the loss of its mother's breast; its first hatred entered its soul when some one not fit to nourish it tried to replace its mother. This, in effect, was the young child René's first account of his hatred for his father ("s'il est arrivé. que cet aliment ait manqué, l'âme en a eu de la tristesse. Et s'il en est venu d'autre en sa place, qui n'ait pas été. propre à nourrir le corps, elle a en pour lui de la haine"). The forbidden fruit, the melon, the maternal symbol, is the knowledge in which he finds the maternal comfort and sustenance against his father. The exemplar of all love, wrote Descartes, is love for the nourishing mother; even before birth, "aliment", as it entered the infant's body, excited "warmth"; hence, warmth always later accompanied the experience of love in all its varieties ("de là vient que maintenant cette chaleur accompagne toujours l'amour, encore qu'elle vienne d'autres causes fort différentes").[10] The nourishment of the foreign melon is the forbidden science, which becomes the agency for union with the mother he lost.

The first dream, in short, re-enacts the trauma and resolution of Descartes ' decision to give himself to the sciences rather than to the established institutions and their modes of thought. The others in the college, more conventional in their ways, remain firm on their feet, erect. They have no anxiety concerning sexual potency. He, on the other hand, troubled and beset by anxiety, is *bent and unsteady.* Castration-fears, born of his own intense revolt against his father, have rendered him impotent. He is offered the fruit of knowledge through Mersenne's good offices; it was indeed an exotic foreign plant, for its pursuit had already taken Descartes from his native land to make of him a physical as well as an intellectual exile. The snubbed man in the street can bring to bear all God's authority against Descartes. Thus the dream mirrors a conflict so full of pain for Descartes that in anguish he wonders whether some evil genius isn't seducing him, whether indeed, the fruit of knowledge is Satan's device to destroy him.

Analysis of the Second Dream:

He hears a sharp and piercing noise which he takes for a peal of thunder. When the fright awakens him, he sees many sparks of fire scattered in the chamber. The key to this dream seems to me provided by Descartes' verbal association with "sparks of fire" ("étincelles de feu"). During his own interpretation of his third dream, Descartes spoke "of the seeds of wisdom", emitted by the divinity of Enthusiasm and the Imagination, and comparable to "the sparks of fire in flint stones" ("les semences de la sagesse ... comme les étincelles de feu dans les cailloux"). The divine enthusiasm, in other words, emits ("fait sortir") "the seeds of wisdom" in a process similar to a sexual emission; the seminal ideas emitted by God are as intellectual sperm. The sparks of fire symbolize the eroticized character of seminal ideas. Freud has observed that primitive men regarded fire as "analogous to the passion of love—we should say, as a symbol of the libido. The warmth radiated by fire evokes the same kind of glow as accompanies the state of sexual excitation, and the form and motion of the flame suggest the phallus in action. There can be no doubt about the mythological significance of flames as the phallus."[11] The "sparks of fire" are by an extension of the metaphor the "seeds of wisdom", the intellectual sperm, emitted by the "stones" ("les cailloux"), that is, the testicles.

We can now begin to see how the second dream continues the first. The latter ended with Descartes in a state of anxiety, seeking the forbidden fruit of knowledge, but unsteady, unable to maintain himself erect, in an anxiety of impotence. The second dream represents an achievement of sexual potency, to which is conjoined the eroticization of intellect. C. P. Oberndorf has remarked: "Nearly always, when the child turns to thinking as a goal, it is because of an intense reaction to deprivation of love (pre-Oedipal or at the Oedipus level) by one or both parents—a wound to the child's narcissism."[12] With Descartes henceforth, thinking was to have the full glow of the sexual liberation and potency which he had that night of November 10, 1619. Indeed, the emotion which union with God inspired in him as he later was to describe it was precisely the replica of sexual love: "and the very idea of this union suffices to excite warmth around the heart, and to cause a very violent passion" (et la seule idée de cette union suffit pour exciter de la chaleur autour du coeur et causer une très violente passion).[13] The liberation of thought, of the intellect, was thus tied to the liberation of sexual feeling, potency, and fulfillment. The Cartesian revolution in philosophy thus had a sexual basis which provided the energy for the individual's new self-assertion. Free thought, fruitful, seminal thought, was precisely thought endowed with sexual power, eroticized, de-asceticized, repotentated. The sharp and piercing noise, like a peal of thunder, was the full force of phallic ejaculation, which brought forth fright, as to an adolescent's first experience.[14] There had been other occasions when Descartes had awoken in a state of sexual excitation. *This had often happened to him before ... But this time he desired philosophical reasons.* The sharp, piercing sexual experience suddenly gave to his thought its full eroticization. He was liberated from the fears, the self-aggression, the castrational de-manning anxiety which he had imbibed from father and clerical surrogates. With this liberation from self-aggression which had filled him with the neurotic doubt of existence and self came the full, thick experience of the overwhelming reality of his own existence. It was thinking which was eroticized which affirmed in fullest measure that he existed. Thinking which was de-sexualized, castrated, had about it a neurotic unreality; the thinking which was animated with emotional, sexual affirmation bore the conviction of existence.

Analysis of the Third Dream:

He finds a book on the table, and on discovering it's a dictionary, is delighted. At the same instant, another book appears, he knows not how; it is the CORPUS POETARUM, an anthology of classical poets. He opens it, and chances on the verse "Quod vitae sectabor iter?" A man he doesn't know then presents him with a verse beginning "Est et Non" which the man praises. Descartes tells him it's one of the Idylls of Ausonius which can be found in the anthology. This was the dream which finally convinced Descartes that he had God's blessing in the vocation he was undertaking, that of the independent scientist, the thinker taking his own road. It is important to understand the significance of the dreamwork's use of Ausonius as a vehicle for a declaration of emotional and intellectual independence. There had been a revival of interest in Ausonius a few years before Descartes became a student at La Flèche.[15] Ausonius, a Latin poet of the fourth century, and governor of Gaul, had been at most a nominal Christian. When a friend of his began to base his life upon Christian precepts, Ausonius was "totally unable to understand his friend's attitude" and could only believe that he was "crazed".[16] Ausonius himself conformed decorously to Christianity as a matter of expediency, to avoid collision with the dominant cult.[17] Thus, Descartes chooses, to begin with, as a philosophic oracle a classical dissenter who knew too how to dissimulate and conceal his private views. Descartes' tactic in life was to be that of Ausonius; he would wear a mask, as he said, agree with the authorities, but carry on his ordained work in a strategy of subversion.

The verse by Ausonius " Quod vitae sectabor iter " is the beginning of an eloquent poem concerning the anxieties of choice. It is "a Pythagorean Reflection on the Difficulty of Choosing One's Lot in Life":

"What path in life shall I pursue? The courts are full of uproar; the home is vexed with cares; home troubles follow us abroad; the merchant always has fresh losses to expect, and the dread of base poverty forbids his rest; the husbandman is worn out with toil; frightful shipwreck lends the sea a grim name; the unwedded life has its sore troubles, but sorer is the futile watch and ward which jealous husbands keep; to serve Mars is a bloody trade; the tarnished gains of interest and swift-mounting usury slaughter the needy. Every stage of life has its troubles, and no man is content with his own age; . . . With one accord we all scorn our present lot: . . ."[17a]

Immortality and mortality, chastity and promiscuity, friendship and misanthropy, ambition and status, learning and ignorance, patron and client, fatherhood and childlessness, thrift and spendthrift, all the alternatives of the "conflicting wishes" which "beset and distract our hearts" lead to frustration and disillusionment: "All paths in life confront you with unfavorable issues. Therefore the opinion of the Greeks is wisest; for they say that it is good for a man not to be born at all, or, being born, to die quickly."

This then is the anxiety of choice which Descartes faces in early manhood. At this juncture he is presented by a man with another verse of Ausonius which the latter recommends, "Est et Non". It is the Pythagorean "Yea" and "Nay":

" 'Yes' and 'No': all the world constantly uses these familiar monosyllables. Take these away and you leave nothing for the tongue of man to discuss. In them is all, and all from them. Sometimes two parties both use one word or the other at the same time, but often they are opposed . . . From these arises the uproar which splits the air of the courts, from these the feuds of the maddened Circus and the widespread partisanship which fills the tiers of the theatre, . . . They are the instruments with which the schools fit for peaceful learning wage their harmless war of philosophic strife. On them the whole throng of rhetoricians depends in its wordy contests 'You grant that it is light? Yes? Then it is day?' 'No, the point is not granted; for whenever many torches or lightning-flashes give us light by night, . . . that is not the light of day.'—'Countless squabbles.' "[17b]

It is clear from the content of this poem that the interpretation which Descartes himself is said to have proposed can scarcely hold. The "Est and Non" stands for empty scholastic disputation, wrangling over words, unlike truth and falsehood in the secular sciences. It is noteworthy that Descartes, after having found in his dream the "Quod vitae sectabor iter", was unable to find the "Est et Non". In back of the two editions of the CORPUS OMNIUM VETERUM POETARUM, which might have been used by him, those of 1603 and 1611, as Norman Kemp Smith tells us, the two poems "stand face to face, under the eyes simultaneously."[18] In the 1603 edition, they are on the same page, while in the 1611 edition, though on different pages, they stand opposite each other. Clearly then we have operative here a tremendous motive for the repression of the "Est et Non", the symbol for scholastic disputation. The man whom Descartes doesn't know, we may infer, is his father who abandoned him in his earliest years (he "was left to the management of women for what at that time seemed the unusually long period of eight years"). This was the father who had made René, much against the boy's wish, bear the name "du Perron", after a small property he owned. Such place-names were not unusual in the seventeenth century. Descartes, however, evidently seemed to take this deprivation of the parental name as a symbolic rejection by his father. And it was Descartes' father who had enrolled him in 1604 as a student at the Jesuit college at La Flèche where the scholastic method of disputation was very much used.

The significance of the third dream now begins to emerge. It is an enactment of the conflict in Descartes between enthrallment to tradition, to his father, and the assertion of individual intellectual independence, the way of science. This conflict was carried out against a background in which Descartes experienced great guilt; therefore he always wore a mask, to hide his inner self. His philosophy began in doubt which enveloped, to begin with, all existence. Such a universal doubt is the characteristic of a philosophy which harbors a tremendous aggressive feeling, sometimes externalized against the outer world and its constituents, sometimes internalized in despair against one's own existence. This aggression struck back at the whole scholastic ascetic outlook, which, symbolized by his father, had threatened to castrate thought as well as the body, and to dehumanize the person. In later years, when he mourned the death of his beloved illegitimate daughter, Francine, Descartes was attacked for his alleged immorality: in reply, "he contented himself with laughing at it; and to answer to the reproach made by his enemy, that not having taken a vow of chastity, and not being exempt from the frailties natural to man, it would not be difficult to confess them if he had them." He was pleased, he said, "not to pass as a big saint", especially since the Calvinist minister who criticized him "had no great opinion of the continence of the ecclesiastics of the Roman church who live in celibacy." Descartes' dream reflected this process of recovery of hedonistic, anti-ascetic, sexual feeling. But against this recovery were arrayed all the forces of the traditional super-ego, the father and the cultural surrogate, his church. The Cartesian revolution involved an act of intellectual parricide. No revolutionist, however, is strong enough as an individual as to be able to do without the support of a super-ego. And the significance of the third dream was that in it God Himself seemed to speak to Descartes directly, and to countermand the will of the earthly father and church. Every revolutionist creates a new god in his own image, and in the third dream, the Cartesian God gave his unmistakable blessing to Descartes' choice of the vocation of the freethinking, independent scientist. *Both the Dictionary and Anthology appeared mysteriously. The Dictionary then disappeared and re-appeared mysteriously.* Here indeed was the sign that God Himself was speaking directly, without intermediary, to Descartes. Unable to find the "Est et Non", the way of scholastic theology, he discovers that *the Dictionary is not complete as it had been previously.* This incomplete Dictionary is the unfinished inventory of science, and it will be his mission to complete it as it was when it first came from God's hand. *He has not been able to find "Est et Non", but he tells the man he knows a more beautiful poem by the same author "Quod vitae sectabor iter?"* He is abandoning the fixities of scholasticism, and is prepared to undergo all the risks and uncertainties of choice which Ausonius delineated; he will follow the pagan adventure of experience rather than the traditional path of words. *On the person's request, he looks for the poem; he finds several small copper-plated engravings that are very beautiful, but did not belong to the original book.* What then is the significance of the copper-plated engravings which had rendered this new edition of the poets more beautiful? Now as Norman Kemp Smith informs us, there were no portrait engravings in either of the editions of the CORPUS POETARUM which might have come to Descartes' hand. The engravings, in other words, represent something new which is being added to the old which make it even more beautiful. What then is this something new? To explain this, we must call attention to the original French expression which Descartes used, "les petits Portraits de taille-douce." The crucial word is "taille", for in a simple play of words, Descartes' dream is indicating that the addition of experimental science will raise the old poetic aspirations to the still higher experimental level. Dreams often thus play with words, and the word "taille" had come to have an especial significance for Descartes. "Taille" denotes any cutting of glass or diamonds or material, and Descartes a few years later made himself " a great master in the art of cutting glasses" ("tailler les verres").[19] This was of central importance for him in his work on dioptrics. Since mathematicians were hampered by the lack of skilled workmen, Descartes himself undertook to train several cutters. His own teacher in these matters was Claude Mydorge, known as "the first French mathematician" of the day, and next to Mersenne, Descartes' best friend. Mydorge in 1627 and 1628 cut ("fit tailler") for Descartes a multitude of specimens. "Nothing seemed to him [Descartes] more useful than these glasses for knowing and explaining the nature of light, vision, and refraction. Mydorge made for him parabolics, hyperbolics, ovals, and elliptics."[20] Descartes had already met Mydorge in Paris when he went there after graduating from La Flèche. The older Mydorge had been "the most important of the new friends" he had made, and the young Descartes " found in this new friend something which pleased him extremely, whether it was his humor or character of mind." This then was the "taille douce", literally, "the sweet cut", the grinding of lenses and glasses for investigation into optics. That this was indeed the latent content of this symbol is indicated by the fact that after he was visited by an Italian painter the day after the dream, Descartes was no longer troubled by the significance of these symbols. For the mathematician who was most celebrated in Descartes' time for achievements made possible by cutting glass was the Italian, Galileo Galilei, who with his telescope a few years earlier had opened a whole new world to view. That Galileo's identity is thus glossed over is intelligible when we remember that Descartes was always reluctant to make any acknowledgment of debt to Galileo.

Descartes, thankful for the emotional liberation which came to him in his dreams, evidently thenceforth regarded his dream-life in general with an affection and warmth which no philosopher has ever shown. Medieval ascetics like Jerome used to fear their dreams for the temptations they brought to their unconscious. But Descartes derived from his dreams an immeasurable strength. He wrote to his friend Balzac from Amsterdam in 1631 that he was sleeping ten hours a night without any care awakening him; "after sleep has long led my spirit through woods, gardens, and enchanted palaces, where I experience all the pleasures imagined in the Fables, I mix insensibly my reveries of the day with those of the night"; when he awoke, it was only to have his senses participate in his happiness. Fourteen years later he wrote proudly to Princess Elizabeth that he had never had a bad dream ("ainsi je puis dire ques mes songes ne représentent jamais rien de fâcheux"). The night of November 10, 1619 had given him the blessing of manhood self-confidence, and vocation.

A word finally concerning the history of the interpretation of Descartes' dreams. Freud was asked by Maxime Leroy to comment on them, but his own lack of interest in philosophy precluded the kind of speculative venture which he had made in *Leonardo Da Vinci*. Freud evidently had no knowledge of the circumstances of Descartes' life or the self-probing character of Descartes' philosophy. He was content to say that Descartes' dreams are of the kind known as "dreams from above", that is, "formulations of ideas which could have been created just as well in a waking state as during the state of sleep." The content of Descartes' dreams was presumably, according to Freud, close to his conscious thought, and he was therefore ready to accept Descartes' own interpretation of them. "The philosopher interprets them himself and, ... we must accept his explanation, but it should be added that we have no path open to us which will take us any further."[21] Freud perceived that Descartes' dreams mirrored an internal conflict, but he made no effort to determine the elements of that conflict. He regarded the "bizarre" symbols of the dreams, the "almost absurd" melon from a foreign land and the little portraits as "unexplained". The melon was perhaps associated with a state of sin, Freud conjectured, with a "sexual picture which occupied the lonely young man's imagination." Clearly, however, there were paths open in the interpretation of Descartes' dreams which Freud did not take. They bore on Descartes' struggle to liberate himself from his father's authority and that of tradition, his struggle with the guilt which comes with original thought. Clearly, moreover, the content of the dreams proceeded from deep within Descartes' unconscious. For otherwise, they would not have aroused such tremendous emotions, and constituted for Descartes the landmark of his life's work. One cannot help feeling that some resistance-mechanism was awakeed in Freud by the content of Descartes' dreams, perhaps the trauma of his own long quest for intellectual and emotional independence.

"It is undoubtedly very annoying to find at the origin of modern philosophy a 'cerebral episode'," writes Jacques Maritain, one "which would call forth from our savant, should they meet it in the life of some devout personage, the most disquieting neuropathological diagnosis."[22] Several such analyses have been forthcoming, notably those by J. O. Wisdom, Stephen Schönberger, and Iago Galdston.[23] The unusual insights of these writers have not, however, it seems to me, been free from serious error. Both Schönberger and Galdston feel there was a strong homosexual ingredient in Descartes; Galdston avers as evidence the fact that Descartes was so concerned about the welfare of his servants. Freud probably was the first to argue that homosexual tendencies underlie democratic feelings, the "social instincts", the feelings of comradeship, and the "love of mankind in general," especially in those who "set themselves against an indulgence in sensual action." One can only say in reply that the a-social, anti-human, anti-mankind drives of Nazis, for instance, have been much more associated with homosexualism than the democratic emotion of Franklin D. Roosevelt or Thomas Jefferson. Descartes was indeed a spontaneous, instinctive democrat in his dealings with his servants, treating them kindly as his equals, and helping several of them to become scientists themselves. Descartes, however, unlike the medieval ascetics, Catholic monks, or Calvinist clerics, defended his human emotions. To those who railed against him as the father of an illegitimate child, he replied with pride in his human status: "Et nunc adhuc homo sum" (And now, after all, I am a man).[24] Descartes' ethic was part of the emotional hedonistic liberation on which the scientific revolution was founded. Galdston believes that Descartes' dreams represent a renouncing of sexuality, a kind of self-castration, in which he dedicated himself to knowledge as a substitute for sexuality. The evidence seems to me precisely the opposite, that in his dreams, Descartes overcame the fears of castration or impotence occasioned by fear of his father and traditional authority. The liberation of the senses and intellect for scientific investigation was associated with a liberation of sexual feeling. Freud mistakenly interpreted the pursuit of scientific knowledge as the product of the repression of sexual energies; the Scientific Revolution was rather born from a Psychological Revolution against medieval asceticism.

At times, the significance of the symbolism in Descartes' dreams seems to me to have been misapprehended. J. O. Wisdom, in his acute analysis, sees the third dream as based on a "fundamental conflict between the desire for a vivid life and a desire for knowledge." But this is because he regards *Est et Non* as referring generally to the poetic life, instead of noting the significance of the specific title as a symbol for scholastic disputation, for traditionalism of a repressive kind. He thus does not allow for the linkage between Descartes' desire for knowledge and his emotional liberation. The "melon from a foreign land" is altogether lost sight of in Wisdom's analysis; the mysterious "Monsieur N." likewise vanishes, and Descartes' college, the Jesuit classical school of La Flèche, is mistakenly described as a military academy. If Descartes' mother is absent from Wisdom's analysis, she appears in Schönberger's essay as the latent meaning of the copper-plated little portraits. No item of evidence, however, supports this interpretation.

Criticism of details, however, can easily make us overlook the contributions of these writers in opening the way to a psychoanalytical understanding of Descartes' character. The next step remains: to show how the specific ideas of Descartes' philosophy, his unique emphases and standpoint, were linked to the character of the underlying anxieties suggested by his dreams. To this task I shall return in a subsequent essay.

Meanwhile, however, we have tried to use the dreams of Descartes to see partially the anxieties which were hidden behind what Descartes called his "mask".[25] The young Descartes wrote down in January 1619, the year of his dreams, his earliest and most intimate reflections. Among these fragmentary COGITATIONES PRIVATAE, or PENSÉES, the first was his declaration of character: "As an actor puts on a mask in order that the color of his visage may not be seen, so I who am about to mount upon the stage of that world of which I have as yet been a spectator only: I appear masked upon the scene."[26] When we have understood what lay hid behind the Cartesian mask, we shall have grasped the latent meaning of the Cartesian philosophy.

Notes

1. Adrien Baillet, *Vie de Monsieur Descartes,* Paris, (ed. 1946), p. 38.
2. G. Milhaud, "Une Crise Mystique chez Descartes en 1619," *Revue Metaphysique et de Morale,* Vol. 23, (1916), pp. 612–613.
3. The text of Descartes' dream is reprinted in *Oeuvres de Descartes,* ed. Charles Adam and Paul Tannery, Vol. X, Paris, 1908, pp. 180–188. A complete English translation of them will be found in Norman Kemp Smith, *New Studies in the Philosophy of Descartes,* London, 1952, pp. 33–39.
3a. Also cf. Ezekiel 1:4: I Kings 19:11, Psalms 18: 9, 10.: Anthony and Miriam Hanson, *The Book of Job,* New York, 1962, p. 113.
4. Adrien Baillet, *Vie de Monsieur Descartes,* pp. 16, 21–22, 277.

5. Adrien Baillet, *Vie de Monsieur Descartes*, p. 21. Elizabeth S. Haldane, *Descartes: His Life and Times*, London, 1905, pp. 19, 35.
6. W. M. Stewart has suggested that "melon" is to be interpreted by a phrase in Ausonius' third Idyll, "the globe of the world." But this leaves out of account the force of its being "from a foreign country." William McC. Stewart, "Descartes and Poetry", *The Romantic Review*, Vol. XXIX, (1938), p. 215.
7. Letter to Elizabeth, May or June, 1645, Descartes: *Lettres*, ed. Michel Alexandre, Paris, 1954, pp. 117–118. Hans Pullnow, "La Psychologie Infantile chez Descartes", *Travaux du Xe Congrès International de Philsophie*, Etudes Cartésiennes, Paris, 1937, Vol. I, p. 164. *Oeuvres de Descartes*, ed. Charles Adam and Paul Tannery, Paris, 1899, Vol. III, pp. 120–121.
8. *The Philosophical Works of Descartes*, Transl. Elizabeth S. Haldane and G. R. T. Ross, New York, 1955, Vol. I, p. 398.
9. Charles Adam, *Descartes: Ses Amitiés Féminines*, Paris, 1937, p. 11.
10. Descartes, *Lettres*, p. 166.
11. Sigmund Freud, "The Acquisition of Power over Fire", *The International Journal of Psycho-Analysis*, Vol. XIII, (1937), pp. 407–408.
12. C. P. Oberndorf, "Depersonalization in Relation to Erotization of Thought", *International Journal of Psycho-Analysis*, Vol. XV, (1934), p. 273.
13. Descartes, *Lettres*, ed. Michel Alexandre, Paris, 1954, p. 169, Letter to A. Chanut, Feb. 1, 1647.
14. Stekel discusses a dream in which a thunderstorm denotes mental agitation, and lightning the flash of self-awareness. In Descartes' case, the fortunate association of "les semences" which he provides for "les étincelles de feu" provides a firmer basis for interpretation. Wilhelm Stekel, *The Interpretation of Dreams*, transl. Eden and Cedar Paul, New York, 1943, p. 332.
15. Sister Marie Jose Byrne, *Prolegomena to an Edition of the Works of Decimus Magnus Ausonius*, New York, 1916, p. 82. John Edwin Sandys, *A History of Classical Scholarship*, Cambridge, 1908, Vol. II, p. 201.
16. *Ausonius*, transl. Hugh G. Evelyn White, New York, 1919, Vol. I, p. xiii.
17. H. J. Rose, *A Handbook of Latin Literature From the Earliest Times to the Death of St. Augustine*, London, 1954, p. 528, Sister Marie Jose Byrne, op. cit. p. 20.
17a. *Ausonius*, op. cit. pp. 162–169. Leibniz, who had read the text of Descartes' dreams, later used the question, "quod vitae sectabor iter", to denote the anxieties of choice in religion and vocation. St. Augustine, he observed, after repeating the phrase, resolved the question as to which Christian sect to join by opening at random the book of the Holy Scriptures. Cf. G. W. Leibniz, *Theodicy*, transl. E. M. Huggard, New Haven, 1952, pp. 55, 178.
17b. *Ausonius*, op. cit. pp. 170–172.
18. Norman Kemp Smith, *New Studies in the Philosophy of Descartes*, London, 1952, p. 35.
19. Adrien Baillet, *Vie de Monsieur Descartes*, p. 66.
20. Adrien Baillet, *Vie de Monsieur Descartes*, p. 20, 65, 66. Elizabeth S. Haldane, *Descartes: His Life and Times*, p. 37.
21. Sigmund Freud, "Some Dreams of Descartes," in *The Complete Psychological Works of Sigmund Freud*, transl. and ed., James Strachey, Vol. XXI, London, 1961, pp. 203–204.
22. Jacques Maritain, *The Dream of Descartes*, transl. Mabelle L., Andison, New York, 1944, pp. 15–16.

23. John O. Wisdom, "Three Dreams of Descartes", *The International of Psycho-Analysis,* Vol. XXVIII, (1944) pp. 118–128. Stephen Schönberger, "A Dream of Descartes: Reflections on the Unconscious Determinants of the Sciences," *The International Journal of Psycho-Analysis,* Vol. XX, (1939), pp. 43–57. Also cf. Robert Fliess, *The Revival of Interest in the Dream,* New York, 1953, pp. 118–121. Iago Goldston, "Descartes and Modern Psychiatric Thought", *Isis,* Vol. XXXV, (1944), pp. 118–128.
24. Charles Adam, *Descartes: Ses Amitiés Féminines,* Paris, 1937, p. 87.
25. Descartes' "state of mind was almost morbid", writes Elizabeth S. Haldane. Cf. *Descartes: His Life and Times,* pp. 48–49.
26. "Ut comoedi, moniti ne in fronte appareat pudor, personam induunt: sic ego, hoc mundi theatrum conscensurus, in quo hactenus spectator exstiti, larvatus prodeo". *Oeuvres de Descartes,* ed. Charles Adam and Paul Tannery, Vol. X, p. 213.

13

Anxiety and Philosophy: The Case of Descartes

The basic problems of modern philosophy were set in the thought of René Descartes. Its central themes, motives and arguments were those which likewise became central in the consciousness of modern man. His philosophy sought to resolve anxieties which in large measure were representative of those which perdured in the unconscious of the seventeenth century Scientific Revolutionists. The answers which Descartes gave are a congeries of ideas which seemed to him to arrange themselves in a system. They present a unique problem for our psychoanalytical understanding. As we try to isolate those ideas the analysis of which would take us into the psychological core of Descartes' philosophy, four principal tenets suggest themselves for study:

(1) that animals are incapable of thinking,
(2) that God's goodness is the guarantee that the external world exists,
(3) that God necessarily exists,
(4) that because I think, I am certain that I exist.

The philosophical ideas of Descartes, we shall find, issued from a deep struggle in his unconscious with the authority of a father whom he blamed for the death of his mother, and who had abandoned him during his childhood years. From his view that his father was heartless, soulless, sprang his doctrine that animals are soulless. From the severance of parental affection in his earliest years, there issued the mistrustful solipsistic mood in which Descartes wondered whether he could put faith in his perceptions of an external world. Deserted by his father as a child, he blamed himself, the rejected one, as unworthy of love, and with this self-aggression came a loss in the conviction of his own existence, with which his philosophy had later to cope. From the guilt of his own intellectual parricide, his own revolt, came the necessity to prove that God, a Surrogate Father who blessed his scientific vocation, necessarily existed,—that mathematics, the weapon of revolt against the natural father, itself guaranteed the existence of the Surrogate Father, who endorsed the scientific revolt of his earthly son.

The conflict within Descartes' unconscious brought into sharp focus certain basic conflicts which were central in the emotional perspective underlying the Western Age of Reason and Revolt.

I shall begin with Descartes' notion of animal automatism, not because it is a primary tenet of his philosophy, but because its unique character affords us a strategic vantage-point for the beginning of a psychoanalytical study.

Animal Automatism ("Contra-totemism")

(1) *That animals are incapable of thinking.* This doctrine of animal automatism, writes Norman Kemp Smith, was the view which Descartes held persistently from the years in which he was composing the *Regulae* up to his last utterances. "We are left guessing," he says, "what can have been Descartes' reasons for so extreme—we may indeed say so monstrous a thesis."[1] The purely scientific grounds for this doctrine, Descartes acknowledged, were unconvincing. Descartes conceded that it was undemonstrable whether animals were unthinking or thinking beings. He granted that common sense ("all minds from infancy") are persuaded that brutes think because their bodily organs are so much like our own. Nevertheless, he insisted, this belief was "the greatest of all the prejudices we have retained from infancy."

Descartes adduced one important scientific consideration for his tenet of animal automatism. Animals lack language, he noted, and "the word is the sole sign and the only certain mark of the presence of thought." To this argument Descartes' contemporary Pierre Gassendi, made the obvious reply: that the difference between men's intelligence and that of beasts was one of degree, of more and less, rather than of kind. Descartes' answer then left the field of science for that of theology: "I have nothing to reply, except that, if they could think as we do, they would have an immortal soul as well."[2] He refused to open the gates of immortality to a host of "imperfect" organisms "such as oysters, sponges, etc." This shift of the gravamen of the argument to the theological fitness of things clearly however opened the door even further to undemonstrable possibilities. The animal lovers, the theriophilists, could portray a heaven in which all living beings would enjoy the immortalities proper to their kinds, and share in the final community which Isaiah foretold for the last of days.[3]

Evidently behind Descartes' doctrine of animal automatism, there were strong emotional, non-logical determinants. For, in his own terms, he chose between two undemonstrable views, neither of them a certain, clear and distinct idea. When he departed from scientific grounds to make a theological appeal, he only emphasized the non-logical component of his belief. What then was the character of these non-logical determinants? A sociological explanation was proposed briefly by Karl Marx, who advanced the view that Descartes' theory was an expression of the new industrial age: "Descartes, in defining animals as mere machines, saw with eyes of the manufacturing period, while to the eyes of the middle ages, animals were assistants to men."[4] Such an explanation is congenial to those who seek socio-economic accounts for both the genesis and reception of ideas. The fact of the matter is, however, that Descartes' notion of animal automatism was quite unrepresentative of the new bourgeois society. It won adherents only for a brief period. The poet, La Fontaine, made his animals into spokesmen for the bourgeois ethic, and believed that animals had souls; he was typical in this respect of almost all his fellow poets. The playwright Molière shared La Fontaine's view, and the literary Mlle. de Scudéry expressed the preponderant opinion in 1683: "the tiniest monkey by its industry and intelligence destroys all of Descartes' doctrines." By the end of the seventeenth century, writes the scholarly historian, Leonora Cohen Rosenfield, "a revulsion of feeling set in against the theory of the beast-machine."[5] The ordinary person sensed a streak of cruelty and hatred in Descartes' doctrine. Malebranche, in Cartesian fashion, might kick a pregnant dog, and say it had no feeling. But the ordinary person could not regard his animal pets as feelingless extensions; rather he found a less distorted standpoint in Locke's view which granted to animals a capacity for feeling, memory, and simple ideas.

A highly individual emotion, not a sociological pattern, was the controlling factor in determining Descartes' doctrine of animal automatism. We shall call it Descartes' "contratotemic" emotion.[6] What is this contra-totemic element in Descartes' philosophy? Freud has illumined how the totemic device enables a child to vent upon a surrogate being its hostility against its father; "the child displaces a part of its feelings from the father upon some animal."[7] In Descartes' case, because of unique circumstances, the resentment against the father issued in a contra-totemic feeling; he denied to animals, the surrogates for his father, the existence of souls. They were soulless and unfeeling as his father had been; they were automata even as his father had been in his infancy.

Ordinarily, a child attributes a "full equality to animals," and feels himself more akin to them than to the mysterious adults. He does not set a dividing line between his nature and that of the animals. When animal phobias do arise, it is because the child has displaced its fear of its father upon an animal. In one case, a small child regarded the slaughtering of poultry as a festive occasion because it ambivalently hated its totem animal, the chicken. And Descartes indeed in manhood made the pursuit of animal dissection into a festive occasion. Baillet, the early biographer, writes of the "ardor" which led Descartes to "go almost everyday to the butcher to see him kill the beasts." Accused of "going about the villages to see pigs killed," Descartes replied that it was no crime to be interested in anatomy; "I was in Amsterdam one winter, and I went almost everyday to a butcher to see him kill the beasts, and had brought thence to my house the parts of the animals which I wished to anatomize at leisure; which I did again several times in all the places where I was, and I do not believe any intelligent man would blame me for it."[8] The animals which he watched being killed with such pleasure were, Descartes assured himself, mere machines, without consciousness of thought. In his contra-totemism, he deprived animals of any common lot with his own humanity. His contra-totemism was not only a projection of his father's feelinglessness toward animals; it was also a rationalization for the repetitive re-enactment of his own tremendous parricide, his own revolt.

The contra-totemic theme was strong in Descartes from his earliest reflective years. "His sentiments on the *soul of beasts,* or *Automatons,*" writes Baillet, his earliest biographer, were found in the "works of his youth, . . . twenty years before he published his principle on the distinction between thinking and extended substance."[9] He had read no author from whom he might have absorbed this idea. From the logical standpoint, the doctrine of animal automatism might have provided Descartes with a rationale for the eating of animal food. Actually, however, Descartes advocated a vegetarian diet of roots and fruits, "which he believed more suitable for prolonging the life of man than the flesh of animals."[10] And the vegetarian unconscious seems often to be characterized by severe feelings of hostility and guilt toward one's father.[11] Descartes' doctrine issued from such an unconscious anxiety, not from his spontaneous attitudes toward animals. He was evidently personally fond of dogs, and kept at least one pet dog called "Monsieur Grat" whom he once lent to his friend, the Abbé Picot, for breeding purposes.[12] At odds both with the popular common sense and the scientific philosophers of his time, Descartes however clung tenaciously to his rigid intellectualistic stance for animal automatism.

Descartes began practicing dissections in 1628, the year in which he made his final break with father, family and friends in France, and settled in Holland. In 1639, he wrote to Father Mersenne, his old friend, that he had been doing dissections often on various animals in the last eleven years. ("C'est un exercice ou je me suis souvent occupé depuis onze ans.")[13] His zealous dissections and his contra-totemism both had a common emotional source. They stemmed, the facts suggest, from those same childhood anxieties which underlay Descartes' questions of his own existence. The infant René, orphaned of his mother when he was a year old, could not love the father who deserted him to find another wife. The infant's first sadness, wrote Descartes in later years, comes with its loss of its mother's breast; its first hatred entered its soul when someone not fit to feed it tried to replace its mother.[14] Thus in effect, Descartes described the trauma of his mother's death and the genesis of his resentment toward his father. Nor would he accept as sincere his father's sorrow for his lost mother. With a distinct note of personal experience, Descartes analyzed the "secret joy" of a bereaved husband: "when a husband laments his dead wife whom (as sometimes happens) he would be sorry to see brought to life again, . . . he yet feels a secret joy in the inmost parts of his heart, . . . "[15] What do we have here but the son's critical observation of his own father whom, in essence, he now blames for his mother's death? From his childhood trauma arose the distinctive Cartesian anxieties and sense of guilt,—anxiety at the severance of a mother's love which left him deprived of the emotional warmth which sustains a child's sense of existence,—fear and hatred toward the father who disrupted its emotional haven, and coldly left him in strangers' hands,—guilt for the inner hatred which he bore his father, and which led him to mask his character. A child will harbor Oedipal resentments towards its father when the mother is alive, but when its mother is dead, the resentments for the deprivation of maternal love turn all the more against the father.

The inner guilt in Descartes awoke such strong resistances that he refused to discuss the question as to whether the doctrine of animal automatism lightened the burden of man's guilt. Cardinal de Polignac, Descartes' sole defender among the poets, said that unless animal automatism were true, we would be guilty every day of murder.[16] But Descartes refused to consider such suggestions of guilt; they were akin "to the extravagances of Pythagoras, which attached to those who ate or killed them (the animals) the suspicion even of a crime."[17] Evidently his own parricidal motive and inner conflict with his guilt were so strong and deep-seated that he tried to block their entry into consciousness by refusing to discuss questions which would come dangerously close to the root of his guilt.

Always hiding and fleeing throughout his life, Descartes seems indeed, like the children in their animal phobias, to have cast his father in the role of a wolf. When he looked for an example of the complexities of an animal machine, he spoke of a wolf which frightens a sheep into flight: "why should we marvel so greatly if the light reflected from the body of a wolf into the eyes of a sheep should be equally capable of exciting in it the motion of flight?"[18] Descartes' contra-totemism, the evidence indicates, was a defence mechanism for the guilt of his intellectual parricide.

The Cartesian Anxiety

We have alluded to the Cartesian anxieties which made for distinctive tenets in his philosophy. We wish to answer questions such as why Descartes began his philosophy with a search for assurances that his existence was not an illusion. Why did he worry that perhaps God was a deceiver? Let us therefore enquire briefly into the salient facts of Descartes' character in order to perceive more clearly the roots of his basic anxiety.

The pattern of the Cartesian anxiety has one component familiar to child psychiatrists; it is that of the child deprived in its early years of its mother. Even the partially deprived child, who receives, as Descartes did, a loving nurse to replace its mother, experiences "acute anxiety, excessive need for love powerful feelings of revenge, and, arising from these last, guilt and depression." The child sustaining this blow at the age of one has an especially severe emotional response; often it rejects a substitute mother, and usually it undergoes "a regression to primitive functioning and increased difficulty in learning afresh." Descartes, seeking to reconstruct with his method of doubt the childhood inference from personal experience to an external world, was later to re-enact this regression and re-learning. His fondness for remaining in bed ten to twelve hours a day, a habit he had all his life, was, we may surmise, a persistent regressive longing and substitute for a mother's warmth. The deprived child, moreover, has violent fantasies against a parent who deserts it; René's mother deserted him in death, and his father deserted him in life. An acute conflict and anxiety arise in the case of such a child; "so far from idolizing their parents and wishing to become like them, one side of them hates them and wishes to avoid having anything to do with them." The child also begins to withdraw from people so as not to suffer pain again: "To withdraw from human contact is to avoid further frustration and to avoid the intense depression which human beings experience as a result of hating the person whom they most dearly love and need... Thenceforth he becomes a lone wolf, pursuing his ends irrespective of others..."[19] Withdrawal from almost all social existence too became the law of Descartes' life.

The young Descartes grew up in a strange aloofness from the world; he and his fellow-beings moved in shadowy, unreal scenes. The earliest of his writings depicts his existence as that of a spectator masked from the world: "As an actor puts on a mask in order that the colour of his visage may not be seen, so I who am about to mount upon the stage of the world of which I have as yet been a spectator only: I appear masked upon the scene."[20] This strange youth, as Elizabeth S. Haldane observes, felt himself to be in the world without being part of it.

Now one wears a mask only when one wishes to hide one's real character from his fellow-men. And the masked Descartes was always hiding. An inner guilt was never altogether to leave him; as Baillet writes, Descartes believed he "may not have been as innocent before God as before men." His celebrated "mask" hid the nature of that guilt from men. He had a passion all his life, as Baillet tells, for "une vie Cachée," a hidden life. His favorite motto was a verse from Ovid: *Bene vixit, bene qui latuit,* he who has lived well, has hid well. He went to enormous precautions while living in Paris to keep his whereabouts secret from friends and family. When he was nineteen years old, he managed to seclude himself for almost two years, but when a friend found him by accident, Descartes soon fled to the curious anonymity of a soldier's life in the Dutch Army of Prince Maurice of Nassau,—a kind of T. E. Lawrence quest for identity in non-identity. Eleven years later he was again in Paris taking the most elaborate measures to keep his address unknown, and was so dismayed when a worried relative located him that he fled first to the siege of La Rochelle, and subsequently to Holland.[21]

The direction of his sexual desire indicated a man who inwardly feels himself maimed emotionally by some guilt, some unworthiness. His first affection was for a cross-eyed girl, and long thereafter, he felt a special attraction for persons with this physical defect. "When I was a child," he wrote, "I loved a girl my age, who was a little squint-eyed (louche); it made such an impression on my brain, that when I looked at her cross eyes (ses yeux égarés), it joined with that which stirred in me the passion of love, and I felt more inclined to love them more than others only for having this defect, but nevertheless I did not know that it was for this reason. On the contrary, after I had reflected on it, and recognized that it was a defect, I was no longer affected by it." For he himself bore an injury within, a defect within himself, so that in his choice for whom to love he felt most secure with some one who externalized, as it were, the inner defect; and from this pattern of choice in love, Descartes only liberated himself by a self-analysis of his unconscious.[22]

From whence came these anxieties in Descartes of detachment and concealment and the sense of being maimed? "A child who has lived deprived of its mother, at a home without warmth, a scholar who liked to dream alone in his room where doubts assailed him,"—in such words Maxime Leroy has portrayed the early years of Descartes.[23]

Existence acquires a hollow, insubstantial character for those whose emotional lives are disrupted in their earliest years. Deprived of his mother when he was a little more than a year old, Descartes felt himself nonetheless linked to her by common physical features and a common sickness. But where his mother had succumbed, Descartes, in his early manhood, overcame what he regarded as their common illness; his health was achieved during that same period when his three famous dreams brought him a liberation from the disabling antipathy of his earthly father, and the blessing on his vocation from the Heavenly One.[24] The Cartesian optimism and independence of spirit matured at that time: "For, having been born of a mother who died a few days after my birth, of an illness of the lungs, . . . I had inherited from her a dry cough and a pale complexion, which I kept until more than twenty years old, and which made all the doctors who saw me before that time, condemn me to die young. But I believe the inclination which I have always had to see things from the point of view which would make them most agreeable to me, and to see that my principal happiness depended on me alone, is cause why this indisposition, which was as if natural to me, passed away entirely little by little."[25] It was in his early twenties evidently that Descartes experienced a psychological liberation from some neurosis which had haunted his childhood and youth.

The origin of love, according to Descartes, reaches back to the nourishment we receive in our pre-natal period; its nature is difficult to understand because of the "confused feelings of our infancy." Love originates, he said, as an emotion directed to the person who provides the child with nourishment, and when that nourishment is lacking, sadness first arises. Thus Descartes in his theory of emotions depicted the history of his own emotional experience and the trauma of his own orphanhood. Sadness had supervened when the loved mother who fed him had died. Father-hatred had then entered his life: "And if another came, who was unsuited for feeding its body, it [the child] had hatred for him." (Et s'il en est venu d'autre en sa place, qui n'ait pas été propre à nourrir le corps, elle a eu pour lui de la haine.)[26] It was at this juncture in Descartes' life that his relation to the external world was sundered; the loss of his mother upset the smooth, loving, stable equilibrium, in which he found the security of nourishment and love. Suddenly, the mother's breast and body, the secure assurance of the external world, was not there. For our first relationship to the external world is with our mother's breast as the source of "aliment." In her place was the father who deceived one, for he was not suitable to nourish one; the external world was shaken,—it was somehow an illusion, for it left one insecure and unfed, and one hated the deceiver.

Descartes' sense of rejection by his father persisted throughout his life. After the mother's death, the father, Councillor Joachim Descartes left the child behind in Poitiers, where the maternal grandfather lived, under a nurse's care. Joachim Descartes went to Brittany where he subsequently re-married, and settled down with his second wife. He assigned to René the surname "du Perron" to use instead of "Descartes"; such place-names were customary, but René bitterly resented it, evidently regarding it as a symbol of rejection, and he resumed "Descartes" when he left his home. The father wished René to devote himself to the study of law, and to become a Councillor, a civil servant. René did in fact take a degree in law at Poitiers in 1616, but afterwards managed to evade the parental wish. He noted sadly six years later, after a brief visit with his father that "nobody in his family could give him good advice (de bonnes ouvertures) on the way of life he should choose."[27] Although his father had called him as a child "his philosopher, because of the insatiable curiosity with which this child asked him the causes and effects of everything which passed its senses," the rest of his family "despised him for the odious title that he bore of *philosophe,* and tried to banish him from their memory as a family disgrace." Descartes made it a point to insist during his life on how little his father's guidance had meant to him. "He told his friends readily that even if his father hadn't made him study, he would nevertheless have written in our language the same things he wrote in Latin," and that if he had been born a workingman and been taught a trade, he would have succeeded just as well.[28] And when his father was dying in 1640, Descartes found himself too involved, though indirectly, in a petty controversy between two Dutch mathematicians to make him a farewell visit.[29] All his life, furthermore, there was an utter dislike between Descartes and his elder brother, Pierre, who followed their father's legal career.

Here at manhood's outset then, was the young Descartes seeking to affirm his masculinity and independence, and hoping to find a kind of Surrogate Family, an alternative brotherhood, which would support him in his intellectual parricide, and thereby enable him to recover his sense of existence. Descartes was led into two unusual steps, his enlistment in the Dutch Army and his joining the secret society of the Rosicrucians. He was twenty-three years old when he joined the army of Prince Maurice of Orange seeking, he wrote, to read "the great book of the world" and to spend his time "in collecting various experiences, in proving myself in the various predicaments in which I was placed by fortune." Biographers have been puzzled by Descartes' decision to become a soldier. Baillet wrote that he joined an army to be free from distractions, and to develop his thoughts. "A rough camp of soldiers," however, was, as Elizabeth Haldane notes, scarcely the setting one would choose for philosophical meditations; "there must have been some strong impelling power to make a young man of three-and-twenty devote his time to living the life of a soldier in a climate as cold and bleak as that which he encountered on the Danube."[30]

What then was the "strong impelling power" which moved Descartes to a soldier's existence? For the military profession itself he never seems to have had in later years any respect; "he had always spoken of his military profession with so indifferent and so cold a manner, that one easily judged that he regarded his campaigns as simple trips." The soldier's life, wrote Descartes, is one which "accustoms men to see their best friends die unexpectedly."[31] The last work of his life, *La Naissance de la Paix,* a poetic ballet, written in Stockholm for Queen Christina in 1649, was virtually a pacifist tract. It opened with a paean to Pallas, the goddess of peace:

> Pallas is right in thinking that war,
> Even the best one can have
> Takes always much beauty from the land,
> And that to give us Peace,
> This is the greatest of her benefits.
>
> (... Pallas a raison de penser que la guerre,
> La meilleure qu'on puisse avoir,
>   Oste toujours beaucoup des beautez de la terre,
> Et que de nous donner la Paix
> C'est le plus grand de ses bienfaits)

It presented a spectacle of characters: Mars, who covered the most beautiful fields with dead, Panic Terror, "fille de la nuit," boasting that she dominated war, and then in an autobiographical note, a troop of Volunteers, seeking Victory and Glory, but followed by maimed soldiers reciting:

> Who sees us as we are now made
> And thinks that war is beautiful
> (Qui voit comme nous sommes faits
> Et pense que la guerre est belle)[32]

Then there were Pillagers and ruined, despoiled Peasants. Pallas, however, appealed to the peoples: Mars was defeated, and the Goddess of Wisdom was supreme. Justice and peace reigned with her. Such were Descartes' anti-militarist views in later years, when he could affirm: "I profess nothing now but poltroonery." In his youth, however, a certain "warmth in the liver" (chaleur de foie), to use his words, had made him love weapons.[33]

Behind Descartes' enlistment as a soldier was the impelling drive of a revolt against the residual bonds which linked him emotionally to his father and the Jesuit upbringing he had given him. Years later when Descartes' freedom was threatened by Calvinist fanatics at the University of Leyden, Descartes told of his motive in becoming a soldier. He, a Frenchman, had taken up arms for the Dutch in order to help drive the Spanish Inquisition from Holland. He hoped therefore that he would now be spared an Inquisition by Dutch ministers.[34] Descartes had joined an army which was fighting against Catholic authoritarianism. Its commander, Prince Maurice. was the sceptic who had mocked at theology: "I know nothing of predestination, whether it is green or whether it is blue."[35] Descartes' decision revealed the spirit of revolt which lurked behind his mask; he volunteered for an army which was fighting against his own cultural superego. Most important, however, was the fact that in this setting. he could seek his own full masculinity away from his father's maiming influence, in this rough camp, where in anonymity, he could achieve a sense of his own vocation. It was a time of crisis, searching, and vacillation for Descartes. He worried as to his path of life. Was some evil genius misleading him? He left the Dutch Army after two years, and enlisted in that of Maximilian, Duke of Bavaria, chief of the Catholic League. Then came the three liberating dreams of November 10, 1619 which brought a new strength into Descartes' life.

Meanwhile, however, Descartes believed himself for a time to have found a brotherhood in which he could find sustenance in his intellectual parricide. This was the secret brotherhood of the Rosicrucians. They were rumored to have discovered a new method, a new science, and Descartes, excited by this news, tried to seek out its members. The truth concerning Descartes' relations with the Rosicrucian Order will never be altogether known. In 1623, when Descartes returned to Paris, he was exceedingly worried by the story which spread everywhere that he had joined the secret confraternity ("l'on commençait à faire courir le bruit qu'il s'était enrolé dans sa confrérie"). Strange stories were current in Paris concerning the Rosicrucians; the *Invisibles,* as they were called, were said to have worked out a detailed plan for the secret intellectual infiltration of Paris. According to Baillet, Descartes had been unable during the fateful winter of 1619 to find a single member of the Society, so secret were they. Descartes, indeed, replying to rumors, said that because the Order was invisible, he had never been able to discover any members in Germany.[36] On the other hand, the ineluctable fact remains that Descartes dedicated a mathematical treatise which he wrote in 1619-20 "to the illustrious Brethren of the Rosicrucian Order in Germany." One hardly dedicates a treatise to a society of which one knows not a single member. Descartes' motto, *bene vixit bene qui latuit,* was also that of the Rosicrucians.[37]

Descartes, like Leibniz later, as well as Francis Bacon and Sir Kenelme Digby, evidently went through a stage of enthusiasm for the notion of a secret society of scientists working for the common cause of reason in a Europe riven by religious intolerances and wars.[38] The publication in 1615 of two books on the history of the Rosicrucians and their aims had brought the society to people's attention and caused a great sensation. "Secret societies at that time had an evil reputation," remarks Elizabeth Haldane. They have always aroused fear in times of political and intellectual crisis. A secret society is characteristically a child's pattern of defiance of its father; the new brotherhood is a substitute family to which the father cannot belong; from it the son in revolt derives an added strength to assert himself against the father. The secret society lightens the guilt of its individual members by having them all share in the parricide. Descartes, in his brief association with the Rosicrucians, had allowed his mask to drop away too much. Thenceforth the very word "secret" disturbed him. He wrote in 1629 to Mersenne: "As soon as I even see the word arcanum (secret) in some proposition, I begin to have a bad opinion of it."[39] The secret of his own Rosicrucian relations became part of the Cartesian mask of guilt.

In the end, there was no brotherhood with which Descartes could alleviate his anxiety,—only such rare daring spirits as Constantin Huygens, to whom he would avow the inner isolation which he lived and sought. He confessed himself a man who wished passionately "to avoid even the shadows of all who could trouble him" (éviter même les ombres de tout ce qui pourrait le troubler), and when these shadows did cast themselves even tangentially on his life, he compared their people hatefully to "flies which fly around the face of a man who is asleep in the darkness of a woods to rest himself" (mouches qui volent autour du visage d'un homme qui s'est couché à l'ombre dans un bois pour s'y reposer).[40] And he told Huygens how much more he preferred to live expatriate among the Dutch people, non-Catholics though they were, than among his own. Thus he lived, secluded in Holland, hoping in strange self-negation that nobody would think of him.

For many years, Descartes withheld publication of his ideas. In 1635, however, an event took place which awoke in him a warmer feeling than he had ever known; it moved him out of his anchorite's self-concealment and rejection of the world. A child was born to him of a Dutch woman, Hijlena; the child was baptized in a Reformed church and christened Fransintge (Francine). Descartes acknowledged himself the father. Probably the child was illegitimate though its birth was registered in the book for legitimate children.[41] Possibly in the name "Francine" Descartes expressed a longing for the love he had never known in France, but could now fulfill. He was much concerned with the child's education, and, unlike his own father, insisted that the child live with him. Francine died when only five years old, and Descartes mourned her deeply, showing as Baillet wrote pedantically that "the true philosophy cannot extinguish natural feelings."[42] The birth of Francine seems to have freed Descartes in large measure from the strong resistance he felt to publication.[43] At the age of forty, in full maturity, he had asserted his manhood as against both Catholic celibates and Calvinist clerics. When he was assailed for having had an illegitimate child, and even accused of having had others, he retorted mockingly that he had never taken a vow of celibacy, and that not being free from the natural human weaknesses, he would avow them if he had them. "But even if there weren't any, he nevertheless agreed not to pass himself off as a big saint," especially in the eyes of a Protestant minister who had such a low opinion of the celibacy of Roman priests.[44] This successful emotional assertion of Descartes set the ground for a public act of intellectual revolution; Descartes began to bring to completion his *Discourse on Method* which was published in June, 1637. A year after Francine's death, Descartes referred to himself as "so useless" (si inutile), but he had the enduring will to complete the *Meditations*.[45]

The central motif of Descartes' philosophy was joy as the theme of man's existence. The Cartesian revolution, in its emotional significance was the affirmation of an optimism concerning man's status. Like all philosophers, Descartes too began with anxiety; but whereas many philosophies also ended with anxiety, Descartes succeeded largely in overcoming anxiety. The birth of modern philosophy in Descartes' rationalism had this as its underlying psychological foundation:—that man, beginning his thinking with a sense of guilt, could emerge from that ordeal; that he could liberate himself from the past, and seek his own way of life, confident that in doing so, he had God's blessing; that he could have faith in his human reason and its limitless promise of progress. Energies consumed in the guilt of self-aggression turned outward to conquer the world.

"There are always more goods than evils in this life," Descartes could write, even in a Europe ridden by religious wars and hatreds. Joy, he said, confers a "secret force," elevating one's life above circumstances: "I have often noticed that the things which I have done with a gay heart, and without any internal repugnance, have generally succeeded happily, even when, in games of chance, where only fortune reigns, I have always experienced it more favorable when having joyous thoughts than when having sad ones."[46] The spirit is most satisfied, he wrote, when it has the utmost of gaiety, and the very beatitude of the soul was linked to the gaiety and ease of the body, aided by such "exercises of the body as hunting, tennis, and the like." He took as his morality: "to love life without fearing death" (d'aimer la vie sans craindre la mort). He hadn't been sick in thirty years, he wrote contentedly in 1639, and contemplating his teeth still strong and good, ("les dents si bonnes et si fortes"), he thought himself further from death than even in his youth.[47] The Cartesian philosophy had been a successful revolution against anxiety and guilt. To its tenets we therefore now return.

(2) *That God's goodness is the guarantee that the external world exists* ... The *Meditations* of Descartes trace the road by which he surmounted anxiety. Brooding upon the reality of his existence and God's, he wondered too how he could guarantee the veracity of his sensations which testified to the existence of bodies. For the soul was entirely and absolutely distinct from the body; how then could its ideas validate the existence of the body and physical objects generally? Then the fear arose in Descartes that perhaps our senses were deceptive, and that God, a Supreme Deceiver, took delight in deceiving us: "But how do I know that He has not brought it to pass that

there is no earth, no heaven, no extended body, no magnitude, no place, and that nevertheless they seem to me to exist just exactly as I now see them? . . . I shall then suppose, not that God who is supremely good and the fountain of truth, but some evil genius not less powerful than deceitful, has employed his whole energies in deceiving me; I shall consider that the heavens, the earth, colours, figures, sound, and all other external things are nought but the illusions and dreams of which this genius has availed himself in order to lay traps for my credulity. I shall consider myself as having no hands, no eyes, no flesh, no blood, nor any senses, yet falsely believing myself to possess all these things; I shall remain obstinately attached to this idea. . . . " He was determined, wrote Descartes, to avoid "being imposed upon by this arch deceiver, however powerful and deceptive he may be. But this task is a laborious one, and insensibly a certain lassitude leads me into the course of my ordinary life . . . , and I dread awakening from this slumber, . . . "[48]

Again and again Descartes referred to this haunting fear that "there is some deceiver or other, very powerful and very cunning, who ever employs his ingenuity in deceiving me." Jeremy Bentham once invented the word "cacotheism" to signify belief in a malevolent God.[49] I shall use a less harsh sounding word, "maledeism," to designate the belief in an evil God.

Descartes' tentative maledeism derived from the anxiety of his own experience with an unloving father. A child's conception of God, as Freud illumined, is generally a projection of its father. But in Descartes' case, the conception of God, linked as it was to an evil father, was directly connected with a traumatic doubt as to the existence of the external world. And Descartes, struggling with philosophical means to issue from the thinking self to the external world, is abreacting a traumatic experience of his own early life. For as we have mentioned, the death of Descartes' mother when he was one year old disrupted his whole relationship to the external world. The child's first experience and verification of the external world is associated with its mother's breast; there it seeks its nourishment; there the environment's existence as an external source of food is confirmed. Love, as Descartes said, originates as an emotion directed to our nourisher. Instead, there came another, "unsuited for feeding his body," and the child "had hatred for him." This was the Evil Power who somehow had removed the mother, a Deceiver, for he was "unsuited" for feeding; the Deceiver fooled him at first by presenting himself in his mother's place, but the mother's milk, the nourishment, that is, the reality of the external world as the source of aliment, was lacking. The child was like the Pavlovian dog in whom a nervous collapse is induced by not giving it food after it has been stimulated to expect it. God the Arch Deceiver was a projection of Descartes' early experience of his father's deception.

It is important to perceive how much the experience of a mother's and father's love is related to the conviction of the existence of the external world. The child, as Freud tells us, moves from attachment to its self to an interest in external objects. The love of elders supervenes however to guide and support the child in this transition; the nursing child is caressed, and receives the reassurance of love as it does its nourishment. But what if some traumatic experience of deprivation intervenes at this time? Where love is withdrawn or denied, uncertainty and doubt arise, the expression of the child's panicky, terrified insecurity. A malign genius is then, indeed, playing with the child's helplessness, Descartes in adulthood, in self-analysis, copes with the unloving father in the guise of the unloving God. The therapy came with the optimistic conviction of the existence of a loving God, the Surrogate Father.

Descartes' struggle for an external world illumines our contemporary conception of the physical external world as an "intersubjective order." For if the road to belief in the existence of the external world depends at the outset on our emotional linkage to other subjects, other persons, then what is primary is that we have a community of subjects to begin with. In psychological terms, this means that the physical world is one concerning which the child has the loving assurance of mother and father that it's really there, that it exists. The love of an adult thus guides the child to its next libidinal attachment to physical things. But that avenue to an intersubjective order is disrupted when the primary co-subject withdraws his or her affection, and menaces you with hostility. The emotional community of co-subjects is the psychological precondition for the intersubjective physical order. In this sense, Martin Buber is right when he says that an I-Thou relationship precedes the I-it.

The very examples which Descartes used to show what he meant by doubt in his senses suggest an underlying castration-fear in relation to his father, and his projective counterpart, God the Arch Deceiver.[50] He notes, to begin with, that persons with a "part" which has been "amputated," "whose arms and legs have been cut off" still think they feel pain in that part: "I could not be quite certain that it was a certain member which pained me, even although I felt pain in it." Then he goes on to make an unusual statement that pain (not pleasure), is the most intimate experience: "for is there anything more intimate or internal than pain?" Since in his own theory of emotions, joy and love precede sadness in the order of experience, the primacy which he informally assigns to pain, giving as an example a pain of whose origin he could not be quite certain, suggests that an unconscious amputative fear was strong in Descartes. In the same paragraph, he gives a specific example of the deception of the senses which revolves around a failure in the perception of height, of erection: "colossal statues raised on the summit of these towers appeared as quite tiny statues when viewed from the bottom."[51] This "intimate pain" of uncertain origin suggests that experienced by a man who has strong sexual desire, but finds himself impotent. This "intimate pain," deriving from fear of his father (God the Arch Deceiver), would then coincide with the anxiety which was underlying to all the doubt which Descartes felt in the veracity of his senses. Cure came when the Perfect Father's love replaced the earthly father's hatred.

Descartes' procedure was indeed a therapeutic one. He reenacted the process of the child's groping inference from its own psychic life to the existence of an external world. Above all, he abreacted the traumatic experience of his father, an arch deceiver, when his mother died. But with a difference. The trauma was now surmounted by the assurance of the existence of a Perfect Father.

It is when he recognizes the existence of a Perfect God that Descartes feels that he is liberated from the compulsion "to doubt them all (the matters which the senses seem to teach us) universally."[52] And God "cannot be a deceiver, since the light of nature teaches us that fraud and deception necessarily proceed from some defect." For "the desire to deceive doubtless testifies to malice or feebleness, and accordingly cannot be found in God." So Descartes fashioned for himself a Heavenly Father who, unlike his natural father, "could not desire to deceive me."[53] Every deprived, deserted child seeks in its fantasy the assurance and comfort of the loving parent it has never known. God, unlike his father, would be an unfrustrated being, engaging in no aggression. He would not deceive anyone. God's perfection was necessary for

Descartes to overcome the internalized anxiety of his father's hatred which had maimed his powers and his own access to the external, intersubjective world. The possibility of maledeism had to be denied for Descartes to overcome his anxiety. And Descartes' substitution of the Perfect God for his imperfect father now went so far that his words denied in a metaphysical sense the parental role; his mother and father had not created him, he said; God was his Creator. "Possibly, . . . I am created either by my parents or by some other cause less perfect than God. This cannot be, . . . "[54] For he had within him the idea of God, and this, said Descartes, could have come from no earthly parents but only from a Perfect God, "like the mark of the workman imprinted on his work." Descartes shied away from any genetic analysis of the idea of God. That was the only other alternative for dealing with his anxiety. He chose instead to make the monumental effort to construct a whole projective world in which his anxiety would be appeased, a world which with an "omnipotence of thought" (in Freud's phrase), he would say, necessarily existed.

(3) *That God necessarily exists*—In examining the idea, writes Descartes, "which I had of a Perfect Being, I found that in this case existence was implied in it in the same manner in which the equality of its three angles to two right angles is implied in the idea of a triangle; . . . Consequently it is at least as certain that God, who is a Being so perfect, is, or exists, as any demonstration of geometry can possibly be."[55] God simply could not be conceived as not existing; the very definition of God, according to Descartes, entailed that He necessarily existed.

Now Descartes was well aware of the objections which philosophers from Kant on have made to the so-called ontological argument for the existence of God. Indeed, it was Pierre Gassendi who reminded Descartes in his own words that just as from the idea of a mountain, it does not follow that the mountain exists, "so from the fact that I think of God as existing, it does not follow that He exists." To those who tried to formulate this objection in linguistic terms by saying that "exists" is not a predicate of an object in the sense in which its color and texture are, Descartes had the ready reply: "nor do I see why it (existence) may not be said to be a property as well as omnipotence, taking the word property as equivalent to any attribute or anything which can be predicated of a thing."[56]

Now it must be observed that Descartes' argument is essentially a psychological one, not a logical one. He appeals to the fact that psychologically he cannot conceive of God as not existing: "For it is not within my power to think of God without existence (that is of a supremely perfect Being devoid of a supreme perfection) though it is within my power to imagine a horse either with wings or without wings "[57] Here we have the crux of the argument: Descartes finds that he does not have the psychological power to dissociate existence from the idea of God. He is indeed reporting a psychological fact, not a logical one.

A few scholars have surmised that the so-called ontological argument was psychological in character. Professor Harry Austryn Wolfson wrote thirty years ago that "in both Descartes and Spinoza the ontological argument is really psychological, resting as it does upon the view that God is a direct object of our knowledge."[58] The question, however, arises: if the ontological argument was psychological in character, why did it seem so convincing to persons such as Descartes, and why did it seem so unconvincing to thinkers in the eighteenth century? What was it in Descartes' psychology which made the ontological argument appear as cogent as any geometrical demonstration?

The Cartesian anxiety and guilt, we shall answer, provided the inarticulate, unconscious premises which conferred on the ontological argument its high validity. An emotional perspective of anxiety and guilt was a necessary condition for the acceptance of the ontological argument. With respect to the question of the existence of God, Descartes was in a psychological state characterized by the pattern of "the omnipotence of thought."

In all neurosis-conditioned experience, it is not the reality of the experienced object, but the reality of its being thought which is important. The neurotic lives in a universe in which the criterion of reality is the intensity of emotion with which an object is invested; the neurotic's criterion of truth is not the correspondence of his belief to external reality, but the coherence of external reality with his emotions. The neurotic's guilty conscience, for instance, is not founded on real misdeeds: "And yet his sense of guilt is justified: it is based upon intensive and frequent death wishes which unconsciously manifest themselves towards his fellow beings . . . Thus the omnipotence of thought, the over-estimation of psychic processes as opposed to reality, proves to be of unlimited effect in the neurotic's affective life and in all that emanates from it."[59]

Now we have traced the guilt which was hidden behind Descartes' mask,—his hostility to his father, his potency-fears. This guilt gave rise to the typical overestimation of "psychical reality" which is part of the neurotic's experience. The compulsive neurotic is always engaged in ontological arguments. "Thus," writes Freud, "if he happened to think of a person, he was actually confronted with this person as if he had conjured him up." The ontological argument is the formal expression of a projective response to his guilt. From the emotional perspective of the Cartesian anxiety: if Descartes thinks of God, then He necessarily exists. In Descartes' case, however, unlike that of the compulsive neurotic, the ontological argument applies only to God's existence; the neurosis has been localized to the existence of the Father-Surrogate object.

Precisely to the extent that Descartes harbored strong unconscious death wishes against his father, to that extent did the ontological argument seem unassailable. God was a projected Surrogate Father whose assured existence diminished somewhat the guilt which Descartes felt for his own death-wish against his father. Descartes' father was an old and dying man when the son undertook the writing of the *Meditations.* God dominated its pages as He never had the earlier *Discourse on Method* or *Regulae.* As Descartes said, he had only "slightly touched" on the question of God in the *Discourse.*[60] The guilt which grew with his own satisfaction in his father's prospective death seemed to heighten the "omnipotence of thought." Affected by guilt, Descartes saw the Father Surrogate reinstated in existence by the processes of critical logic itself. The ontological argument is a regressive mode of reasoning; it is a regression to childhood responses. The child cannot imagine its father as not existing; he seems so all-capable, all powerful. Other entities might not exist, but not so the father. When hatred against the father ensues, and the guilt of revolt is strong, the self-aggression is lightened by the reinstatement of the Surrogate Father whose non-existence cannot be conceived. Descartes prided himself on the Method which would overturn all prejudice and make man the master of Nature; now God came back as the Guarantor of the Method.

But why did the ontological argument win such a following in the seventeenth century? In our time, Albert Camus writes: "I have never seen anyone die for the ontological argument."[61] The will to martyrdom was certainly something which Spinoza and Leibniz would have regarded as irrational; on the other hand, they too, like Descartes endorsed the ontological argument. Since they presumably did not share the Cartesian anxiety, how shall we account for their firm adherence to the Cartesian argument?

There was a common pattern of guilt which ran through the great seventeenth-century thinkers. The seventeenth century was the century of scientific revolt and the religious wars. The European scientists labored with the sense of guilt which affects all revolutionists. Their rigorous, mathematical method carried within it a tremendous destructive power against the received religions, against the world of their fathers. Guilty in their possession and use of this method, they sought the aid of secret societies, small circles, deviant sects; these were their surrogate brotherhoods to share the guilt of their intellectual parricide. Leibniz, like Descartes, sought out the Rosicrucians; Spinoza went among the Mennonites, and later had his own trusted inner circle. Under the pressure of their unconscious guilt, they tried to make amends by reinstating their fathers' supremacy in the guise of the Perfect Father. Mathematicians all, their form of the "omnipotence of thought" expressed itself in the ontological argument. This is the argument of a generation of mathematicians anxiety-haunted by the guilt of their revolt and their death-wishes against religious forebears. In the less guilt-ridden, more optimistic and benevolent eighteenth century, the ontological argument dropped away, to be replaced by the argument from design, which looked at the goodness of things, and saw the evidence of God's works.

The ontological argument thus acquired a vogue in an era of scientific, mathematical rebels. A great guilt was upon them as they turned their method against tradition. The ontological argument was a way of assuaging this guilt. God would be shown after all to exist by geometrical necessity; the mathematical rebellion would not destroy Him.

The unconscious content of the ontological argument may thus be stated: I feel tremendous guilt when I harbor death-wishes against my father, when I revolt against his religion and way of life, when I use a mathematical method which is ominous with portent against the existence of his God; I must lighten the guilt I feel with what my method has done to my earthly father; my guilt is too great and unbearable if I were to destroy God logically as well. Therefore, to alleviate my guilt, God must necessarily exist. I do not have the power to dissociate God's existence from his essence.

Thus the ontological argument brought a psychological easement;

its projection-mechanism resolved an emotional conflict.[62] The Method of Revolt was conciliated with the existence of the Perfect Father, and revolution merged with a Higher Counterrevolution. The sons destroying the Primal Father were said by Freud to have fashioned for themselves a Surrogate Totem; the seventeenth century scientific revolutionists, in the anxiety of their Parricide, made for themselves a Mathematical Totem.

To the person with the Cartesian anxiety, existence is perceived with such an interpenetration and determination by emotional needs that the Kantian critique simply is felt to be without force. To the person without such an anxiety, the ontological argument is utterly implausible. We shall not be led, however, into a socio-psychological relativism according to which equally valid different universes are defined by different anxieties or emotional bases. Karl Marx, for instance, in his youthful discussion of the psychological character of the Cartesian argument went on to derive a psychological relativism in which the distinction between illusion and reality tended to get lost:

> In this respect the critique of Kant is also without value. If someone imagines he has a hundred dollars, if this presentation is not something purely subjective for him, if he believes in it, then the hundred imagined dollars have the same value for him as a hundred real ones . . . Kant's example could have, on the contrary, strengthened the ontological argument. Real dollars have the same existence as imaginary gods. Does a real dollar exist elsewhere than in the imagination, although it be also a general imagination or one rather common to men? Introduce paper money in a country where they don't know the usage of paper, and everybody will laugh at your subjective representation. Come with your gods to a country where other gods are honored, and they will prove to you that you suffer from hallucinations and abstractions."[63]

Real dollars, however, do not have the same existence as imaginary gods; it is not enough to know the usage of paper money, for there has to be a backing of goods and products in the economy. Persons living under inflationary conditions learn only too well the distinction between notes with and without value. It is not a matter of "subjective representation." The distinction upon which Kant in good bourgeois fashion insisted between one hundred real dollars and a hundred imaginary ones is the ultimate one between external reality and a purely psychical occurrence. Freud speaks of the "neurotic standard of currency" in which the intensity of psychical desire rather than objective perception determines the constitution of "realities."[64] The young Marx was indeed putting all currency on a neurotic standard. The distinction, however, between hallucination and perception is not a matter of sociological customs; the peoples which once shared the fantasies of the Children's Crusade did not by their community in neurosis transform their illusions into realities. Metaphysics can be said to be a definition of the cosmic situation; the ontological argument defined the cosmic situation as perceived in terms of the needs of the Cartesian anxiety.

(4) *I think, therefore I am.* The doubt that one exists is characteristic of all persons who have experienced much self-aggression. In Descartes' case, the self-aggression, the self-blame or guilt, had two sources; it was the internalized form of his father's hostility to himself, and it was also his guilt for nurturing death-wishes against his father. Descartes explores the philosophical expression of his anxiety to its last consequence; his self-aggression leads him unconsciously to a final assault on the sense of his own existence; thinking under the aspect of the Cartesian anxiety has a self-destructive, outcome; its "dialectic" is a suicidal one.

The "I think, therefore I am" is what we might call a *"psychological existence-argument."* Such an argument appeals to some experience of the person which alleviates the underlying anxiety, and tends therefore to restore to him the feeling that he exists. In our time, there have been a variety of such psychological existence-arguments. I shall mention several of them:

(1) I revolt, therefore I am.   (Camus)
(2) I fight, therefore I am.   (M. Beigin)
(3) I have sexual intercourse, Therefore I am.   (D. H. Lawrence)
(4) I work, therefore I am.   (Freud, Marx)
(5) I choose, therefore I am.   (Jaspers, Sartre)
(6) I suffer, therefore I am.   (Graham Greene)

The doubt as to self-existence which these arguments are meant to resolve is not the kind of doubt which arises in ordinary life or in scientific situations. It is an anxiety-induced doubt which infects the sense of your self with unreality, so that you seem phantom-like, hollow, insubstantial to yourself, and all scientific evidence and considerations whatsoever share in this same unreality. Descartes was well aware that the "doubt" which he had of his self's existence and the world's were radically unlike ordinary doubt. His, he wrote, was "that supreme kind of doubt which, I have insisted, is metaphysical, hyperbolical and not to be transferred to the sphere of the practical needs of life by any means." Readers, he said, "who know nothing of the metaphysical doubt, referring it to the practical life, may think that I am out of my mind."[65] All psychological existence-arguments are designed to reduce self-aggression or guilt. The idle, functionless, or unemployed person who feels himself superfluous and guilty for his purposelessness, *de trop,* or who finding no channel for aggressive energies outwards, expends them against himself, or who blames himself because society has no place for him, discovers that the chance to work alleviates his self-aggression, and restores the sense of meaningfulness to his existence. We may refer to this sense of the reality of one's existence as the "ontological sense." Existence-arguments try to strengthen it by emphasizing some guilt-reducing experience. Even Graham Greene's masochist existence-argument, "I suffer, therefore I am," has this aim; the sufferer feels less guilty because suffering will admit him into the Christian community, and thereby reduce his guilt and loneliness "You once said that where one suffers, one begins to feel part of the human condition on the side of the Christian myth, do you remember? 'I suffer, therefore I am.' "[66]

Descartes' "I think, therefore I am" was much similar in its significance to "I revolt, therefore I am." For the crucial, ultimate mode of thinking which always came to his mind was doubting: "I saw from the very fact that I thought of doubting the truth of other things, it very evidently and certainly followed that I was."[67] In principle, any kind of thinking could have sufficed as well as doubting for showing that we exist; for the "I think, therefore I am," as Descartes insisted was no syllogistic argument, but a "simple act of mental vision," and this simple act could be found equally in every act of thinking: "But when we become aware that we are thinking beings, that is a primitive act of knowledge derived from no syllogistic reasoning. He who says, 'I think, hence I am, or exist,' does not deduce existence from thought by a syllogism, but, by a simple act of mental vision, recognizes it as if it were a thing that is known *per se.*"[68] But it was the act of doubting which was the preeminent, indubitable mode of thought. Arguments such as "I walk, hence I exist," or "I breathe, therefore I am," said Descartes, were unconvincing because one might always doubt their occurrence, and presume the walking and breathing to be as in a dream.[69] Again, we might object to Descartes that "the simple act of mental vision" of our breathing or walking should convey as much certainty of our existence as that of doubting. Descartes said that doubting itself could never be challenged, for to doubt that one was doubting, one would still be thinking. This argument, however, was a syllogistic one, and was leaving behind the appeal to simple, mental vision. The central point was indeed an unstated psychological one; doubting fulfilled the role of an experience of challenge and revolt. Thereby it restored the maimed sense of reality of one's self.

Doubting liberated Descartes from self-aggression because it turned his aggressions outwards against the received opinions of school, nation, society, and father; it was his weapon against the traditional, cultural super-ego. He was aware that it would appeal especially to the younger generation, and that it was potential with subversion: "All young people," he wrote to Father Dinet, "seek truth when first they apply themselves to the study of philosophy."[70] And he was aware that if "dubious and controversial questions shall be agitated," it raised the possibility that subjects would become "more contentious, more refractory, and more opinionative, and thus less obedient to their superiors and more likely to become seditious." But doubting, we might say, was Descartes' "proto-experience;" it released within him the utmost libidinal sense of reality. Doubting, he could destroy as it were his father; doubting removed the obstacle which stood in his way to a fuller, libidinous sense of reality. His own self alone would emerge unscathed in doubting; he would be re-born, and only his act of credence would save his father from nullity.

As he pursued his doubt to the last consequence, the symbolic imagery for his new birth came spontaneously to Descartes' mind. It was "as if I had all of a sudden fallen into very deep water, I am so disconcerted that I can neither make certain of setting my feet on the bottom; nor can I swim and so support myself on the surface."[71] Freud tells us: "Birth is almost invariably represented by some reference to water: either we are falling into water or clambering out of it, saving someone from it or being saved by them, i.e. the relation between mother and child is symbolized."[72] And curiously, during the period of his mental liberation, Descartes was taking to the sea in voyages designed to prove to himself his masculinity. To Isaac Beeckman, the Dutch mathematician who for a brief period was something of a father-figure to the young man, Descartes wrote in 1619 of the pleasure he was having in risking his life: "Let us talk then of my voyages. The last passed well, all the better because it seemed to offer more danger, especially on leaving your island. The first day, at Flessingue, the winds forced me to return to port; but the following day, embarked on a ship of the smallest kind, I sustained a tempest still more furious. I had however more satisfaction than fear: it was an occasion to prove myself (c'etait une occasion de m'éprouver); I had never attempted a sea crossing and I have done it without nausea. Here I am become braver for undertaking a longer voyage."[73] Three years later, he undertook another curious voyage, the outcome, as Elizabeth Haldane writes, of "a strange desire to visit at leisure the sea-coasts of Germany a dreary enough expedition, one would say, in a small boat in winter." Although Descartes almost lost his life at the hands of murderous, dishonest boatmen, he proved himself again a man with his sword. The search for masculinity was both a physical and philosophical one. Philosophically, the aim was to be reborn anew, in the conviction of his own existence, freed from the unnerving doubt in himself and the world which issued from a father's aggression and his own guilt.

The danger of Descartes' method of philosophizing was that it would fail in those persons, unlike himself, in whom self-aggression remained strong. They would become mired in a neurotic solipsism. This was what happened to his good friend and disciple, the Princess Elizabeth. She conscientiously followed Descartes' method of detaching his mind from his senses, and she found she could not arrive at any notion of the interaction of mind and body. Then in a remarkable letter in 1643, Descartes warned against his own method. He allowed himself to use it in a philosophical direction, he wrote, only a few hours a year.[74] He advised the Princess to base her philosophy on everyday experience, not on his method of self-withdrawal: "and finally, it is only in making use of life and ordinary conversations, and in abstaining from meditating and studying things which strain the imagination, that we learn to conceive the union of the soul and body."

For the Princess Elizabeth, melancholy-ridden as her father had been, estranged from her mother, and living in a povertied, lonely exile, was not easily rid of her self-aggression. She asked Descartes why if the soul were so much greater than the body should we not seek death, and escape the body's ills and passions?[75] Descartes' philosophical abreaction began with what we might call a "libidinal detachment" from things, a cultivation of the narcissistic stage. But his vitality and optimism were then strong enough so that he went on to "libidinal attachments," to belief in a Benevolent God and the joys of the external world. He could destroy in thought and totem the malevolent earthly father and the Arch-Deceiver. The Princess, however, remained sunk in Calvinist gloom and ill-health. "Libidinal detachment," Descartes saw, was no therapy for her. When he advised her to have recourse to ordinary life for her therapy, he was confessing the limitations of his method. Detach your emotions, withdraw from all bodily involvement, he acknowledged, you will lose your sense of reality. Allow your emotions to function normally, and the sense of reality will be restored; you will then perceive the psychophysical unity of mind and body. In short, the Cartesian method would not always resolve anxiety; it had its limits as a therapy which Descartes, in his own terms, could never adequately explain.

In his therapist's perplexity, Descartes advised the Princess too to cultivate his actor's mask, to see the world as a theatre; great souls, he said, look at unfavorable events as we do "those of comedies. And as the sad and lamentable stories which we see represented on the theatre, give us as much recreation as the gay, although they draw tears from our eyes, so the greatest souls, of whom I speak, have the satisfaction in themselves of all the things which happen to them, even the most painful and insupportable."[76] The Princess wisely did not see detachment as the path to any high contentment. And Descartes had resolved his anxieties in a way he could never explicate; he was not consciously aware of the psychological function of his philosophical tenets. How successful he was in his own case can be seen when we remember that this was a man who could boast in 1631 that he never had a bad dream, and who welcomed sleep for the endless joy which it brought him: "I sleep here ten hours every night, and without any care ever awaking me, after sleep has long led my mind through woods, gardens, and enchanted palaces, where I experience all the pleasures which are imagined in the Fables, I insensibly mix my daytime reveries with those of the night; . . ."[77]

Anxiety had maimed the young Descartes into emotional passivity and fear; he had hidden his guilt of death-wishes behind his mask. In the act of doubting, he found an agency for revolt against the superego, for a manifestation of his sense of power; it created a socially acceptable channel for the liberation of energies of guilt and self-aggression against something much closer to the original object of hatred. The child, who had suffered in sorrow, could now strike back against its father in socially constructive ways. In Holland Descartes could feel an affection for people restored.[78] As anxiety was subdued, the conviction of existence returned.

The Cartesian argument, "I think, therefore I am," has been found unconvincing in later times in two principal ways. First, there are those who doubt there is an "I"; second, there are those who, granting that there is an "I", doubt that it thinks. Bertrand Russell with his view that the "I" is a "grammatical convenience," J. B. Watson with his view that there are no such mental occurrences as thinking, Oxonians with their polemic against the "ghost in the machine," all these are variants of a sado-masochistic anxiety which recurs with a new revolt against the human consciousness. One might conclude each age has its anxiety, and every age is an age of anxiety. Every age, however, one trusts, will also have those who will share the Cartesian optimism, and will refuse to accept anxiety as the central fact of life. For Descartes did not make of philosophy a descriptive rendition of anxiety; his aim was always to raise the human status above it.

Notes

1. Norman Kemp Smith, *New Studies in the Philosophy of Descartes,* London 1952, p. 136.
2. Descartes, *Selections,* ed. Ralph M. Eaton, New York, 1927, pp. 357–360. Descartes, *Lettres,* ed. Michel Alexandre, Paris, 1954, p. 39. *The Philosophical Works of Descartes,* tr. Elizabeth S. Haldane and G. R. T. Ross, New York, reprinted 1955, Vol. II, p. 144.
3. William James many years later explicitly proposed a hypothesis of animal immortality. William James, *Human Immortalilty,* Second ed., Boston, p. 35. The term "theriophily" was invented by George Boas. He uses it to denote an attitude in which men regard animals as their models or superiors. I am taking the meaning of "theriophilist" simply as "animal-lover." George Boas, *The Happy Beast in French Thought of the Seventeenth Century,* Baltimore, 1933, p. 1.
4. Karl Marx, *Capital: a Critique of Political Economy,* tr. Samuel Moore and Edward Aveling, Chicago, 1906, p. 426.
5. Leonora Cohen Rosenfield, *From Beast-Machine to Man-Machine: Animal Soul in French Letters from Descartes to La Mettrie,* New York, 1941, pp. 50, 112–113, 181–182, 191, 202. George Boas, *The Happy Beast in French Thought of the Seventeenth Century,* pp. 102–105.
6. Descartes fluctuated in his contra-totemism probably in accordance with differences in his mood. At one time, he held that his doctrine was indemonstrable. At another time, he maintained that he had given a most stringent proof that "brutes do not possess thought." To those who argued that if animals were solely mechanical creatures, then man too might be regarded as devoid of consciousness, Descartes replied that we cannot divest ourselves of our consciousness. He had no such ultimate reply to the "Pythagorean" view that animals have minds distinct from their bodies. *Selections,* p. 358. *Philosophical Works,* Vol. II, pp. 244–245.

7. *The Basic Writings of Sigmund Freud,* ed. A. A. Brill, New York, 1938, pp. 904–906. Also cf. Melanie Klein, *The Psycho-Analysis of Children,* Sec. Ed., London, 1937, trans. Alix Strachey, p. 220. Joachim H. Seyppel, "The Animal Theme and Totemism in Franz Kafka", *The American Imago,* Vol. 13 (1956), pp. 79–84.
8. Descartes, *Lettres,* p. 57. Adrien Baillet, *Vie de Monsieur Descartes,* Paris, ed. 1946, p. 84. Elizabeth S. Haldane, *Descartes: His Life and Times,* London, 1905, pp. 126, 280. It remains a further problem in the psychological understanding of Descartes to trace why at certain times, as in 1623, he developed a dislike for mathematics, and turned to the study of morals or the practice of anatomy. Cf. Elizabeth S. Haldane, op. cit., pp. 86–87, 103.
9. Adrien Baillet, *op. cit.,* p. 27.
10. *Ibid.,* p. 276.
11. Cf. Hyman S. Barahal, "The Cruel Vegetarian," *Psychiatric Quarterly Supplement,* vol. 20, (1946), pp. 3–13.
12. Leonora Cohen Rosenfeld, *op. cit.,* p. 70; cited from a marginal note in Baille, *La Vie de Monsieur Descartes,* Paris, 1691, Seconde Partie, p. 456. Also cf. John Dewey, "The Ethics of Animal Experimentation," *Atlantic Monthly,* Vol. CXXXVIII (1926), pp. 343–346.
13. *Lettres,* p. 52.
14. *Lettres,* p. 166.
15. *The Philosophical Works of Descartes,* Vol. I, p. 398.
16. Leonora Cohen Rosenfeld, *op. cit.,* p. 52.
17. Descartes, *Selections,* p. 360.
18. *The Philosophical Works of Descartes,* Vol. II, p. 104.
19. John Bowlby, *Maternal Care and Mental Health,* Geneva, 1951, pp. 12, 23–24, 32, 56–57.
20. Elizabeth S. Haldane, *op. cit.,* pp. 48–49.
21. Descartes, *Lettres,* p. 23. Adrien Baillet, *Vie de Monsieur Descartes,* pp. 22, 68, 282. Elizabeth S. Haldane, *op. cit.,* pp. 37, 104–105.
22. According to Baillet, however, this strange attraction to cross-eyed women remained with Descartes all his life: "by an effect of this bizarre inclination, there had remained strong within him during the course of his life an inclination to affection for cross-eyed people which came from the impression of his childhood, when in early age he loved a little girl who was a bit cross-eyed." Baillet, *op,. cit.,* p. 290. *Lettres,* p. 179.
23. Maxine Leroy, *Descartes: le philosophe au masque,* Vol. II, p. 165.
24. Lewis S. Feuer, "The Dreams of Descartes," *The American Imago,* Vol. 20, (1963), pp. 3–26.
25. Descartes, *Lettres,* pp. 117–118.
26. *Lettres,* pp. 165–167. Elizabeth S. Haldane, *op. cit.* pp. 3–5.
27. Baillet, *op. cit.,* pp. 19, 69, 119. Elizabeth S. Haldane, *op. cit.,* pp. 95–96.
28. Baillet, *op. cit.,* pp. 6–7, 18–19. E. S. Haldane, *op. cit.,* pp. 34, 208.
29. *Lettres,* p. 73. E. S. Haldane, *op. cit.,* p. 202. *Philosophical Works,* Vol. I, p. 86.
30. E. S. Haldane, *op. cit.,* p. 59.
31. Baillet, *op. cit.,* p. 45. *Lettres,* p. 77.
32. William McC. Stewart, "Descartes and Poetry," *The Romanic Review,* Vol. XXIX, (1938), pp. 234–240. (The translations are mine.)
33. *Lettres,* p. 49.
34. "C'est ce qui m'oblige à vous (Servien) supplier d'intercéder pour moy auprès de M. le Prince d'Orange, à ce qu'il luy plaise comme chef de l'Université de Leyde . . . Car je suis assuré qu'ils n'approuveront pas qu'après tant de sang que

les François ont répandu pour les aider à chasser d'icy l'Inquisition d'Espagne, un François, qui a aussi portè autrefois les armes pour la même cause, soit aujourd'huy soumis à l'Inquisition des Ministres de Hollande." *Oeuvres de Descartes,* ed. Charles Adam and Paul Tannery, Paris, 1903, Vol. V, pp. 25–26. Charles Adam, *Vie et Oeuvres de Descartes,* Paris, 1910, p. 347. Also, *Lettres,* p. 176.
35. Lewis S. Feuer, *Spinoza and the Rise of Liberalism,* Boston, 1958, p. 73.
36. Baillet, *op. cit.,* pp. 39–41, 50–51. E. S. Haldane, *op. cit.,* p. 72. *Oeuvres de Descartes,* Vol. X, Paris, 1908, pp. 176–177.
37. Maxime Leroy, *op. cit.,* Vol. I, p. 76.
38. G. C. Guhrauer, *Gottfried Wilhelm Freiherr v. Leibnitz,* Breslau, 1846, Vol. I, pp. 46–47. John Theodore Merz, *Leibniz,* Edinburgh, 1884, pp. 23–24. Arthur Cadbury Jones, "Rosicrucians", *Encyclopaedia of Religion and Ethics,* ed. James Hastings, New York, 1920, Vol. X, pp. 856–857.
39. *Lettres,* p. 4.
40. *Oeuvres de Descartes,* Vol. II, Paris, 1898, pp. 586, 350. Vol. III, Paris, 1899, pp. 101, 158. Baillet, *op. cit.,* p. 283.
41. Maxime Leroy, *Descartes: le philosophe au masque,* Vol. I, Paris, 1929, p. 142. E. S. Haldane, *op. cit.,* pp. 162–164, 207.
42. Baillet, *op. cit.,* p. 163.
43. E. S. Haldane, *op. cit.,* p. 164.
44. Baillet, *op. cit.,* p. 163.
45. E. S. Haldane, *op. cit.,* p. 208. *Oeuvres de Descartes,* Vol. III, p. 101.
46. *Lettres,* pp. 144, 156.
47. *Ibid.,* pp. 133, 117, 49, 54. Baillet, *op. cit.,* p. 281.
48. *Philosophical Works,* Vol. I, pp. 147–149.
49. *Ibid.* p. 150. David Baumgardt, *Bentham and the Ethics of Today,* Princeton, 1952, p. 483. Norman Kemp Smith, *op. cit.,* p. 296.
50. When Gassendi objected to Descartes: "you consider yourself not as a complete human being," he was indeed approaching the underlying anxiety in Descartes' philosophy. *Philosophical Works of Descartes,* Vol. II, p. 138. François Meyer, "Gassendi et Descartes," *Actes du Congrès du Tricentenaire de Pierre Gassendi,* Paris, 1957, p. 222.
51. *Philosophical Works,* Vol. I, p. 189.
52. *Philosophical Works,* Vol. I, p. 190.
53. *Ibid.,* pp. 171–172.
54. *Ibid.,* p. 169.
55. *Ibid.,* p. 104.
56. *Ibid.,* Vol. II, pp. 186, 228.
57. *Ibid.,* Vol. I, pp. 181–182.
58. Harry Austryn Wolfson, *The Philosophy of Spinoza,* Cambridge, Mass., 1934, Vol. I, p. 170.
59. *The Basic Writings of Sigmund Freud,* ed. A. A. Brill, New York, 1938, p. 873.
60. *Philosophical Works,* Vol. I, p. 137. The *Meditations,* furthermore, unlike the *Discourse on Method,* were written in Latin, and dedicated to the Dean and Doctors of the Sacred Faculty of Theology in Paris.
61. Albert Camus, *The Myth of Sisyphus,* tr. Justin O'Brien, New York, 1959, p. 3. For Descartes' views on martyrdom, cf. Henry A. P. Torrey, *The Philosophy of Descartes,* New York, 1892, p. 340.
62. *The Basic Writings of Sigmund Freud,* pp. 857, 878.
63. Karl Marx, *Fruehe Schriften,* Erster Band, ed. H. J. Lieber and P. Furth, Stuttgart, 1962, p. 75.

64. *The Basic Writings of Sigmund Freud,* p. 784.
65. *Philosophical Works,* Vol. II, p. 266.
66. Graham Greene, *A Burnt-Out Case,* New York, 1961, p. 232.
67. *Philosophical Works,* Vol. I, p. 101.
68. *Ibid.,* Vol. II, p. 38.
69. *Ibid.,* Vol. II, p. 207. *Lettres,* p. 38.
70. *Philosophical Works,* Vol. II, pp. 357–358.
71. *Philosophical Works,* Vol. I, p. 149.
72. Sigmund Freud, *a General Introduction to Psychoanalysis,* tr. Joan Riviere, New York, 1938, p. 137.
73. *Lettres,* p. 3. Elizabeth S. Haldane, *Descartes: His Life and Times,* p. 78. Adrien Baillet, *Vie de Monsieur Descartes,* pp. 47–48. Maxime Leroy, *Descartes: le philosophe au masque,* Paris, 1929, Vol. I, p. 63.
74. "The Principal rule which I have always observed in my studies, and that which I believe has served me most to acquire some knowledge, has been that I have used it only for a very few hours a day for those thoughts which occupy the imagination, and very few hours a year, for those which concern the understanding alone, and that I have given the rest of my time to the relaxation of the senses and to the repose of the mind." *Lettres,* pp. 100–101.
75. E. S. Haldane, *op. cit.* p. 263, pp. 254–255.
76. *Lettres,* pp. 115, 144.
77. *Lettres,* p. 17.
78. In a letter to Huygens in 1640, Descartes referred to "this life, with those of this country (Holland), with whom I have shown indeed that I like to live more than in my own," (cette vie, avec ceux de ce pays, avec lesquels j'ai montré par effet que j'aimais mieux vivre que dans le mien proper). *Oeuvres de Descartes,* Vol. II, p. 158.

14

Spinoza's Thought and Modern Perplexities: Its American Career

I

Spinoza's place in the history of thought is rather anomalous. He is the only great philosopher who founded no school. There have been Cartesians, Leibnizians, Lockeans, Humeans, Kantians, Hegelians, Nietzscheans, even Berkeleyans, but in this sense, there have never been Spinozists.[1] For no one recorded in the history of ideas seems ever to have accepted all of Spinoza's major tenets.[2] Different men have selected for their use some particular doctrine of Spinoza's while rejecting the others; nobody seems to have found it possible to accept all the central ones. Many a Kantian could quite whole-heartedly affirm all of Kant's chief propositions concerning the synthetic *a priori*, the categories of the understanding, the antinomies of pure reason, and the imperatives of the practical reason. And indeed, Jewish philosophers in the nineteenth century were the chief proponents of Kant's standpoint.[3]

Why was it then that a Spinozist school never arose as did, for instance, a Kantian one? The answer seems to me to lie primarily in the fact that Spinoza's philosophy, unlike Kant's, was much more evidently a system of propositions with profound inconsistencies, a system in a logically unstable equilibrium, vulnerable at several *loci*. Here was a thinker who said God was Nature, and that the laws of His nature were to be grasped by the geometrical method, but who also said that numbers were only aids to the imagination, and not the basis for the truths of reason: how then could any law of nature be regarded as part of our rational understanding of God's nature?[4] Spinoza the mystic was incompatible with Spinoza the scientist. Here was a thinker who, because he affirmed a psychophysical parallelism—that the order of mental modes corresponded to that of the physical modes—was consequently obliged to assert that the human mind can be no more immortal or eternal than its corresponding brain tissues and occurrences. Nonetheless, contravening his own propositions, Spinoza declared: "Nevertheless, we feel and know by experience that we are eternal." But how could Spinoza introduce a feeling into his philosophical geometry? Here was a thinker who bade us accept the doctrine of determinism, and to love the God from whose nature "infinite numbers of things" followed in "infinite ways," the God who could fashion the varieties of men from Albert Einstein to Adolf Hitler. The same thinker told us that "the mass of mankind remains always at about the same pitch of misery, . . . [and] is always best pleased by a novelty which has not yet proved illusive," a trait that "has been the cause of many terrible wars and revolutions."[5] But would not these infinite modes of horror lead us to an intellectual hatred of God rather than to an intellectual love of Him, for God Himself would, despite Spinoza's dialectical efforts, be the cause of evil.

Spinoza himself could scarcely take delight in the varieties of human evil and folly as exhibiting the infinitude of God's power. His own moral intuitions were at odds with his meta-physics. No contemporary event more deeply affected Spinoza than the brutal lynching in 1672 of John de Witt, the Grand Pensionary of Holland, whom Spinoza admired as the rational statesman of his time. As he told Leibniz, Spinoza wished to go out into the street bearing a placard with the words *ultimi barbarorum* (lowest of barbarians), and to confront the murderous mob.[6] Spinoza's landlord, knowing that Spinoza would likewise be torn to pieces, barred the door. Yet Spinoza, re-reading his own *Ethics,* would have had to say: "God's power shows itself in infinite ways, realizing all things that can be conceived by infinite intellect, and one of them is the lynching by barbarians of a free spirit." In Spinoza's own words: "Because to Him material was not wanting for the creation of everything, from the highest down to the very lowest grade of perfection."[7] As a naturalistic political philosopher, furthermore, Spinoza would have had to say: "The mob has more power, hence more right, as well as more of God's power than the free man, de Witt; hence, it is right that they lynch him if they would." But the moral convictions of Spinoza, the free man, scarcely accepted such a proposition, If he felt despite his geometrical metaphysics that "we feel and know by experience that we are eternal,"[8] his action also seemed to say: "Despite my naturalistic equation of right with power, I feel and know by experience that such actions are wrong, and I cannot believe that the infinite nature of God requires an infinite variety of human cruelties." Spinoza's concept of divine perfection, and the mandate to admire the infinite variety of all things, demands a kind of human masochism; we feel that the universe would have been more perfect without a finite mode such as Adolf Hitler, that God would have done better to actualize a second Einstein instead.

Imposing though the architectonic of his geometrical system was, Spinoza himself harbored doubts concerning the logical cogency of his demonstrations. The next to the last proposition of the *Ethics* is the most unusual for a book composed according to the mathematical method: "Even if we did not know that our mind is eternal, we should still consider as of primary importance Piety and Religion, and absolutely everything which in the Fourth Part we have shown to be related to strength of mind and generosity."[9] What would one think of a thorem in the textbook of geometry that said that even if an entire successive batch of theorems proved to be invalid, that would not affect the validity of the theorems already demonstrated in previous sections? Such a theorem would indeed be equivalent to the assertion that a whole batch of propositions was logically independent of its predecessors, for the proof of the logical independence of two propositions, *a* and *b,* consists precisely in showing that both the proposition *a* and its contradictory are equally consistent with *b.* To this extent, Spinoza's system is not one whose parts are completely interdependent; its "unity" is usually overestimated, for, as Spinoza says, all that he writes about the intellectual love of God might be false without affecting the truth of his psychological propositions concerning the hedonistic, free man.

One wonders if other components of Spinoza's system are similarly a-systematic. Here is Spinoza, a determinist, who yet writes a book on the ethics of the free man, as well as a treatise and a half advising him as to what political measures might advance the lot of free men. Presumably the free man can then act upon the prescriptive advice, although the multitude lacks this power of free action: "Experience, however, teaches us but too well, that it is no more in our power to have a sound mind, than a sound body."[10] Free men have the power to act freely, and to adopt Spinoza's ethical and political principles; if so, this act seems to be a choice on their part, for if they are simply determined in accordance with psychophysical laws to do so, they should be no more admired than the wretch who inevitably destroys himself according to those same laws. Might God's infinite power, we suggest, manifest itself in the existence of a variety of men who have varying degrees of freedom from the governance of strict determinist laws? Is it an "inadequate idea" on Spinoza's part that he can define God's power as expressing itself only in a determinist order rather than in one with domains of indeterminacy and freedom?

The unity of Spinoza's system was thus only an apparent one; it conjoined a variety of disparate themes and ideas. Each of Spinoza's themes might at some time, under specific historical and personal circumstances, be appropriated by some group of thinkers; at the same time, they would reject much else that a disciple would have regarded as equally essential to Spinoza's philosophy. Spinoza said he did not wish to create a school "named after him."[11] This wish was fulfilled. But if he never encouraged that species of mental bondage that characterizes the master's disciples in a "school" of thought, Spinoza was of that company such as Locke and John Stuart Mill who helped men in diverse ways to achieve their own freedom.

Thus, Spinoza's philosophy has been used in the most contrary ways in the history of ideas. American transcendentalists drew on Spinoza's pantheism, his notion of an all-encompassing God, but they found his determinism repellent. Justice Holmes found congenial Spinoza's view of Nature as a system of contending forces, but he had no use for Spinoza's logical machinery, his postulates concerning God, substance, and modes. Horace M. Kallen was attracted to Spinoza as the theorist of the dynamic of nature's energies, but he could never ground his faith in the free individual in Spinoza's determinism. By contrast, though Walter Lippmann saw a deep wisdom in Spinoza's conception of the free man, his theology was repugnant to Lippmann's agnostic spirit. The sociological careers of Spinoza's ideas were the outcome of a socio-historical selection in which strangely enough the character of Spinoza, the excommunicate rebel, was far more potent than his formal arguments.

II

It is a curious fact, to begin with, that Spinoza failed to influence in any way the intellectual spokesmen of the American Revolution. Not that his works were unknown. The philosophic Quaker merchant of Philadelphia, James Logan, possessed, for instance, in his library during the first part of the eighteenth century, a copy of Spinoza's *Tractatus Theologico-Politicus*.[12] Jefferson himself moreover had gone to the trouble of purchasing the original edition of Spinoza's *Opera posthuma,* published in Amsterdam, in 1677, as well as the original translation into English of Spinoza's *Tractatus Theologico-Politicus,* published in London in 1689. They were among the volumes he later sold to the Library of Congress.[13] Thomas Jefferson thus knew Spinoza's ideas, but Jefferson felt evidently there was nothing in common between his conception of "inalienable rights" and Spinoza's notion of "natural right." As Jefferson wrote to his old friend and political opponent, John Adams, in 1823:

> ... I can never join Calvin in addressing *his God.* He was indeed an atheist, which I can never be; or rather his religion was daemonism ... Now one-sixth of mankind only are supposed to be Christians; ... This gives completely a *gain de cause* to the disciples of Ocellus, Timaeus, Spinoza, Diderot and D'Holbach. The argument which they rest on as triumphant and unanswerable is, that in every hypothesis of cosmogony, you must admit an eternal pre-existence of something ... They say then, that it is more simple to believe at once in the eternal pre-existence of the world, as it is now going on, ... than to believe in the eternal pre-existence of an ulterior cause, or Creator of the world, ... On the contrary, I hold ... that when we take a view of the universe, ... it is impossible for the human mind not to perceive and feel a conviction of design, consummate skill, and indefinite power in every atom of its composition.[14]

Jefferson in this passage, perhaps the only one in which he explicitly discussed Spinoza's philosophy by name, associates Spinoza's thought with that of the eighteenth-century French materialistic atheists, Diderot and D'Holbach. The latter, disbelievers in the argument from design for the existence of a Creator, simply accept the existence of the world as a *causa sui.* Such a philosophy could never provide Jefferson with a basis for his own theory of "inalienable rights," for men, according to Jefferson, are "endowed by their Creator" with the inalienable rights to life, liberty, and the pursuit of happiness. And if a Creator were not to exist, then the notion of Creator-endowed inalienable rights would collapse. Only on a theistic basis does the notion of "inalienable rights" make any sense. Otherwise all "rights" are sociologically alienable; rulers have decreed that men should be deprived of their lives, liberties, and possibilities for happiness. Whenever these actions, even if recognized as legal, are said to violate inalienable human rights, that is because some higher law has presumably endowed the human individuality with rights inseparable from it.

Spinoza was the most eloquent advocate of freedom of thought and speech in the seventeenth century. His argument was, however, essentially a pragmatic, utilitarian one. It would be the state's greatest misfortune, he declares, for it to try to punish its most "enlightened" citizens, and destroy "its highest examples of tolerance and virtue." Like John Stuart Mill, Spinoza argued that "such freedom is absolutely necessary for progress in science and the liberal arts," and that it is contrary to the state's own interests for it to drive "upright" men into becoming conspirators against itself.[15] Proudly Spinoza, a Dutch patriot, adduced his empirical evidence for the utility of freedom of thought and speech:

> The city of Amsterdam reaps the fruit of this freedom in its own great prosperity and in the admiration of all other people. For in this most flourishing state, and most splendid city, men of every nation and religion live together in the greatest harmony, ... His religion and sect is considered of no importance: for it has no effect before the judges in gaining or losing a cause, ...[16]

The Jeffersonian radical, on the other hand, finds pitfalls in Spinoza's theory of "natural right." For Spinoza makes right not only coextensive but synonymous with power. Hence Spinoza is obliged to concede that the state "has the right to rule in the most violent manner, and to put citizens to death for very trivial causes ... ."[17] It might seem that men have at least a natural right to freedom of speech, for as Spinoza says, "it is impossible to deprive men of the liberty of saying what they think."[18] Yet this is precisely what a totalitarian state undertakes to do, and indeed, they have done so with considerable success. Thus, on Spinoza's premises, it would follow that no natural right of Soviet dissidents to dissent has been suppressed because, as a matter of fact, that suppression was for many years successful.

Spinoza's own moral and political feelings cry out against this equation of "right" with "power," and its consequences. Nonetheless, his formal political doctrine makes it impossible for him to deny that a totalitarian regime has the "right" to suppress the majority. Doubtless he would argue that it was following a mistaken and self-defeating policy in so doing, but no "right to revolution" would devolve upon the suppressed majority."It is true," Spinoza acknowledges, "that the sovereign has the right to treat as enemies all men whose opinions do not, on all subjects, entirely coincide with its own; but we are not discussing its strict rights, but the proper cause of action."[19] Hardly a stirring argument, it would have contributed no moral force to the *Declaration of Independence.* Jefferson sought for an ethically ultimate "right to revolution," not for admonitions to the British king on the expedient exercise of power. Jefferson affirmed: "that whenever any form of government becomes destructive of these ends [life, liberty and the pursuit of happiness], it is the right of the people to alter or abolish it . . ."

Spinoza, by contrast, is a scientific democrat, not an ideological one. An ideological democrat is one such as Jefferson, who finds a natural virtue in the people, and who regards their revolutionary act as justified when such is their will. But Spinoza could never bring himself to any high admiration for the "people." The masses of men were, in Spinoza's eyes, irrational. "[M]an's natural passions are everywhere the same," writes Spinoza. "[T]he mass of mankind remains always at about the same pitch of misery, . . . Men are more led by blind desire than by reason."[20]

John Locke, Spinoza's younger contemporary, has been called the philosopher of the "Glorious Revolution" of 1688, and Jefferson's Declaration was indeed, as distinguished scholars have shown, a Lockean document.[21] To Spinoza, however, every revolution was inglorious. A successful revolution, to Spinoza's mind, would contravene the laws of social psychology: "peoples have often changed their tyrants, but never removed them or changed the monarchical form of government into any other." Of this truth, the English revolution led by Oliver Cromwell in the mid-seventeenth century was "a terrible example." "They [the English] sought how to depose their monarch under the form of law, but when he had been removed, they were utterly unable to change the form of government, and after much bloodshed only brought it about, that a new monarch should be hailed under a different name . . ." Furthermore, Spinoza felt that revolutions had led the peoples who made them to provoke European wars. For men brooding in guilt over their king whom they had killed would be apt to alleviate their self-directed aggression by re-directing it toward an external foreign people. As Spinoza stated it, to "divert" the people's mind "from brooding over the slaughter of the king," Cromwell's government had provoked a foreign war, thus accomplishing "nothing for the good of the country."[22]

Among our contemporaries Solzhenitsyn's outlook toward revolution is probably closest to Spinoza's. For in the judgment of the courageous Russian novelist, the Bolshevik Revolution replaced one tyranny by another that was worse, and then by aiming to establish Communist dictatorships everywhere, contributed to the collapse of European reason and the liberal democratic governments, that made the triumph of Hitler inevitable. Probably, however, there was also a strong Jewish ingredient in Spinoza's antipathy to revolution. The Jewish people and their philosophers have usually stood for stability, and dreaded the instability of mass movements. A mass people's movement generally portended for the Jews violence, massacres, lootings, persecutions, ruinations, and expulsions. No doubt Spinoza himself, at the age of sixteen, had witnessed the arrival at the Amsterdam Synagogue of the refugees from the Chmielnicki Massacre of 1648; at least one-third of Poland's Jews had perished therein. Stable authority guaranteed at least the minimal conditions of survival, whereas a movement of the masses harbored all varieties of unreason.

Spinoza's democratic political standpoint thus bore a Dutch-bound impress; provincial in its setting, it could be adopted elsewhere only under highly favorable frontier conditions; it authorized no mission for international democratic revolutions. The Netherlands fortunately had long been ruled in the towns and provinces by a commercial class without the interposition of a king. Dutch democracy thus happily arose under favored circumstances, as did later the American democracies in the colonial frontier settlements. He therefore intended his constitutional advice for those countries where a "free multitude" existed, that is, where non-democratic institutions were not established: "For a multitude that has grown used to another form of dominion will not be able without great danger to pluck up the accepted foundations of the whole dominion, and change its entire fabric."[23] Where democracy did not already exist, Spinoza counselled only that the existing political system be so directed as to ensure the welfare of the people, and the safeguarding of its "free men." Piety and obedience to God, said Spinoza, require that a man "obey the sovereign power's commands in all things;" moreover, those agitators who promote schism or revolution are probably power-seekers: "he who strives to deprive the sovereign power of such authority is aiming (as we have said) at gaining dominion for himself." The state, wrote Spinoza, "will go much faster [to ruin] if private citizens seditiously assume the championship of Divine rights."[24] According to Spinoza, one does best to work within the framework of the system one has, monarchical, aristocratic, or democratic, trying to adapt its constitution to serve the welfare of its people and the liberties of its free men.

III

According to the great French social observer and thinker, Alexis de Tocqueville, the pantheistic philosophy exerts an unusual attraction upon the citizens of a democracy. "Among the different systems by whose aid philosophy endeavors to explain the universe," he wrote, "I believe pantheism to be one of those most fitted to seduce the human mind in democratic times." "[S]uch a system, although it destroys the individuality of man, or rather because it destroys that individuality, will have secret charms for men living in democracies."[25] Certainly this generalization was validated for the generation of transcendentalists who emerged in the decade of the eighteen-thirties to exercise a profound influence on American thought. And in their pantheon of philosophers Spinoza held well-nigh the highest place. "Spinoza has come to be revered," noted Ralph Waldo Emerson in his lecture on *Character* in 1864.[26]

Neither Spinoza's determinism nor naturalism, nor indeed his geometrical method appealed to the transcendentalists. Nor were they drawn to Spinoza's hedonism. Rather the young transcendentalists were tired of the stale Lockean-Jeffersonian materialism and epicureanism. What drew them to Spinoza was his vision of an all-inclusive, all-comprehensive God, in whom every object, every creature, from the lowliest insect to the lordliest man, was a mode, a manifestation, each equally partaking of and necessary to the unity and simplicity of Substance. According to Spinoza, man above all had the privilege of achieving blessedness in an intuitive knowledge and love of God. The transcendentalists were eager eclectics; they did not hesitate to conjoin a Kantian ethic to a Spinozist base. Furthermore, if every man shared in God's power and intellect, it seemed to follow that slavery was an indignity to man's nature as a manifestation of the divine. Even if

Southern spokesmen could argue plausibly that slavery was in the interests of the greatest happiness of the greatest number, slavery nonetheless stood in clear violation of the intuition of the divine ingredient in every man.[27] These "democratic philosophers," recalled Charles A. Dana in his later years, "this party of transcendental philosophers" proposed that "equality and democracy should characterize our social relations," and sought to accomplish "the reform of society . . . And that was what inspired the socialist movement which began about 1835 or 1838.,"[28]

George Ripley, the organizing spirit in 1840 of Brook Farm, the famed Massachusetts transcendentalist colony, was much moved by Spinoza's all-embracing God. "On Sunday afternoons" at Brook Farm, "during the earlier years, Ripley elucidated Kant and Spinoza to those who cared to listen, . . ."[29] In 1839 Ripley was already embroiled in a public philosophic controversy in which he defended Spinoza, Schleiermacher, and De Wette against the charge of atheism. The well-known Rev. Andrews Norton had declared that "the celebrated atheist Spinoza" was "the first writer" to hold that miracles were impossible, and to maintain that God and Nature were the same. Although Spinoza had an "affectation of religious language," Norton charged his doctrine was atheist, and the "latest form of infidelity," rampant in Germany, that was spreading to the United States: "The commotion of men's minds in the rest of the civilized world, produces a sympathetic action in our own country. We have indeed but little to guard us against the influence of the depraving literature and noxious speculation which flow in among us from Europe." Revolutions were threatening in Europe: "Long-existing forms of society are giving way,"[30] Religion, Norton argued, could not be founded on Spinoza's notion of intuitively apprehended and certain truths, but only on the authority of recorded miracles. To which Ripley replied in an anonymous pamphlet that man's soul was indeed conscious of the highest reality, and that the aspersions on Spinoza's character were cast unjustly on a "devout, sweet, unselfish, truth-seeking" man.[31]

Spinoza's thought exhilarated too the most gifted woman in America, Margaret Fuller. In 1840 Margaret Fuller, together with Emerson and Ripley, founded the transcendentalist monthly magazine *The Dial,* devoted to religion, literature, and art; it was indeed America's first genuine philosophic, critical, literary journal, the organ of an intellectual movement. Margaret Fuller in her youth had discovered Spinoza;

she was 27 years old when she later discussed Spinoza's ideas with the outstanding Unitarian transcendentalist, Theodore Parker. During her editorship of *The Dial*, the famous *Conversations* that she conducted in Boston as her own unique enterprise in adult education included Spinoza among her topics, and his philosophy was a theme for her circle of friends.[32] Conservative New England thinkers, however, fearing the determinism advocated in Spinoza's philosophy, were alarmed that such a man as George Ripley was Spinoza's proponent. The leader of Vermont transcendentalism, James Marsh, president of the University at Burlington, felt that Ripley, a follower "of Spinoza" despite himself, was losing "the idea of a free personal agent . . ." "I must write to Mr. Ripley about this matter," he informed a friend in March, 1838.[33]

The idealistic American transcendentalists thus diverged notably from their European socialist contemporaries, the German left Hegelians such as Karl Marx, Moses Hess, and Friedrich Engels. The American transcendentalists were as much socialists as the German group, but the Americans rejected materialism. George Ripley read Ludwig Feuerbach's materialist writings, and rejected them, saying in 1852 that Feuerbach was "crabbed and dogmatic in his atheism"; Marx and Engels, on the other hand, extolled Feuerbach as their materialist mentor. The American transcendentalists based their socialism on a Spinozist pantheism; Marx and Engels turned instead toward a materialist conception of history.[35]

In pre-Civil War America, numerous audiences in Eastern cities and frontier towns learned of Spinoza's import from Ralph Waldo Emerson. That greatest of American teachers, riding indefatigably in the railway cars and stagecoaches along the trail of the lecture lyceums to far-flung points, drew in his famous lecture, "The Over-Soul," the contrast between two types of philosophers, those like "Spinoza, Kant and Coleridge," and on the other hand, those like Locke, Paley, and Stewart; the "one class speak from *within*, . . . and the other class from without, as spectators merely, . . ."[36] As far as Emerson was concerned, the latter were of "no use" to him; only those spoke to him who were "parties and possessors of the fact," who felt "that the Highest dwells with him," who felt the "influx of the Divine mind into our mind," experiencing "that Unity, that Over-Soul, within which every man's particular being is contained and made one with all other . . ." This was indeed the pantheistic creed that de Tocqueville thought so congenial to democratic nations. But as the smoke of the Civil War battlefields and the ideological fervor receded, even transcendentalists began to feel a certain hollowness in their creed, and to perceive that the harsh reality of sin and suffering were glossed over in the Over-Soul.

IV

The transcendentalists found themselves dissatisfied, as the Civil War ended, with their pre-Civil War metaphysics. Before the war, enthusiastic with Spinoza's concept of an all-inclusive God, they had heard critics charge that Spinoza's views made the fact of Evil as well as man's free choice unintelligible; such objections seemed trivial and puny against their mystical conviction. But not so after the war. It was as if the transcendentalists went through their own "end of ideology," for this phenomenon in the history of ideas was not unlike that which took place after the Second World War when intellectuals of the thirties who had been ardent Marxist determinists decided in the latter forties that an existentialist recovery of the individual's free choice was in order. The war was almost an occasion for the oscillation of the meta-physical wave.

Thus, when the transcendentalist assembled during the post-Civil War years in their Boston Radical Club to hear one of their founders, Frederic H. Hedge, discourse on pantheism, they heard him say: "Spinoza not only denies freedom of will to man, but denies to man substantial existence. He considers the human mind to be part of the infinite intellect of God; . . . In other words, there *is* no such entity as the human mind or soul; . . ." Moreover, since "Spinoza's ontology supposes a single and whole substance, comprising all that is, . . . he is said to have turned the Devil out of the world. There is no room for his Satanic Majesty in a universe which is an expression of God." Consequently: "The great weakness of Pantheism, as expressed by Spinoza, consists in the relaxation of the moral sense consequent on referring all action, good or bad, to God as the one immediate and direct cause of all." Hedge acknowledged "the quickening sense" that Spinoza's philosophy had previously given of "the all-pervading" presence of God, thereby revivifying "a cold and unmeaning dogma."[37] But the distinction of good and evil, in his judgment, likewise vanished if all events were necessitated in God or Nature. The later historian of the movement, Octavius Roy Frothingham, while replying that "Americans are more pantheistic than other peoples because their sympathies are more general and quicker," still felt likewise that "the antithesis between good and evil remained essential to all mankind."[38] The poet Henry Wadsworth Longfellow reminisced that one of his professors had justly warned him and his fellow students many years earlier that Pantheism was "more Dangerous" than Atheism "for it takes away the sense of moral responsibility." The famed Abolitionist militant, Wendell Phillips, veteran of a hundred platforms, added: "Spinoza gives no theory which explains away the fact of suffering, and he had seen suffering which he felt sure was unmitigated evil."[39] In short, it seemed now that life could not be earnest on a Spinozist foundation. One might still admire the character of Spinoza, as Thomas Wentworth Higginson did, for his determination not "to forego his appointed work" though he had to live "on five cents a day."[40] But even the abolitionist colonel, and the discoverer of the belle of Amherst, the poet Emily Dickinson, seemed to have left aside the Over-Soul. Precisely at this juncture a new variation of Spinoza's thought was developed by the emancipated spirit of Oliver Wendell Holmes, Jr.

V

The young Oliver Wendell Holmes, Jr. joined the Union Army in 1861 at the age of twenty. At the age of twenty-three, thrice wounded, and mustered out in 1864, his outlook on man and the universe had been recast on the battlefields. He had entered the war as an adherent of the abolitionist ideology ("in the emotional state not unlike that of the abolitionists in former days, which then I shared and now much dislike—as it catches postulates like the influenza," he wrote in 1918). When he emerged, he had henceforth had his "belly full of isms." As a wounded Lieutenant, he had begun to mistrust the facile moralizings of the militant reformers.[41] As a Captain, wounded in the heel, Holmes had prayed that he might lose his foot so that he might not have to return to duty at the front for a third time.[42] During his convalescences he read Lewes' *Biographical History of Philosophy* and above all Spencer's *First Principles* with its conception of an evolving world governed by an all-encompassing law.[43] Good and evil were tested in the crucible of a soldier's experience; the transcendentalist moral absolutes seemed a feeble fortification in the carnage of confusion, cowardice, and courage at Ball's Bluff: "I doubt if the intellect accepts or recognizes that classification of good and bad," wrote the wounded youth.[44] He tried to make his notion of "good" into a synonym for the "general law" of the universe. Recovering from the dysentery after the battle of Fredericksburg, he doubted whether the Union's arms could overcome "the unity or determination of the South," feeling that "civilization and progress" would conquer better in peace than in war.[45]

Probably it was in 1883 that Holmes may have first studied Spinoza's philosophy through the agency of the book by his friend Frederick Pollock, *Spinoza: His Life and Philosophy,* the most influential work in drawing the attention of scholars and philosophers to Spinoza's thought.[46] More than 40 years later in 1927, Pollock wrote Holmes: "You are, I think, the kind of reader B. d. S. desired. He says expressly that it is not a wise man's part to found a school to be called after his name; and there is nothing to show that he regarded his own work, as final or even finished...."[47] Holmes indeed valued Spinoza precisely because he regarded human good and evil as strictly human responses with no bearing on the meaning of the cosmos as a whole. The transcendentalists had derived from Spinoza an idealistic pantheism in which all living beings shared in the divinity. Holmes washed all such concepts including "natural right" in what he called "cynical acid";[48] the residue that was left of irreducible reality was the notion of contending psychophysical entities, of men, pursuing their fighting faiths, ignorant of any cosmic scheme that underwrote any of them. Spinoza's naturalism was a philosophy congenial to the unsentimental philosophical pharmacologist.

Holmes read and re-read Spinoza's *Ethics.* "If you leave out his logic chopping and theological machinery," he wrote his young friend Harold Laski in 1923, "his view of the cosmos seems to me better than any other I know in the past."[49] With the Sacco-Vanzetti case straining the American legal system, Holmes stated tersely his attitude to systems of philosophy: "All that any of the philosophers has to contribute is a small number of insights that could be told in two minutes... In 50 years, more or less, the system goes to pot; posterity doesn't care for it—but you have to read the book to get the author's aperçus—... I care more for Spinoza's than for the other old ones, but I don't believe his postulates or yield to his logic. What I care for is an attitude and a few truths that are independent of his machinery."[50] Holmes saw a world of competing men, classes, nations, and races, struggling against the background of a planet's dwindling resources; civilization resolved itself into a fortunate enclave of "polite manners." It was Spinoza's world divested, however, of God and *amor intellectualis dei.*

The echo, however, of Spinoza's phrases on behalf of freedom of speech can be heard in at least one of Holmes's classical dissenting opinions. In the year 1925, the United States Supreme Court considered the appeal of a New York Communist, Benjamin Gitlow, who had been convicted of "criminal anarchy"; Gitlow had circulated two publications advocating that a dictatorship of the proletariat be established. Holmes had been reading two years earlier "Santayana's charming introduction" to the Everyman's Library edition of Spinoza's *Ethics* that cited the basic proposition of Spinoza's liberalism: "If acts only could be made the ground of criminal prosecutions, and words were always allowed to pass free, sedition would be divested of every semblance of justification, and would be separated from mere controversies by a hard and fast line."[54] This was the doctrine indeed that Holmes was formulating: that freedom of thought reached a limit only when it posed a "clear and present danger." Thus Holmes argued in Gitlow's case that "there was no present danger of an attempt to overthrow the government by force on the part of the admittedly small minority who shared the defendant's views."[52] No doubt the Spinozist influence on Holmes was largely unconscious, dating from his reading of Pollock. "Spinoza has had no conscious influence upon me," he once wrote, though he added, "when I find myself sympathizing with him, the probability of an influence, even if indirect, is great."[53]

The old Civil War veteran in the last 25 years of his life found his chief intellectual joy in the companionship of young Jewish intellectuals. He wondered at this phenomenon himself: "When I think how many of the younger men that have warmed my heart have been Jews I cannot but suspect it," that "loveableness is a characteristic of the better class of Jews," wrote Holmes to Laski in 1921.[54] Holmes also told that he had heard say that his ancestors, the Wendells, were Dutch Jews, who had been originally Vondells.[55] Doubtful though this hypothesis of Jewish descent may be, there can be little doubt that Holmes was the most distinguished American thinker to draw sustenance from Spinoza; he imbibed not a theory of God but rather of Nature, the *Natura,* not the *Deus,* the vision of creatures and objects functioning, and man pursuing "an unknown end."[56]

VI

Unquestionably the group to whom Spinoza appealed most were the young Jewish intellectuals, children of the first generation of immigrants to America. Rebels against Jewish orthodoxy, they felt poignantly the division that grew between them and their elders. The young student of philosophy, Morris R. Cohen, twenty-one years of age, wrote that the deepest tragedy in New York's East Side was its conflict of generations affecting the youth with cynicism and pessimism.[57] To these young rebels searching for a master-thinker, the figure of Spinoza, the courageous rationalistic philosopher whom the Jews excommunicated, was a dramatic model.[58] Spinoza seemed to have coped with and solved their own problems; he had united Science and God, combining the deterministic causal laws they learned at college with an infinite God who was Nature; even in studying one's textbook of physics one was engaged in the worship of God, and achieving an intellectual love of Him. Each laboratory experiment became a prayer to God for knowledge of His essence. Cohen in later years thought there was a duplicity in this solution to the generational conflict between science and religion. On the East Side, "to mask our unorthodox ideas," the rebellious intellectuals would "use the word 'God' with Spinoza, to mean what scientists call the system of nature, . . . Every impulse of filial piety . . . drove us to this hypocrisy," he said.[59] Yet at the height of his own intellectual powers, Cohen saw no such duplicity, and found a resolving power in "the intellectual love of God." He declared in an essay in 1922: "I undertook to defend the validity of the Spinozistic ideal." For: "Of all philosophers, it seemed to me that Spinoza had most clearly developed the rational and tolerant attitude to the values of religion for which I had been searching." Spinoza left the most "deeply religious" impress, if religion signified "a sense of infinite powers beyond our scope," combined with "a sense of the mystic potency in our fellow human beings." Spinoza, he wrote, "showed me the path to that serenity which follows a view of life fixed on those things that go on," which is imbued with "a sense of the limitations of all that is merely material."[60] From Spinoza, Cohen derived a sense of existence with its unknown infinite attributes, not unlike the Unknowable of Herbert Spencer. At the same time, Spinoza's conviction of the eternity of the human mind was lacking in Cohen's thought; the fact of death for

Cohen could not be philosophically overcome as it was for Spinoza: "I cannot agree with Spinoza that the free man thinks of nothing less than of death"; rather the thought of death "is ever present in my mind."

The first Jewish generation to confront death without the sustenance of a religious faith in immortality was all the more drawn to the personality of the excommunicate Spinoza. Perhaps it was a masochistic, self-immolating declaration of loyalty to truth even if it should slay one. But it was peculiarly true that insofar as Spinoza's philosophic reputation was concerned, the fact that he had been excommunicated was the most fortunate thing that could have happened to him. The unconscious of the young Jewish rebels had its own *a priori*: because Spinoza was excommunicated, what he said must be true. Newspaper articles, essays in the *Jewish Daily Forward,* and in such magazines as *The Jewish Tribune,* periodically recounted the exciting tale of the rupture between Spinoza and the Amsterdam Jewish community. The play *Uriel Acosta* was a perennial favorite of the young Jewish intellectuals of the East Side; portraying sympathetically the ill-fated precursor of Spinoza, his bitter battle with Jewish orthodoxy and his final self-destruction, its last act depicted the crushed Acosta transmitting his freethinker's mantle to the child Spinoza.[61]

English agnostics of the nineteenth century had been similarly moved by the character of the excommunicate Spinoza. George Henry Lewes, the life-long companion of the novelist George Eliot, and the author of an influential *Biographical History of Philosophy,* narrated that when he was just over twenty years of age, he hungered for "some knowledge of this theological pariah, partly, no doubt, because he was an outcast, for as I was then suffering the social persecution which embitters all departure from accepted creeds, I had a rebellious sympathy with all outcasts, and partly because I had casually met with a passage, quoted for reprobation, in which Spinoza maintained the subjective nature of evil, ..."[62] Lewes encouraged George Eliot in her translation of Spinoza's *Ethics.* Francis Bacon once wrote of the sources of human error—The Idols of the Tribe, the Idols of the Marketplace, the Idols of the Cave, and the Theatre; in strict logic, we might add the Idols of the Pariah, the attraction to the rebellious, the "underdog," the defiant, the truth of the defeated: not "Magna est veritas et prevalebit," but "Magna est veritas et perdet."

Many a young East Side Jew shared in an emotional identification with Spinoza. When Will Durant, later the celebrated author of *The Story of Philosophy,* began his career in 1913 as a popular lecturer at the Labor Temple on Second Avenue and Fourteenth Street, his opening lecture on Spinoza drew "some five hundred new Americans."[63] A Jewish auditor of Durant's versified in a "little magazine":

ON A TALK ON SPINOZA

Durant spoke of Spinoza yesterday
And I sat list'ning, feeling, meditating,
And now and ever afterwards will feel
And live and think more deeply than before,
For having heard Durant speak of Spinoza.

Spinoza! what a mighty, mighty name!
All Alexanders, Caesars and Napoleons—
Mere specks of dust upon a polished lens,
Compared to this poor polisher of lenses.
He polished lenses for myopic eyes; . . .

The World's myopic eyes have need of them—
And long will need them—poor myopic world.
My own sight seems improved since I heard
Durant speak of Spinoza yesterday.[64]

Durant's lecture transcribed into a booklet, sold widely as one of the most influential series of cultural publications in American history, the famous Haldeman-Julius Blue Books, retailed at five cents a copy; later it provided the most moving chapter in the still unsurpassed and oft reprinted best-seller *The Story of Philosophy.*[65]

The first American to write a doctoral thesis on Spinoza, Gabriel R. Mason, was indeed one of the young East Side Jews who was in rebellion against the orthodoxy of his parents. Gabriel, himself born in Russia in 1884, was brought by his parents to the United States at the age of eight in 1893. During the year after his Bar-Mitzvah, as he was shedding "the ritual and the superstitions of the Jewish religion," he sought for some sort of intellectual support, or more accurately the reassurance of a tradition within Jewish history of his philosophical rebellion against orthodoxy. To his delight, he came in his senior year at college upon Spinoza, learning that "for these same views Spinoza was excommunicated by the Jews of the Amsterdam Synagogue." Thus, he writes, "I was attracted to this road [Naturalism] by the lovable personality and profound philosophy of Spinoza, . . ."[66] Mason had a notable career as an educator, serving for many years as a principal of a public elementary school and high school, when such posts were the highest academic places to which Jewish scholars, except for unusual exceptions, could aspire. But Mason also was for 44 years a member of the Socialist Party during periods especially before and during the First World War when such a membership brought one close to the permissible boundaries of academic freedom. And if Spinoza lived close to the Collegiant sect, Mason found fellowship in the Ethical Society.

Horace M. Kallen on the other hand, held fast to his ties with the Jewish community, striving rather to re-shape its philosophy to a "Hebraism" that was much influenced by Spinoza. Born in 1883, Kallen was among the very first of the Jewish sons to find in Spinoza a guidepost for liberation. Horace's father, after serving as a Rabbi in Prussian Silesia, had migrated to Boston, Massachusetts; Horace was five years old at the time.[67] Great strains arose between the Rabbinical father and the freethinking son. As Kallen described it in retrospect: "the internal devaluation of Jewish heritage was continuous and progressive. My father was a strict man; I didn't like him; . . ."[68] In later years, Kallen would still recall with some acerbity how his father would humiliate him publicly before his pupils. Horace felt adrift, without intellectual moorings., Then, by chance or divine incursion, as Kallen narrates:

> One day I chanced upon a German rendering of Spinoza's *Theological-Political Tractate*. It set me free. I began to read English versions of his works and English commentaries on these. It was in the year of my graduation from high school that I became enamored of the man and convinced of his philosophy. His image, his thought, and his story became the point of no return in the ongoing alienation from my father and the ancestral religion. I identified with Spinoza.[69]

For a brief period Kallen seems to have shared Spinoza's estrangement from Judaism. As a Harvard undergraduate, however, he was blessed by the friendship of two great teachers, Barrett Wendell and William James. From Wendell, "a Tory Yankee with Puritan heritage," Kallen acquired a heightened sense of the significance of his Hebraic heritage, while James attracted him toward pragmatism and "radical empiricism." Upon this metaphysics of plural individual realities Kallen based his unique contribution to American thought, the notion of a "cultural pluralism" as America's democratic alternative to the diversity-annihilating concept of the melting pot. Kallen described himself as "a libertarian who had been a determinist, a temporalist who had been an eternalist, ... a pluralist who had been a monist, ... The sum of it—a Spinozist who became a pragmatist."[70] Yet if Kallen came to reject Spinoza's determinism, eternalism, and monism, his vision of existence, as set forth in his remarkable *The Book of Job as a Greek Tragedy,* a neglected classic, was embedded in a Spinozist matrix. An impersonal natural order, whose processes and objects forever in flux provide the setting in which man affirms his own unique excellence, was delineated by Kallen: "God is immanent in the movement of events: each is an aspect of him, each reveals him, yet he transcends each and all such ... God is the dynamic of the universe, and the range of his power is co-extensive with it ... In human terms he cannot be thought; being omnipotent, he is self-sufficient, absolute, consequently altogether incommensurable with human nature, ... From any point of view that is human that you may choose, God has no preferences nor can his will and interests be defined in terms of preference."[71] The God that Job finally envisaged, the Infinite Existence of substance and its modes transcending all human standards of good and evil, was for Kallen identical with the "God or Nature" of Spinoza. Although as a Jamesean empiricist, Kallen abandoned altogether the rationalistic theory of knowledge and its pretense of geometrical demonstration from unchallengeable axioms and postulates, the naturalism of Spinoza was retained as a basic tenet.

In Kallen's view, however, Spinoza's "inevitably necessary world of mathematical physics was replaced by the new one of freedom and contingency of physical mathematics." Still, as he estimated Spinoza's significance in 1932 on the three hundredth anniversary of his birth, he declared that though Spinoza's determinism was superseded, "His ethic remains as appropriate to modern conditions of life as it was to his own personal life; nay, far more appropriate."[72]

A striking division of opinion thus characterized the spiritual descendants of Spinoza as to the value of his determinist tenet. Einstein, most notably, quite unlike Kallen, felt that the deterministic tenet was essential for retaining one's sanity in a world consumed by the passions of unreason. "The spiritual situation with which Spinoza had to cope peculiarly resembles our own," wrote Einstein. "The reason for this is that he was utterly convinced of the causal dependence of all phenomena ... In the study of this relationship he saw a remedy for fear, hate and bitterness, the only remedy to which a genuinely spiritual man can have recourse."[73]

Cohen and Kallen, both selective Spinozists, likewise selected and discarded in different ways from their philosophic source. Cohen, as a youthful Marxist and an adherent of the Socialist Labor Party, had valued Spinoza the cosmopolitan, who altogether rejected Judaism, and was indifferent to its continued communal existence. Zionism, Cohen later maintained in a provocative article, was a species of tribalism, and as such, incompatible with the liberal standpoint. To this Kallen replied in his most powerful controversial essay: no, the foundation of Zionism was rather the liberal conviction that "all nationalities are created equal and endowed with certain inalienable rights"; as such, it was continuous with the philosophy of Jefferson's Declaration of Independence.[74] The Jews, as a historic people, "neither better nor worse" than others, are entitled equally to an autonomous existence. Only an abstract internationalism, bred on a theory that levelled all men into "economic men" would obliterate their cultural qualities. Liberalism should rejoice in the cultural diversities of men. Kallen's Hebraism attached a greater significance to the cultures achieved in the temporal flux than did Spinoza for whom the "free man" was essentially a member of a trans-national community of scientists and scholars.

During the post-World War I years until the advent of Hitler, it was the kind of Spinozism that Morris Cohen advocated rather than Kallen's which attracted Jewish college students. During that age of scientism, universalism, and assimilationism, the Menorah Society, for instance, dedicated as it was to Jewish culture, declined in numbers at the City College of New York. Few Jewish students took Zionism seriously, and the feeling was that as liberal democracy took hold, Eastern European Jews as well would find Zionism an obsolete regressive nationalism. Spinoza became the prophet of science and liberalism; the knowledge of science was the knowledge of God. They could regard themselves as disciples of Einstein, the world's greatest Jew, whose "cosmic religion" was the modern equivalent of Spinoza's "intellectual love of God."

Spinoza, indeed, was more often cited in the nineteen twenties as a source of inspiration for Americans than he has ever been since. Indeed, both in political philosophy and metaphysics his name now rarely occurs. His rationalism is out of fashion, and likewise, his freedom from ideology and subservience to the adoration of the masses. His appeal was principally to the second generation young Jews in transition from religion to science. To that generation Morris Cohen and Walter Lippmann were proponents of Spinoza's wisdom.[75]

Cohen and Lippmann, both like Kallen nurtured by William James, and writing their first articles in their organ *The New Republic,* admired Spinoza's conception of life rather than his cosmology. Cohen in 1927 wrote of Spinoza as the "prophet of liberalism" who would have spared America the pathetic humiliations of the era of the Prohibition Amendment;[76] Spinoza had indeed warned that "such attempts have never succeeded in their end," that men will "outwit laws framed to regulate things that cannot be effectively forbidden." Walter Lippmann had listened as a student to his professor, George Santayana, gracefully interpreting Spinoza. Lippmann learned from Spinoza a notion of blessedness that he, a humanist, could accept; blessedness was not the reward of virtue, but virtue itself; the aim of life was not the accretion of things but their enjoyment as objects of reflection; "pure science is high religion incarnate." And Lippmann noted that Spinoza was at one with Freud in his view that "an emotion which is a passion ceases to be a passion as soon as we form a clear and distinct idea of it." On such "disinterestedness," dispassionate understanding, Lippmann felt a humanistic culture could be founded.[77] As America's most influential and analytic political writer, Lippmann brought this insight to bear upon America's problems during five decades of columns, editorials, and books, in which he appraised the time's turmoil of war, depression, and destruction. Lippmann remained exempt, as Santayana noted, from that "form of cowardice peculiarly modern" of "those who feel safe in their ethics and politics if they are swimming with the tide."[78] Spinoza helped keep Lippmann free from historicism, that ideology which endorses the bandwagon with a metaphysics, and looks at all existence *sub specie politiae.*

VII

The name of Spinoza is little invoked in our own time because his political philosophy has no ideological component. Spinoza offers no law of history, no law of progress, no outline of a Utopia to be achieved. There is no sense of directionality in history from the standpoint of Spinoza's philosophy: "I do not believe that we can by meditation discover in this matter anything not yet tried and ascertained, which shall be consistent with experience or practice . . ."[79] He stands with the tradition of wisdom of Ecclesiastes: "There is nothing new under the sun," and dismisses the aspiration of Platonic philosophers to be king: "No men are esteemed less fit to direct public affairs than theorists or philosophers." For political understanding he turns to experienced practitioners in statecraft, to "the most ingenious Machiavelli," to Antonio Perez, the former Secretary of State to the King of Spain, to Pieter de la Court, the economist-collaborator of the Grand Pensionary of Holland.[80] Spinoza's "free man" is not an ideologist seeking a political mandate to fulfill some purported historic mission to realize history's aim. That Spinoza should reject the credentials of philosophers as reliable political advisers might seem strange since he is a philosopher himself. But Spinoza, as we know, ridiculed "the speculations of Platonists and Aristotelians," those who "rave with the Greeks," and he rejected the authority of Plato and Aristotle as not having "much weight with me."[81] He thought of himself as a scientist, a follower of the mathematical method, in the tradition of the ancient materialists.

Spinoza's culminating aim was to provide a set of practical proposals for ensuring the stability of a democratic state. His book of 1670 had extolled democracy as the best form of government, and pointed to the glory of Amsterdam, its liberties and prosperity, as confirming that superiority. The year 1672 had brought the Republic's downfall and the murder of de Witt. He could no more write with assurance about democracy than could a Marxist about Communism after Nikita Khrushchev's speech in 1956 had confirmed the crimes of Stalin's era. In the parts of his *Political Treatise* that were completed before he died in 1677, Spinoza gave a series of specific constitutional provisions for assuring the stability of a monarchy or an aristocracy. What was more important to Spinoza than the form of the government was to insure that, whatever its form, free men would be safeguarded in their liberties. To achieve that goal, there was only one way, and that was to ensure the hegemony of the middle class. He had various empirical models that could guide him in his constitutional theorizing; he used the famed Venice as an example of a stable aristocracy, and Aragon as a model of constitutional monarchy before it was subverted by the absolutist Philip II. But where was he to find an empirical model for a stable democracy? The Athenian history was not one he would emulate; the English colonies in America were small and little known. The Swiss cantons in Spinoza's lifetime were wracked by class struggles and religious civil wars; nor would Spinoza have admired the earlier Genevan democracy, that under Calvin's direction had executed the free man, Servetus.[82] Thus Spinoza's chapters on democracy perforce remained unwritten; the first theorist of democracy in the history of political philosophy could not write the book he wanted to write because he could not find the factual foundation he required. The philosophy of democracy was born in hope overlayed with uncertainty. Spinoza could not base his democratic standpoint on a belief in the virtues of the common man, the Lincolnian populist tenet; he had too often asserted the irrationality of the masses. Spinoza was evidently searching for such mechanisms of checks and balances as the Lockean authors of the American Constitution contrived as guarantees against populist and executive despotisms alike. But more than a hundred years of colonial democratic political experience were required to enable such thinkers as Jefferson, Franklin, Madison, and Hamilton to formulate their analyses of the conditions for democratic stability.

Together with Madison and Hamilton,, Spinoza shared what might be called the principle of cancellation, namely, that in a large democratic body, the selfish aims of opposing groups and persons tend mutually to cancel each other out; thus, the "pressure groups" and lobbies of workingmen, business executives, bankers, consumers, farmers, all presumably arrive at an equilibrium through the mutual veto and correction of their extravagant claims. Therefore, Spinoza could write that in a democracy "it is almost impossible that the majority of a people, especially if it be a large one, should agree in an irrational design."[83] Precisely herein, however, we do have the most difficult problem of modern democracies; the principle cancellation often fails to work; large powerful unions of workingmen can achieve higher wages, large corporations can raise their prices and profits proportionately or more, while the unorganized consumers, middle classes, and poorer workingmen complain of inflation. A cumulative disequilibrium can also be an outcome of democratic process. Probably such questions were central in Spinoza's mind after 1672; he did not pretend that he had solved them. No ideological democrat, Spinoza would probably have found congenial the judicious evaluation and concerns of Alexis de Tocqueville's *Democracy in America*. For de Tocqueville, a discerning empirical observer of the democratic American society, and mindful of the universal laws of human nature, was concerned with the preservation of liberties, culture, and science.

Despite his eloquent plea for freedom of speech, Spinoza had small regard for the "schismatics" (or ideologists, as we would say today); "schisms," he writes, "do not originate in a love of truth, . . . but rather in an inordinate desire for supremacy."[84] To give ballast to the government against the schismatics, he weighted his constitutions toward the gerontocratic side. Whereas Machiavelli was cooly prepared to resort to a periodic terror to restore political equilibrium, Spinoza proposed instead a standing committee of public safety, made up of older, conservative men, "of an age to prefer actual security to things new and perilous," a constitutional device that the Venetian republic in his view had used.[85] Spinoza. though only 44 years old when he died, was above all concerned with obviating seditious and revolutionary violence.

Lastly, Spinoza is the first philosopher who proposes to assure the stability of the state by harnessing to it the energies of men's "avarice," their passion for accumulation. He is not one of those philosophers such as George Berkeley who deplored the rise of the search for commercial and industrial profit-making; Berkeley indeed had perceived the ruin of Great Britain in the rise of its capitalist civilization, and his theme has echoed in our time in R. H. Tawney's tract against *The Acquisitive Society* and its successor pamphlets and books on the alienation of man under bourgeois society.[85] No, Spinoza aims to take the energies of men as he finds them, and to harness them constructively so that they will contribute to the stability of the commonwealth. Societies, he notes, have hitherto decayed because "men in time of peace . . . become soft and sluggish." To counteract the evolution toward decadence, Spinoza proposes that the state encourage, honor, and look for others to emulate the accumulators of capital. "And therefore the chief point to be studied," writes Spinoza, "is that the rich may be, if not thrifty, yet avaricious. For there is no doubt that, if the passion of avarice which is general and lasting, be encouraged by the desire of glory, most people would set their chief affection upon increasing their property, without disgrace, in order to acquire honors, . . ."[87]

Thus, it would follow, according to Spinoza, that only a bourgeois democracy would be a stable democracy. The "decay of capitalist civilization" that has been taking place has not been the outcome of the passion for accumulating capital; rather, as the economist Joseph A. Schumpeter held, it is the spread of the anti-capitalist mentality among intellectuals especially that erodes the self-confidence and enterprising vigor of the capitalist society. When the drive to accumulate is disparaged, the psychological basis of the capitalist economy is undermined. And insofar as liberal democracy is founded on a pluralistic, capitalist economy, with economic power dispersed through a multiplicity of enterprises, it too rests on the freedom of men to venture and accumulate their own capital. Stalin's huge concentration of political power into his dictatorship thus followed on his destruction of private ownership on the farms, his obliteration of the relatively prosperous and more efficient farmers, and his termination of the segments of private commercial and industrial enterprise that Lenin had allowed under his New Economic Policy. The political principles of free thought and speech not only came into existence historically with the advent of Dutch and British capitalism; according to Spinoza's standpoint they were causally grounded in it. As Descartes once wrote, he felt freest doing his writing on the docks of Amsterdam where people were too busy making money to bother about what he was about.[88]

It is strange therefore that Spinoza, the political theorist of bourgeois democracy, has been especially the favorite of so many leftist and socialist-minded youth. Perhaps in part this was because Spinoza renounced trade, became a lens-grinder, and made friends with a religious communitarian group, Collegiant-Mennonites. Then too he was the eloquent advocate of freedom of thought and speech. Nonetheless it is altogether likely that Spinoza would have felt that socialist societies were inherently unstable because they failed to base themselves on the actualities of human nature; such societies became tyrannies precisely because to the extent that their social requirements were inconsonant with the passions of men, they were driven to using compensatory mechanisms of coercion.

VIII

The greatest scientists and philosophers among the Jews in modern times have felt an affinity with Spinoza less as the liberal democrat than as the philosopher of nature. Einstein, Samuel Alexander, and Henri Bergson are men who were stirred by Spinoza's ideas. The fact, however, is that Spinoza might well have repudiated their adaptations of his ideas. He might well have rejected Einstein's concept of the simple order of nature as an "inadequate idea," an all-too-human limitation of God's nature to the capacities of human simple-mindedness; he might have taken umbrage with Alexander's notion of the nisus to Deity as a teleological intrusion upon God or Nature, while Bergson's belief in human freedom he would probably have adjudged as sheer illusion. We would nonetheless suggest that perhaps on a profounder level the tenets of these Jewish thinkers may be found continuous with Spinoza's essential vision.

Einstein's conception was, we might say, that the laws of Nature are noumenal; he once said that wherever he approached a problem as to the structure of Nature, he asked himself: "How would God have solved it?"[89] The laws of physics, Einstein felt, would reveal an order, a beauty, a grandeur of simplicity in Nature. Now Spinoza, in his revolt against the notion of purpose in God, repudiates the notion of an "order in nature"; "order" he states, is not a conception of reason but of the imagination, and men attribute it to God because they "prefer order to confusion," wishing to believe that God "has disposed things in which they can most easily be imagined." But God, says Spinoza, can actualize things that "far surpass our imagination."[90] Nonetheless, Spinoza the scientist in practice sought for laws that were so simple, elegant, and rational that they would appear to be identical with logical necessities. A disorderly world would not share this logical character. A multiplicity of Humean universes is alien to Spinoza's essential conception, which Einstein seems to me indeed to embody. The Jewish philosopher of science, Emile Meyerson, who has interpreted the theory of relativity as a geometrizing of Nature in which causes and effect are seen as identities in substance has translated Spinoza's conception of God for the workings of scientific theory.

This world of order to Bergson, however, is pre-eminently that of physical objects. Einstein holds to Spinoza's standpoint that the human mind as well is in all its activities completely governed by deterministic laws. Bergson, on the other hand, availed himself of Spinoza's notion of intuition, the direct insight into truths, this highest mode of knowledge, called by Spinoza "intuitive science," which surpasses both the data of perception and rational analysis.[91] Bergson maintains that we intuit directly our own freedom, that is, our unpredictable creativity; the creative freedom of the human being, his segment of the *élan vital,* his role in the making of reality, operates exempt from determinist law.[92] Now Spinoza was aware that human beings feel they are free. To Bergson he would respond: a stone falling to earth might well think (if it could) that it was creating its own act of descent; this feeling of freedom, Spinoza would argue, is illusory.[93] To which we might answer: a stone, if endowed with consciousness, might rather regard the external gravitational forces as compelling it to move contrary to its own wish to continue in its previous state of motion; the stone's experience of determinism would be that of an externally imposed obliteration of alternatives. Human beings, moreover, do have the experience that causal forces, both external and internal, are not always exclusively decisive, that in the margin of indetermination, their own choices intervene.

Spinoza argues that such phenomena of non-waking life as dreams and sleepwalking are grounds for the belief in mental determinism, because in dreams, "without the direction of the mind, . . . we do some things which we should not dare to do when awake"; our feeling of a "free decree," illusory in dreams, is, he argues, likewise illusory in wakefulness, This conclusion scarcely follows; determinism may indeed characterize our dream contents, but our waking decisions can nonetheless have their ingredient of freedom. "[T]here is nothing which men have less power over than the tongue," says Spinoza, but not all our speech consists of slips of the tongue, for which Freud tried to provide a deterministic explanation. Spinoza would leap from such instances to the belief that all human works of art are likewise determined, but this is a deterministic leap of faith, perhaps founded, we might assert, on an "inadequate idea."

Of all Jewish philosophers in he twentieth century, Samuel Alexander probably came closest to the spirit of Spinoza's naturalistic pantheism. He took pride, as he said, "in showing his affiliation to such a philosopher as Spinoza, and the more if he is himself a Jew speaking to Jews . . ."[94] Yet clearly Alexander's central notion was deeply inconsistent with that of Spinoza. For a teleological drive characterized Alexander's world of Space-Time; its restlessness, the potent source of new emergent levels, sought the fruition of Deity; beyond present levels of existence, the nisus to Deity pressed for a higher level of qualities surpassing those of man, even as man's emergence had constituted a level rising above its predecessors: "God is the infinite world with its nisus towards deity."[95] Spinoza doubtless chafing at the concept of emergence, would probably insist that the "restlessness" of Space-Time was the outcome of deterministic laws that governed all the emergent levels. Are we then to regard the nisus to Deity as a concept irrevocably incompatible with the directionless, timeless nature of Spinoza's God? It is possible that Spinoza's theology might be enriched by Alexander's conception. For if in the course of time, not all the infinite ideas conceived in God's intellect are realized, then at any given time, the actual universe has as yet not actualized an infinite number of possibilities. Spinoza wrote that "desire (*conatus*) is the essence of man," and joy "man's passage from a less to a greater perfection."[96] Even so, insofar as the unrealized possibilities of existence are those of a greater perfection than those presently realized, one might say that a nisus to Deity obtains in the current segment of existence insofar as it moves towards a more perfect level of reality. Until the highest forms of the finite modes of mind have been achieved, God's mind continues to require them. The nisus to Deity is then a corollary of the world's incompleteness in time.

Spinoza, whose ideas have been adapted selectively in ways evidently incompatible with his philosophy as a whole, might conceivably have found himself much at home, and his thought evolving, in the company of his successors, Einstein, Bergson, and Alexander.

Notes

1. As Sir Frederick Pollock, the greatest English authority on Spinoza, wrote, shortly before his death: "There have been Kantians and Hegelians according to the letter, . . . ; there are no Spinozists in that sense but many according to the spirit." Cf. "Contemporary Appreciations of Spinoza," *The Spinoza Quarterly,* Vol. II, No. 2, (1932), p. 94.
2. Professor Vance Maxwell of Memorial University of Newfoundland informs me that H. F. Hallett, the author of *Aeternitas,* Oxford, 1930 was an exception to my generalization.
3. Noah H. Rosenbloom, *Tradition in an Age of Reform: the Religious Philosophy of Samson Raphael Hirsch,* Philadelphia, 1976, p. 14.
4. "Measure, Time and Number are nothing but Modes of Thought or rather of imagination. Therefore it is not to be wondered at that all who have tried to understand the course of Nature by such notions, . . . should have so marvelously entangled themselves . . . committing even the most absurd absurdities." *The Correspondence of Spinoza,* tr. A. Wolf, London, 1928, pp. 118–119.
5. *The Chief Works of Benedict de Spinoza,* tr. R.H.M. Elwes, London, 1885, vol. 1, pp. 292, 313, 5.
6. Foucher de Careil, *Réfutation Inédite de Spinoza par Leibniz,* Paris, 1854, p. lxiv, *The Oldest Biography of Spinoza,* ed. A. Wolf, London, 1927, p. 180.
7. Spinoza, *The Ethics,* Book 1. Appendix. Cf., Spinoza, *Selections,* ed. John Wild,

New York, 1930, p. 143.
8. Spinoza, *Ethics,* Book V, Prop. XXIII, Scholium, Wild, ed., p. 385.
9. *Ibid.,* Prop. XLI, p. 397.
10. R.H.M. Elwes, tr. *The Chief Works of Spinoza,* Vol. I, p. 293.
11. Spinoza, *Ethics,* Book IV, Appendix, Par. XXV. Also, Book III, The Affects, Def. XLIV, Wild, ed., *op. cit.,* pp. 360, 279.
12. Frederick B. Tolles, *Meeting House and Counting House: The Quaker Merchants of Colonial Philadelphia,* 1682–1763, 1948, reprinted New York, 1963, p. 175.
13. E. Millicent Sowerby, comp. *Catalogue of the Library of Thomas Jefferson,* Vol. II, Washington, D.C., 1953, pp. 16–17. I am grateful to Mr. Steven H. Hochman, of the Jefferson Biography Project at the University of Virginia for bringing my attention to these items.
14. Adrienne Koch and William Peden, eds., *The Life and Selected Writings of Thomas Jefferson,* New York, 1944, pp. 705–706.
15. R.H.M. Elwes, tr., *The Chief Works of Spinoza,* Vol. I, p. 263.
16. *Ibid.,* p. 264.
17. *Ibid.,* p. 258.
18. *Ibid.,* pp. 261, 262, 264.
19. *Ibid.,* p. 258.
20. *Ibid.,* pp. 292, 313, 5.
21. Carl Becker, *The Declaration of Independence: A Study in the History of Political Ideas,* New York, 1922, pp. 27–30. Gilbert Chinard, *Thomas Jefferson: The Apostle of Americanism,* Boston, 1929, p,. 72. Adrienne Koch, *The Philosophy of Thomas Jefferson,* New York, 1943, p. 138.
22. Spinoza, *The Chief Works,* Vol. I, pp. 243–244.
23. *Ibid.,* p. 340.
24. *Ibid.,* pp. 251, 252, 254.
25. Alexis de Tocqueville, *Democracy in America,* tr. Henry Reeve and Francis Bowen, ed. Phillips Bradley, New York, 1954, Vol. II, p. 33.
26. Ralph Waldo Emerson, *Lectures and Biographical Sketches,* Boston, 1895, p. 110.
27. Theodore Parker, "Transcendentalism," in *The World of Matter and the Spirit of Man,* Boston, 1907, reprinted in George F. Whicher, ed., *The Transcendentalist Revolt against Materialism,* Boston, 1949, p. 70–71, 77–78. The American democracy, wrote Parker, was founded on a "transcendental proposition." He had been credited with having been the progenitor of Lincoln's phrase, "government of the people, by the people, and for the people." Cf. Henry Steele Commager, *Theodore Parker,* Boston, 1936, p. 266.
28. Charles A. Dana, "Brook Farm," in James Harrison Wilson, *The Life of Charles A. Dana,* New York, 1907, p. 521.
29. Lindsay Swift, *Brook Farm: Its Members, Scholars, and Visitors,* 1890, reprinted, New York, 1961, p. 59. Charles Crowe, *George Ripley: Transcendentalist and Utopian Socialist,* Athens, Georgia, 1967, p. 157.
30. Andrews Norton, *A Discourse on the Latest Form of Infidelity, Delivered at the Request of the Association of the Alumni of the Cambridge Theological School* on the 19th of July, 1839, Cambridge, 1830, pp. 9–11.
31. Octavius Brooks Frothingham, *George Ripley,* Boston, 1882, pp. 98–104. Also cf. Paul F. Boller, Jr., *American Transcendentalism* 1830–1860: *An Intellectual Inquiry,* New York, 1974, p. 14.
32. R. W. Emerson, W. H. Channing, and J. F. Clarke, *Memoirs of Margaret Fuller*

Ossoli, Boston, 1874, vol. I, pp. 127, 351, Frederick Augustus Braun, *Margaret Fuller and Goethe,* New York, 1910, p. 50. Madeleine B. Stern, *The Life of Margaret Fuller,* New York, 1942, pp. 157, 184, 304.

33. John J. Duffy, ed., *Coleridge's American Disciples: The Selected Correspondence of James Marsh,* Amherst, 1973, pp. 218–219. Lewis S. Feuer, "James Marsh and the Conservative Transcendentalist Philosophy: A Political Interpretation," *The New England Quarterly,* Vol. XXXI (1958), pp. 14, 20–23. Those New Englanders who early had serious misgivings about transcendentalism were critical of Spinoza's philosophy. Channing and Brownson were of this group. Cf. Robert Leet Patterson, *The Philosophy of William Ellery Channing,* New York 1952, pp. 64, 95. Henry F. Brownson, ed., *The Works of Orestes A. Brownson,* reprinted, New York, 1966, Vol. I, p. 436, vol. VI, p. 83 ff. Arthur M. Schlesinger, Jr., *Orestes A. Brownson, A Pilgrim's Progress,* Boston, 1939, p. 154 ff.
34. Frothingham, George Ripley, p. 229. Charles Crowe, *George Ripley: Transcendentalist and Utopian Socialist,* Athens, Georgia, 1967, p. 249.
35. There were indeed at least two Jews who were members of the Brook Farm colony, most notably the ebullient Marx E. Lazarus and "his handsome sister"; the literary contributions of Lazarus to the transcendentalist movement are described by Charles L. F. Gohdes as having "surpassed" in number and extent, "those of any other writer for the periodical save the editor himself." Cf. John Thomas Codman, *Brook Farm: Historic and Personal Memoirs,* Boston, 1894, p. 270. Clarence L. F. Gohdes, *The Periodicals of American Transcendentalism,* Durham, 1931, pp. 203, 135, 112. Lazarus, however, for all his ardent blend of Associationism. Abolitionism, Utopianism, Mysticism, and Vegetarianism does not seem to have been a student of Spinoza's thought. Marx Edgeworth Lazarus, a bizarre personality, was probably the first Jewish socialist in the history of the United States; in later years, he served as a private in the Confederate Army, and died in obscurity. Cf. Edgar E. MacDonald, ed., *The Education of the Heart: The Correspondence of Rachel Mordecai Lazarus and Maria Edgeworth,* Chapel Hill, 1977, p. 328.
36. Ralph Waldo Emerson, *Essays, First Series,* Rev. ed., Boston, 1883, p. 209.
37. Mrs. John T. Sargeant, ed., *Sketches and Reminiscences of the Radical Club of Chestnut Street, Boston,* Boston, 1880, pp. 155–157.
38. *Ibid.,* p. 159.
39. *Ibid.,* pp. 159–160.
40. Thomas Wentworth Higginson, *Cheerful Yesterdays,* Boston, 1899, p. 360.
41. Mark De Wolfe Howe, *Justice Oliver Wendell Holmes: The Shaping Years,* 1841–1870, Cambridge, Mass., p. 111. *Holmes-Laski Letters: The Correspondence of Mr. Justice Holmes and Harold J. Laski, 1916–1935.* ed. Mark De Wolfe Howe, Cambridge, Mass., 1953, Vol. I, pp. 164, 689.
42. Howe, *Justice Oliver Wendell Holmes,* p. 155.
43. *Ibid.,* pp. 112, 156.
44. *Ibid.,* p. 106.
45. *Ibid.,* p. 138.
46. *Holmes-Pollock Letters: The Correspondence of Mr. Justice Holmes and Sir Frederick Pollock 1874–1932,* ed. Mark De Wolfe Howe, Cambridge, Mass., 1941, Vol. I, p. 21.
47. *Holmes-Pollock Letters,* Vol. II, p. 203.
48. Oliver Wendell Holmes, *Collected Legal Papers,* New York, 1920, p. 174.
49. *Holmes-Laski Letters,* Vol. I, p. 478.

50. *Ibid.*, Vol. II, pp. 971–972. Also, pp. 939, 1132–1133, 1135.
51. George Santayana, "Introduction," *Spinoza's Ethics and De Intellectus Emendatione*, New York, 1910, 1922, p. xii.
52. Gitlow v. New York, 268 U.S. 652 (1925).
53. Harry C. Shriver, ed., *Justice Oliver Wendell Holmes, his Book Notices and Uncollected Letters and Papers*, New York, 1936, p. 159.
54. *Holmes-Laski Letters*, Vol. I, p. 304.
55. Leonora Cohen Rosenfield, *Portrait of a Philosopher: Morris R. Cohen in Life and Letters*, New York, 1962, p. 443. Also Cf. Edmund Wilson, *A Piece of My Mind: Reflections at Sixty*, New York, 1956, pp. 97–98. Beryl Harold Levy, "Justice Holmes and the Jews," *Commentary*, Vol. 22 (Dec. 1956), p. 577.
56. Catherine Drinker Bowen, *Yankee from Olympus: Justice Holmes and his Family*, Boston, 1945, p. 416. Oliver Wendell Holmes, *Collected Legal Papers*, Boston, 1945, pp. 305, 315–316. Holmes's animus against humankind generally grew with time even larger than Spinoza's. Once, when the noted historian, Carl Becker, was visiting him, the Justice said: "Becker, do you love the human race?" Becker replied that though his heart was not one "overflowing with human kindness," he wished them well. Whereupon Holmes answered: "I don't, Becker. God damn them all, I say." Cf. Irving Bernstein, "The Conservative Mr. Justice Holmes," *The New England Quarterly*, Vol. XXIII, (1950), pp. 435-436.
57. William Knight, ed., *Memorials of Thomas Davidson: The Wandering Scholar*, Boston, 1907, p. 87, Thomas Davidson, *The Education of the Wage-Earners*, ed. Charles M. Bakewell, Boston, 1904, p. 113.
58. The socialist historian, Max Beer, noted how Spinoza served as a transitional guide for rebellious Jewish youth. Thus, much earlier, Moses Hess, the "communist rabbi" as Engels called him, and the co-worker of Karl Marx, "read the writings of Spinoza, whither Jewish youths were wont to turn when they strayed from parental faith," Max Beer, *Social Struggles and Modern Socialism*, tr. H. J. Stenning, London, 1925, p. 51.
59. Morris Raphael Cohen, *A Dreamer's Journey*, Boston, 1949, p. 99.
60. *Ibid.*, pp. 216–217.
61. Karl Gutzkow, *Uriel Acosta*, ed. S. W. Cutting and A. C. von Noe, New York, 1910, p. 93 ff. Hutchins Hapgood, *The Spirit of the Ghetto: Studies of the Jewish Quarter in New York*, Rev. Ed., New York, 1909, pp. 122, 166. Harry Roskolenko, *The Time That Was Then: The Lower East Side, 1900–1914: An Intimate Chronicle*, New York, 1971, p. 152.
62. The Editor, "Spinoza," *The Fortnightly Review*, Vol. 4, April 1, 1866, p. 387, Lawrence and Elizabeth Hanson, *Marian Evans and George Eliot: A Biography*, London, 1952, p. 178.
63. Will and Ariel Durant, *A Dual Autobiography*, New York, 1977, p. 58.
64. *Ibid.*, pp. 57–58.
65. *Ibid.*, pp. 101–103, 95–96. E. Haldeman-Julius, *The First Hundred Million*, New York, 1928, pp. 107, 130–131.
66. Gabriel Richard Mason, *Gabriel Blows His Horn: The Evolution of a Rebel*, Philadelphia, 1972, pp. 13, 14, 15, 23, 149. Mason's theses, "Spinoza's Idea of God" and "Spinoza and Schelling," secured him the master's and doctor's degrees at New York University, *Ibid.*, p. 16.
67. Horace M. Kallen, *What I Believe and Why—Maybe: Essays for the Modern World*, ed. Alfred J. Marrow, New York, 1971, p. 167. Also cf. Sidney Ratner, ed., *Vision and Action: Essays in Honor of Horace M. Kallen on his 70th Birthday*, New Brunswick, 1953, p. vi.

68. *Ibid.*, p. 181.
69. *Ibid.*, p. 168.
70. *Ibid.*, pp. 166, 182, 169. Cf. M. A. De Wolfe Howe, *Barrett Wendell and his Letters,* Boston, 1924, pp. 183, 185, 273. Robert T. Self, *Barrett Wendell,* Boston, 1975, pp. 140–141.
71. Horace M. Kallen, *The Book of Job as a Greek Tragedy,* 1918, Sec,. Ed., New York, 1959, pp. 66, 68, 70.
72. Horace M. Kallen, in "Appreciation of Spinoza," *The Spinoza Quarterly, Tercentenary Issue,* Vol. II, No. 2, November, 1932, p. 105.
73. Albert Einstein, "Introduction," in Rudolf Kayser, *Spinoza: Portrait of a Spiritual Hero,* New York, 1946, p. xi.
74. Morris R. Cohen, "Zionism: Tribalism or Liberalism," *The New Republic,* Vol. 18, (March 8, 1919), pp. 182–183. Reprinted in Morris R. Cohen, *The Faith of a Liberal,* New York, 1946, pp. 326–333. Horace M. Kallen, "Zionism: Democracy or Prussianism," *The New Republic,* Vol. 18, (April 5, 1919), pp. 311–313. Reprinted in Horace M. Kallen, *Judaism at Bay: Essays toward the Adjustment of Judaism to Modernity,* New York, 1932, 1972, pp. 111–120.
75. During the latter twenties, on December 18, 1927, a society actually was founded in New York City called the Spinoza Institute of America. It organized public lectures on Spinoza at one of which, for instance, at the Stuyvesant High School, Morris R. Cohen was the speaker, and Gabriel R. Mason the chairman. One of its grouplets under the guidance of a Marxist, Harry Waton, tried to synthesize Marx and Spinoza. The Spinoza society published irregularly for a few years a journal called *The Spinoza Quarterly* that enlisted brief statements from Romain Rolland, Albert Einstein, Sir Frederick Pollock, and several Harvard luminaries. The little sect took on all the traits peculiar to an association of crackpots, with a creed called "biosophy," and its creedal leader; its existence became shadowy as the years of depression ended. Harry Waton continued into the early 50's to give lectures at the Labor Temple under the banner of the Spinoza Institute of America; copies of them were deposited in the Archives of the New York Public Library. Waton, with an original flair, argued that Marx's historical materialism defined the physical order of things which in accordance with Spinoza's psycho-physical parallelism, had to be complemented with the historical idealism of mental realities. Cf., Harry Waton, "Why a Spinoza Institute?", in Spinoza Institute of America, *Baruch Spinoza: Addresses and Messages Delivered and Read at the College of the City of New York on the Tercentary of Spinoza,* November 23, 1932, New York, 1933, pp. 65–77. Harry Waton, *Marxism Reconsidered: Should We Go Beyond Marx?,* Brooklyn, 1940, p. 5.
76. Morris R. Cohen, "Spinoza: Prophet of Liberalism," *The New Republic,* Vol. 50, (March 30, 1927), pp. 164–166; reprinted in *The Faith of a Liberal,* New York, 1946, pp. 13–19.
77. Walter Lippmann, *A Preface to Morals,* New York, 1929, pp. 192, 34, 43, 220–221, 239.
78. George Santayana, "Enduring the Truth," *The Saturday Review of Literature,* Vol. 6, Dec. 7, 1929, p. 512.
79. *The Chief Works of Benedict de Spinoza,* tr. R.H.M. Elwes, Vol. I, *Tractatus Theological-Politicus, Tractatus Politicus,* London, 1883, p. 288.
80. *Ibid.,* pp. 287–288, 315, 334, 360.
81. *Correspondence,* p. 290. *The Chief Works of Benedict de Spinoza,* Vol. I.
82. E. Bonjour, H. S. Offler, G. R. Potter, *A Short History of Switzerland,* Oxford, 1952, pp. 193–195.

354   Varieties of Scientific Experience

83. *The Chief Works of Benedict de Spinoza,* Vol. I, p. 206.
84. *Ibid.,* Vol. I, p. 265.
85. *Ibid.,* p. 380.
86. George Berkeley, "An Essay Toward Preventing the Ruin of Great Britain" (1721), in *The Works of George Berkeley,* ed. Alexander Campbell Fraser, Oxford, 1901, Vol. IV, pp. 321–338.
87. *The Chief Works of Benedict de Spinoza,* Vol. I, p. 381.
88. Descartes wrote from Amsterdam on May 5, 1631, to his friend Balzac that in the city of merchants he found perfect solitude and comfort: "en cette ville où je suis, n'y ayant aucun homme, excepté moi, qui n'exerce la marchandise, chacun y est tellement attentif à son profit, que j'y pourrais demeurer toute ma vic sans être jamais vu de personne." The alienation from the cities would have been an incomprehensible fashion to Descartes. "The noise of the traffic interrupts my meditations no more than would the flow of the river." He is filled with pleasure "à voir venir des vaisseaux, qui nous apportent abondamment tout ce qui produisent les Indes, et tout ce qu'il y a de rare en Europe." "Can you tell me another country where so complete a liberty can be enjoyed; where one can sleep more quietly; where there are soldiers ready to guard us; where poisoning, treason, and calumnies are less known; and where more of the innocence of our ancestors remains to us?" Spinoza shared the same enthusiasm for the merchants' democracy. Descartes, *Correspondence,* ed. Ch. Adam and G. Milhaud, Paris, 1936, Tome 1, 189–191. Elizabeth S. Haldane, *Descartes: His Life and Time,* London, 1905, pp. 115–116.
89. Lewis S. Feuer, *Einstein and the Generations of Science,* New York, 1974, pp. 78–80. Leopold Infeld, *Quest: The Evolution of a Scientist,* New York, 1941, p. 267. Also cf. Albert Einstein, *The World as I See It,* tr. Alan Harris, New York, 1934, pp. 264–267. Einstein wrote appreciatively of Spinoza's determinism on the occasion of the 300th anniversary of Spinoza's birth: "Spinoza ist der Erste gewesen, der den Gedanken der deterministischen Gebundenheit allen Geschehens wirklich konsequent auf das menschliche Denken, Fühlen und Handeln angewendet hat. Nach meiner ansicht hat sich sein Standpunkt unter den um Klarheit und Folgerichtigkeit Kämpfenden nur darum nicht allgemein durchsetzen können, weil hierzu nicht nur Konsequenz des Denkens, sondern auch cine ungewöhnliche Lauterkeit, Seelengrösse und—Bescheidenheit gehört." Siegfried Hessing, ed., *Spinoza: Dreihundert Jahre, Ewigkelt, Spinoza Festschrift,* 1632–1932, Den Haag, 1962, p. 196. Also, Albert Einstein, "Tercentenary Message," in Spinoza Institute of America, *Baruch Spinoza* New York, 1933, pp. 28–29.
90. Spinoza, *Ethics,* Book I, Appendix. Wild, ed., op. cit., p. 139.
91. Spinoza, *Ethics,* Book II, Prop. XII., Scholium 3.
92. Henri Bergson, *An Introduction to Metaphysics,* tr. T. E. Hulme, 2nd ed., New York, 1955. Henri Bergson, *Creative Evolution,* tr. Arthur Mitchell, New York, 1911, pp. 238–239. S. Zae, "Les Thèmes Spinozistes dans la Philosophie de Bergson," in *Les Etudes Bergsoniennes,* Vol. VIII, Paris, 1968, pp. 140–147. Cf. Henri Chevalier, *Henri Bergsen,* tr. Lilian R. Clare, New York, 1928, p. 320.
93. Spinoza, *Ethics,* Book I, Appendix, Book III, Prop. H. Scholium.
94. Samuel Alexander, *Space and Time,* London, 1921, p. 79. Samuel Alexander, *Space, Time and Deity,* London, 1920, New Impression, 1927, Vol. I, pp. xiii-xv, Vol. II, pp. 45, 345. Also cf. Samuel Alexander, *Philosophical and Literary Pieces,* ed. John Laird, London, 1939, pp. 62, 67, 385.
95. Samuel Alexander, *Space, Time and Deity,* Vol. II, p. 353.
96. Spinoza, *Ethics,* Part III, Proposition LIX, The Affects, Wild, ed., op. cit., p. 266.

15

John Stuart Mill as a Sociologist: The Unwritten Ethology

In 1843, at the height of his intellectual powers, and with his *System of Logic* published and recognized at once as an intellectual landmark, John Stuart Mill prepared for his next book. Virtually announced at the end of his *Logic,* his aim now was to be to establish the foundations of sociology. Ethology, the science of 'the laws of human character,' was to have been the core; then he proposed to set forth the laws both of social statics and dynamics. For a 'considerable time' Mill tried to write this book. But, as Alexander Bain, his friend, tells us, he 'despaired, for the present time at least' of bringing such a work to fruition. Thereupon Mill turned in the autumn of 1845 to composing instead a volume on *Political Economy*.[1]

Among the great social thinkers of the nineteenth century, Mill was the only one who failed to write a system encompassing the evolution of humanity. Hegel, Comte, Marx, and Spencer felt they could enunciate and derive the law of social progress. Mill too would have wished greatly to prove that an empirical law of progress followed from the basic laws of mind. But Mill, author of 'On the Logic of the Moral Sciences,' the most enduring essay on the method of the social sciences which has ever been written, was aware that their simple 'derivations' collapsed under scientific scrutiny.

Comte, Marx, and Spencer could enunciate laws of historical development because their perception was pre-selected by their categorial schemes; they saw a reality, censored though ideological prisms, which arbitrarily excluded a whole set of possible developmental sequences consistent with observable facts. Comte ruled out the likelihood that religious revivals might occur, a possibility which Tocqueville had documented, and he altogether vetoed the notion that anti-civilizational waves might reinstate astrology and fetishism in people's minds. Marx excluded the possible advent of technocratic, totalitarian societies, characterized by managerial rule rather than by the workers' self-administration. Spencer set aside the possibility that the militant motive in men might manifest itself with a renewed intensity to engulf industrial societies; nor did his law of the differentiated progress of societies allow for the possibility of their decline. Mill alone tried to do justice to all the competing drives and motives of human nature; he would never banish from his consciousness the knowledge of the many-sidedness and many-levelledness of social reality. With his immense learning, practical experience, and logical acumen, Mill was more qualified to write the masterpiece of sociology than any other man in the nineteenth century. To understand why he failed in this design will perhaps bring to light truths of social existence that only great failures make explicit. What intellectual problems arose to make it impossible for Mill to compose his sociological treatise?

Mill in his *Logic* had explicated the character of the social sciences in a manner which has basically withstood all criticism. The inverse deductive method, as he called it, distinguished between the three levels of social analysis. Underlying all social truths there were first the elementary, fundamental laws of mind, the laws of psychology, known to us through introspection and empathetic understanding. These provided the basic premises for sociology considered as a deductive science. The sociologist as a deductive theorist knew his major premises and indeed his empirical conclusions; he then sought the intervening minor premises which were still unknowns. The conclusions to be derived were the empirical laws of sociology, confirmed in statistical studies and surveys—such empirical laws, for instance, as those concerning the frequencies and variations of suicide which Quetelet was investigating in Mill's time.[2] Between the major premises of psychology and the empirical laws of sociology was the domain of the middle principles—the *axiomata media,* the laws of ethology. These were to constitute the science of the formation of national character, or social character as it would be called today. To every system or structure of social institutions, Mill affirmed, there was a corresponding formation of social character. The social institutions were the social initial conditions under which the universal laws of human psychology operated. And they gave rise to the laws of ethology, or social psychology—those forms of human feeling, thought, and behaviour which social circumstances educed from the underlying psychological nature of man. The observed empirical laws would then in turn be derivable from the laws of mind and the ethological premises.

Mill's conception of sociology is today part of its common sense. For instance, it provides the framework for the study of suicide, an example in which Mill was interested. Persons of Calvinist background usually show higher rates of suicide than Roman Catholics; again, for many years, the suicide rates among Negro men were about one-third those of white men in corresponding age groups. To explain these empirical uniformities of suicide, we would avail ourselves in Millite fashion of such psychological laws as we may possess. We might use the psychological law that where frustrations persist or increase, the aggressive energies accumulate; and that where the latter cannot be directed toward causative external objects, they are redirected inward against one's self. As sociologists, we should seek the ethological middle laws which would take us to the empirical uniformities. We might use the ethological law that the Calvinist upbringing made for a more rigorous, severe, unbending conscience, that its reproaches, moreover, were alleviated by no social servomechanism, and that the resultant guilt feelings were more intense. We might note that because the Negro family had for many years a matriarchal pattern, with the fathers of the children unknown or transient, the character-structure of the sons, therefore, was such in which the father's commands had a weaker part. And from these ethological laws we could derive the empirical uniformities of suicide.

As a model of what a science of ethology could do, Mill had before him his father's *History of British India,* a book which he had read in manuscript and which, as he said, guided his thoughts by its analysis of Hindu society and civilization; the son regarded it as perhaps the most instructive history ever written.[3] James Mill had traced the causes of the Hindu national character to their political institutions. His character-sketch of them was scarcely flattering: 'No other race of men are perhaps so little friendly and beneficent to one another as the Hindus.' Their 'listless apathy' was not the outcome of their climate; other nations such as the Chinese had lived under as warm a sun but were 'neither indolent, nor weak.' If the Hindus disliked work, it was for one basic reason—their subjection to a wretched government, under which the fruits of their labour were never secure. Other Britons might find profundities in Hindu religious philosophy, but to James Mill their conception of nature was 'the most grovelling and base' and their writings replete with 'a more gross and disgusting picture of the universe' than any other people could adduce.[4]

There is a simplicity and comprehensiveness in John Stuart Mill's conception of sociology. It is utterly free of the exaggerations of later methodological schools which, fastening on one of Mill's three levels of analysis, have declared it to be the all-exclusive sociological one. Thus, Durkheim's school maintained that no psychological components should enter into a sociological explanation, though its own practice contravened its theory; on the other side, empirical surveyors have wished to pursue their inquiries without regard to underlying causal laws; while phenomenologists have argued that only the inner psychological processes, bracketed from the external world, were the social reality. Mill stood above such academic ideologies.

It was the very comprehensiveness of Mill's inverse deductive method, however, that made it impossible for him to bind together the contrary empirical laws he wished to affirm. To begin with, there was the flat contradiction in Mill between the manifest nineteenth-century optimist and the underlying pessimist. As an optimist, he accepted Auguste Comte's law of the three stages—the evolution from the religious to the metaphysical to the positive stage—as a valid empirical generalization; Comte's 'main conclusions,' he wrote, were sound, and the chain of causation Comte outlined as 'in all essentials irrefragable'; this intellectual movement, Mill said further, was 'at the root of all the great changes in human affairs.'[5] But then there was the pessimist Mill who discerned that the law of the future would be the dominance of mediocrity: 'the general tendency of things throughout the world is to render mediocrity the ascendant power among mankind.' '[I]n the world at large,' wrote Mill, there was 'an increasing inclination to stretch unduly the powers of society over the individual . . .' It was the sociological analogue to the great generalization which William Thomson, later Lord Kelvin, had enunciated at almost the same time on the universal tendency to the dissipation of energy. In not dissimilar words, Mill affirmed: 'the tendency of all the changes taking place in the world is to strengthen society, and diminish the power of the individual . . .' The creative energies of men would become increasingly unavailable.

Most unfortunately, according to Mill, no social class or stratum was exempt from the tendency toward mediocrity. The middle classes, the masses, and the reforming intellectuals were all alike mediocrats and intolerant. In England, said Mill, it was 'chiefly the middle class' which imposed the stamp of its 'collective mediocrity' on social existence;[6] in democratic America, it was the 'whole white population' expressing itself through the force of public opinion;[7] and as for the intellectuals, 'almost all the projects of social reformers of these days are really *liberticide.*'[8] From Saint-Simon to Auguste Comte, their aim had been 'dictatorship,' and with Comte it seemed indeed that the crypto-despot in every revolutionizing intellectual emerged explicit: in his scheme for 'the absolute and undivided control of a single Pontiff for the whole human race—one is appalled at the picture of entire subjugation and slavery . . .'[9]

Here then were two empirical laws which stood as contraries to each other—progress and mediocritization. The contraries in his ethological laws also tore apart the psychological basis which Mill provided for his sociology. In his *Logic,* following in his father's footsteps, Mill asserted that the laws of association were a sufficient foundation for the explanation and derivation of sociological laws. Yet, it became clear to Mill, a far broader conception of human drives was required. He noted in his essay 'Nature' (begun in 1854) that there was 'an instinct for domination' in men, 'a delight in exercising despotism, in holding other beings in subjection to our will'; it was linked to an 'instinct for destructiveness,' 'an instinct to destroy for destruction's sake'; men were 'naturally cruel.'[10] Comte too had observed that there were two 'very powerful instincts,' 'a downright taste for destruction' and a repugnance toward labour, which impelled men toward military rather than industrial societies.[11] If so, however, civilization rested on a precarious 'victory over instinct' through self-discipline. There were 'bad instincts' in men which, said Mill, 'it should be the aim of education not simply to regulate, but to extirpate . . .'[12] But if so, on what psychological ground could an empirical law of progress safely repose? The laws of mind, as Mill set them forth, were consistent not only with mediocritization, but indeed with a decline of civilization. What combination of psychological axioms with middle principles would underwrite the empirical law of progress?

Now Karl Marx, confronted by essentially the same problem, could avail himself of the salvaging motor force of the dialectic. According to Marx and Engels, greed and the lust for power could themselves be enlisted to transform a system beset with 'contradictions' into a higher one. But this conception was explicitly repudiated by Mill, most clearly so in his essay on Guizot's theory of history. Guizot, Marx's forerunner as a historian of class struggles, had affirmed that feudal society, 'by its own nature and tendencies,' evolved toward its dissolution. Mill, however, saw no such dialectic trasmutation of evil to a higher good. 'That is an easy solution,' he wrote, 'which accounts for the destruction of institutions from their own defects; but experience proves that forms of government and social arrangements do not fall merely because they deserve to fall. The more backward and the more degraded any form of society is, the stronger is the tendency to remain stagnating in that state, simply because it is an existing state.'[13] Existing societies, far from being 'rational' or 'functional,' were, from Mill's standpoint, as likely or likelier to be irrational and otiose. Then, how then did the feudal society evolve into a free commercial and industrial one? Progress, according to Mill, took place not because of any dialectical breakdown, but rather the spirit of liberty, the aspiration toward improvement, had found within the feudal order a sufficient support. Given the 'imputed causes of the fall of feudalism, the question recurs,' wrote Mill, 'what caused the causes themselves? ... There can be but one answer; the feudal system with all its deficiencies, was sufficiently a government, contained within itself a sufficient mixture of authority and liberty ... to enable the natural causes of social improvement to resume their course.' The feudal age, in Mill's view, had been wrongly 'vilified,' for 'at no period of history was human intellect more active, or society more unmistakably in a state of rapid advance' than during a great part of it.

Only once did Mill in an ethical fervour allow himself to endorse the notion that an evil institution must perish of economic necessity. That was during the American Civil War when Mill argued in 1862 that the confinement of slavery to the Southern states would mean its 'death-warrant,' its 'nearly inevitable and probably rapid' extinction.[14]

Underlying all progressive change, in Mill's view, was simply a persisting moral aspiration in men which could never be stifled but rather endured through all the 'compressions' of human character, and then availed itself of the rare social circumstances which enabled humanity to resume its linear advance. 'All political revolutions, not effected by foreign conquest, originate in moral revolutions,' wrote

Mill.[15] Revolutions of progress were in his view the consequence of an uprising of the spirit of liberty and improvement against the 'yoke of authority." Mill thus attributed the rise of capitalist society, or in his terms, the rise of 'the principle of accumulation," to conditions which allowed 'the growth of mental activity, making the people alive to new objects of desire.' Under such conditions of a better government and more complete security, foreign arts were welcomed; 'by instilling new ideas and breaking the chains of habit, if not by improving the actual condition of the population, [it] tends to create in them new wants, increased ambition, and greater thought for the future.'[16] If Mill was ranged against any Marxist dialectical conception, he would also have rejected the involuted dialectic of Max Weber wherein Calvinist asceticism somehow gave rise to its precise opposite, the development of new industries and new wants. According to Mill, Calvinism constituted a 'narrow theory of life' making for a 'pinched and hidebound type of human character,' for people 'cramped and dwarfed,' crushing the individual and his will through self-denial, and refusing to conceive of God as a Being who takes delight in every increase of human 'capabilities of comprehension, of action, or of enjoyment.'[17] As such, it would be inimical to the free development of the sciences and technology essential to the rise of capitalism. Calvinist doctrine too easily afforded a justification for what Mill called 'an equal chance to everybody of tyrannizing,' a desire, he said, as 'fully natural to mankind' as the desire not to be tyrannized over.[18]

Nonetheless, a sociological mystery still persisted as to the circumstances in which the spirit of liberty and advancement would prevail over the drives toward enslavement and retrogression.

Mill at one point tried to found an empirical law of progress on two universal human motives—the pursuit of truth and 'the desire of increased material comforts.' The latter, the hedonistic ingredient, was, he wrote, 'the impelling force' to most improvements. But, he went on to observe, the 'progress of industry must follow, and depend on, the progress of knowledge.'[19] Every advance in material civilization has been preceded, wrote Mill, by an advance of knowledge; changes in the mode of thought, in the Comtist pattern, have set the stage for these advances, but these changes in the mode of thought have themselves not arisen from the requirements of practical life but solely from the inner tendency of the previous system of beliefs to evolve. Once again, however, Mill's sociological theory was in straits. For what immanent law prescribed that a system of beliefs had to evolve? If it was dominated by myths, why could not a society stagnate in the mythological mode even as it did in its economy? What gave power to the pursuit of truth so that it could triumph over the contrary will to illusion?

At this juncture Mill tended to shift the causal primacy in social evolution to the character of a people's political institutions. He asserted in his *Logic* as a basic principle 'the necessary correlation between the forms of government existing in any society and the contemporaneous state of civilization . . .'[20] The greatness of the Athenian achievement, he thus affirmed, was derived from their free social institutions.[21] By contrast, the impoverished backwardness of many fertile tracts of Asia received its 'acknowledged explanation' in the tyrannical insecurity of rapacious governments, whose agents could deprive one arbitrarily of the fruits of one's labour. Why did the Roman empire decline? Mill felt that Finlay had explained this phenomenon better than had Gibbon,[22] for Finlay traced the decrease in the Italian population to evils inherent in the political system of the Roman government, its public distribution of grain, its arbitrary mode of taxation.[23] But the relation between social institutions and a people's mode of thought and feelings was also asserted by Mill to be circular, interdependent, and interactive, with neither variable ontologically independent: 'The creed and laws of a people act powerfully upon their economic conditions; and this again by its influence on their mental development and social relations reacts upon their creed and laws.'[24]

Mill, however, could scarcely be satisfied with a theory of the multiple causal interdependence of social institutions and modes of thought. It was adequate for what he (following Comte) called static rather than dynamic situations. Thus an equilibrium was defined by the uniformities between a society's different elements of coexistence; the society's institutions would all be values to interdependent mathematical functions, and the static state of affairs would be the counterpart of mathematical conditions of equilibrium. Yet these static mutual correlations themselves arose out of dynamic processes; they were the terminal points of equilibrium of processes in which one variable might well indeed be both primary and independent. The aspirations for liberty, truth, and improvement seemed never to be confined to particular static forms consistent with their coexistent society; the dynamic variables always had a degree of freedom which resisted their simply being assigned the values appropriate to the existing institutions.

Mill at this point verges on a complete declaration for sociological voluntarism as against sociological determinism. For years he had struggled with the problem of determinism. It weighed on him not only logically, but psychologically, part of that 'nightmare' which (in Thomas Henry Huxley's expression) haunted British thinkers of that era. During the 'mental crisis' of his early manhood, it took the form, as Mill described it, of his being 'seriously tormented by the thought of the exhaustibility of musical combination.' He compared 'this source of anxiety' to 'that of the philosophers of Laputa, who feared lest the sun should be burnt out.' Then in 'later returns of my dejection,' as Mill wrote, 'the doctrine of what is called Philosophical Necessity weighed on my existence like an incubus. I felt as if I was scientifically proved to be the helpless slave of antecedent circumstances . . .'[25] He struggled to remove this incubus all his life. He drew the distinction in later years between two kinds of fatalism, the Asiatic and the modified, which he contrasted with his own doctrine. 'Real Fatalism is of two kinds. Pure, or Asiatic fatalism—the fatalism of the Oedipus—holds that our actions do not depend upon our desires.' In the case of modified fatalism, 'our actions are determined by our will,' wrote Mill, 'our will by our desires,' and the last are determined by our motives and character; our character, however, is supposed to have 'been made for us and not by us, we are not responsible for it . . .' The true doctrine of causation, on the contrary, said Mill, affirmed that our character is 'in part amenable to our will . . .' Yet it scarcely seemed that Mill had escaped the fatalism of the Oedipus. All the varieties of fatalism and his own causation as well reduced to an Oedipal determinism. For our decisions and efforts to improve our characters were all in principle predictable; the behaviour of Mill and the modified fatalist were as predictable as that of the Oedipal subject in whose case a superior power intervened as an added variable. This common predictability pervaded all of Mill's thought with something akin to an Oedipal determinism.[26] His sociological theory tried to make real the power of mankind to choose and be unbound by universal laws. But he could never define a sense of freedom which would liberate him from Philosophical Necessitarianism. Perhaps his choice of the Oedipal metaphor to convey the sense of the extreme of fatalism reveals something of the emotional source of the hold of determinism upon Mill.

It was in his *Political Economy* above all that Mill explicated what he thought was the scientific basis for social choice; he drew the distinction between the laws of production with their necessitarian character and the laws of distribution which were voluntarist. Mankind, rescued from sociological fatalism, was acknowledged to be able to choose the kind of society it wanted:

> The laws and conditions of the production of wealth, partake of the character of physical truths. There is nothing optional or arbitrary in them ...
> It is not so with the Distribution of Wealth. That is a matter of human institution solely. The things once there, mankind, individually or collectively, can do with them as they like ... The rules by which it is determined, are what the opinions and feelings of the ruling portion of the community make them, and are very different in different ages and countries; and might be still more different, if mankind so chose.[27]

Thus, the law of diminishing returns was, according to Mill, essentially a law of chemistry and physics stated with reference to agricultural technology; the law of distribution, on the other hand, bore the stamp of men's varying choices. Mill drew on his own experience and knowledge of Indian affairs to illustrate how human choices could be made among diverse possible social systems. The British authorities had introduced different social systems in India; sometimes they displaced an oligarchy of usurpers and collected the taxes directly; in other cases they decided to create landed aristocracies; and in still others they co-operated with the representatives of village communities to arrest social change. Thus, human choice had been efficacious in deciding among the alternatives to the existent system of land ownership.[28]

If choices were genuine, mankind might then choose to progress rather than retrogress or stagnate. The incubus of sociological necessitarianism would be lifted. Yet, choice remained a kind of surd in Mill's sociological theory. For a sociology of choices always was at hand in his own terms for subsuming them under causal laws. Mill strongly rejected the doctrine, akin to the Marxian, that 'the forces ... on which the greater political phenomena depend, are not amenable to the direction of politicians or philosophers.' According to this doctrine of sociological necessitarianism, 'the government of a country, it is affirmed, is in all substantial respects, fixed and determined beforehand by the state of a country in regard to the distribution of the elements of social power.'[29] Choice, according to such a doctrine, was the experience of a social epiphenomenon; the deliberations of philosophers never liberated the so-called choices from their determinants of social power. 'Whatever is the strongest power in society will obtain the governing authority . . . A nation, therefore, cannot choose its form of government.' And James Mill had long previously analyzed the situation of a country divided into a ruling class and a subject class as one in which the members of the former had sympathies almost exclusively for themselves.

To this doctrine of economic determinism, Mill replied that purely ethical convictions, contravening the material interests of economic and social power, did intervene at critical junctures to transcend the latter: 'It was not by any changes in the distribution of material interest, but by the spread of moral convictions that negro slavery has been put an end to in the British Empire and elsewhere.' The emancipation of the Russian serfs was, he felt, another decision which transcended material interests. Then wrote Mill: 'It is what men think, that determines how they act . . .'[30] Even on his own showing, however, it was very rare that men's thoughts contravened their economic interests. For as he affirmed in *On Liberty:* 'Wherever there is an ascendant class, a large portion of the morality of the country emanates from its class interests . . .'[31]

And, indeed, was that segment of men's ideas which possibly transcended material interests free from causal determination? Were acts which transcended selfishness likewise choices in the sense of transcending causal laws? There were moments in the world's history, as in the February Revolution of 1848 in France, when it seemed to Mill that there appeared 'that almost unheard-of-phenomenon—unselfish politicians'; decisions then seemed to become choices rather than the resultants of polygons of social forces.[32] One might argue that the causal processes of individual psychology would account for these materially transcending actions. But there was no causal account at hand to explain how these idealistic motives, so powerless ordinarily, could have persisted in human history to shape its outlines and demarcate its future development. The instinct of domination was so much more forceful; hatred, selfishness, and brutality were so omnipresent that one asked: How had this puny vector of aspiration to truth and fellow-feeling survived in this welter of barbarian forces? That civilization had risen as far as it had against the evil inscribed in man's animal nature seemed a cosmo-historical fact of such improbable proportions that its actual occurrence defied the categories of sociological understanding. Thus it was that Mill was driven toward a sociological theology.

Mill's last essay, on theism, 'dismayed his disciples,' wrote John Morley.[32a] It was a natural conclusion, however, to his lifetime of sociological reflections. Mill had survived troughs of despair, such times as when he found human beings so abhorrent that he speculated with a certain pleasure upon the conditions under which there might take place a 'universal & simultaneous suicide of the whole human race.'[33] He had reflected on how beautiful the English environment would be but for its people: 'The nuisance of England is the English".[34] Nonetheless, this human race seemed somehow to have surmounted partially and periodically the trammels of its heredity and circumstances. Man had fashioned for himself, wrote Mill, 'a second nature, far better and more unselfish than he was created with.'[35] The virtues—courage, cleanliness, truth-telling—were all conquests of instinct; man evolved not through conforming to nature, but from his resolve to amend it, to challenge the maleficient powers. Whence, however, did he derive the resolve to challenge his own given character and status in animal existence? There was the feeling, Mill wrote, that in such an effort 'we may be co-operating with the unseen Being to whom we owe all that is enjoyable in life.' This God was a Limited God, not omnipotent. To his closest disciples Mill seemed suddenly and inexplicably to have subscribed to a Manichaean theology. Yet Mill found himself drawn to such postulates by processes of thought not dissimilar to those which have moved such scientists as Einstein, Bohr, Heisenberg, Russell. Einstein discerned the counterpart of Spinoza's Substance in the all-embracing simple laws of nature; Niels Bohr was moved in his conception of physical quanta by a vision of Kierkegaardian leaps; Heisenberg and Max Born sought to realize conceptions of free will; while Russell hoped that a Kropotkinite anarchism prevailed at least among the world of logical atoms. In a similar sense, Mill found himself affirming a meta-sociological postulate: 'a battle is constantly going on, in which the humblest human creature is not incapable of taking some part, between the powers of good and those of evil, and in which every even the smallest help to the right side has its value in promoting the very slow and often almost insensible progress by which good is gradually gaining ground from evil . . . '

This postulate in Mill's view was 'the most animating and invigorating thought which can inspire a human creature,' and he allowed that it might be grounded in a hope which reached toward the supernatural.[36]

Thus Mill at the close of the utilitarian cycle moved toward views which his father James Mill had set aside seventy-five years earlier. John Stuart Mill regarded himself as 'one of the very few examples, in this country, of one who has not thrown off religious belief, but never had it . . .'[37] The father had 'after many struggles' painfully shed his Presbyterian creed, and affirmed that "concerning the origin of things nothing whatever can be known.' He had, however, kept open for consideration 'the Manichaean theory of a Good and an Evil Principle, struggling against each other for the government of the universe . . .'[38] And the son now felt that such an inarticulate premise was adumbrated in his own experience; the sociologist's inverse deductive method, the laws of mind, and the empirical sociological laws, required the intervening meta-sociological principle of a Limited God.

Mill's sociological Manichaeanism, moreover, was not a sudden aberration of old age. He had used its vocabulary and metaphors spontaneously at the outset of the American Civil War when he had pleaded that England should not for the sake of cotton render aid to the Confederacy and make 'Satan victorious'; the Southern secessionists, he wrote, were undertaking 'to do the devil's work.'[39] and, indeed, a Manichaean world-view has been a largely unspoken axiom of sociologists; only Mill had the courage to articulate it. Physicists such as Einstein might find in the conception of God a regulative principle leading to the discovery of simple, mathematical laws. Is this the way God would have done it? was the question Einstein always asked. Not so, however, with social reality. The sociologist asks as well: How would social reality have been contrived if Satan had had his share in designing it? If God provides a methodological regulative criterion in physical science, Satan provides a partial one in social science. Malthus seeing disaster latent in every happiness of man, Marx and Engels writing of history as a goddess demanding human sacrifices, Weber describing a rationalization of life which made people ever more disenchanted, Pareto seeing idealists pursuing illusions in an endless circulation of élites, all shared a common standpoint with Mill who brooded in 1848 that 'it is questionable if all the mechanical inventions yet made have lightened the days' toil of any human being.'[40]

Indeed, Mill's Manichaeanism led him to enunciate what was perhaps his most original sociological theorem—his theory of the stationary state. The evolution of society, according to Mill, has an upper limit: 'It must always have been seen, more or less distinctly, by political economists, that the increase of wealth is not boundless: that at the end of what they term the progressive state lies the stationary state, that all progress in wealth is but a postponement of this, and that each step in advance is an approach to it. We have now been led to recognise that this ultimate goal is at all times near enough to be fully in view . . . ' Gone was Condorcet's vision of the indefinite progress of man. Mill's was a law of sociological impotence (to use the term of the physicist Edmund Whittaker): 'This impossibility of ultimately avoiding the stationary state—this irresistible necessity that the stream of human industry should finally spread itself out into an apparently stagnant sea . . .'[41] In essence, Mill derived this thorem from the simple consideration that the practice and development of the industrial arts always involved the depletion of energy-resources. The stationary state was a corollary of the second law of thermodynamics applied to the closed system of the planet Earth. But where other economists and sociologists such as Marx paid homage to the everlasting development of the forces of production and industrial civilization, and would have regarded the stationary state as too remote to enter the sociological purview, Mill avowed himself frankly as not charmed by 'the trampling, crushing, elbowing, and treading on each other's heels, which form the existing type of social life.'[42] He wanted solitude and the preservation of natural beauty. To Marx who wrote of 'the idiocy of rural life' he counterposed the imbecility of urban life, and he spent his most joyous days in botanizing, in long tramps, which led to 'frequent notes and short papers' in the *Phytologist,* an obscure journal for botanical collectors.[43] The Limited God of Mill had His counterpart in the finitude of energy resources. If the human species must finally vanish, its one hope was at least to postpone that not distant event by the wise cultivation of a stationary state, stationary in capital, population, and the productive arts.[44]

Is Mill's theory of the stationary state a viable one? Mill felt that in the stationary state 'there would be as much scope as ever for all kinds of mental culture'; the art of living, the arts and sciences would indeed, he argued, improve even more 'when minds ceased to be engrossed by the art of getting on.'[45] What Mill failed to pursue, however, were the social consequences of a stationary state. For such a society would probably evoke the most intense generational conflicts; bereft of the sense of open frontiers and opportunities, the young would expend their energies of aggression and hatred even more unidirectionally upon the old. In pre-historic stationary societies, the middle-aged men evidently died by violence.[46] The Chinese stationary society, which Mill studied, was characterized by a severe discipline imposed on the young; the sciences and arts of living were virtually transfixed. A stationary society would be one without the experience of renaissances. The condition for healthy social existence is that it be revivified for the young with the breath of fresh new industries and material obstacles; otherwise aggressive energies might turn inward, toward self-destruction.

Moreover, every wave of human improvement has been founded on a contemporaneous expansion of capital. It is sociologically doubtful that the arts of life could progress without such a corresponding material progress. And with the depletion of energy resources, a society would be unable to maintain its stationary position without progress in the use of its industrial capital and available resources.

The author of *On Liberty* feared slackening and stagnation; the theorist of the stationary state, however, regarded it as the best compromise with Manichaean reality that the human race might hope to achieve. Here was still another reason why Mill was unable to write his sociology. Not the contradictions of Victorian England were laid bare in his thought but rather the eternal unresolvable oppositions within all human life.

Some might say that Mill's theory of the stationary state was a reflection of the death-instinct which periodically waxed strongly in him.[47] To which Mill could reply that he alone in the nineteenth century had not flinched from drawing the ultimate consequences of Thomas Malthus's mode of thought.[48] And perhaps the Limited God, if not himself possessed of a death-instinct, was finally incapable of sustaining the human race against the harsh odds of the material universe.

Withal, it is probably true that Mill's sociological pessimism, his pervasive sense of pending exhaustion, had its highly personal sources as well. His upbringing by his father had so constrained him that he wrote poignantly: 'Let any man call to mind what he himself felt on emerging from boyhood ... Was it not like the physical effect of taking off a heavy weight . . .? Did he not feel twice as much alive . . .?'[49] Above all, the strains of the two decades of his sexual repression with respect to Harriet Taylor had exacted their toll. To mitigate these strains, Mill cultivated a mocking attitude not unlike Harriet's toward those who made much of the strength of the sexual drive. He railed at those who said it was difficult to control the sexual appetite.[50] He declared that the possibility of progress itself depended on the reduction of sexuality, that no great improvement in human life could be looked for 'so long as the animal instinct of sex occupies the absurdly disproportionate place it does therein'; and he anticipated confidently that this passion would become with men, as it is already with a large number of women, completely under the control of reason—that is, as disciplined as he was in emulation of Harriet.[51] He persuaded the historian George Grote to delete from the Preface to his *History* the words 'feminine' and 'masculine' in the discussion of the aspects of the ancient Greek character. He showed an unusual sympathy toward the mediaeval institution of a celibate clergy.[52] Mill's *Autobiography* impressed Freud as 'so prudish or so ethereal that one could never gather from it that human beings consist of men and women, and that this distinction is the most significant one that exists.'[53]

Yet, periodically, resentment against the régime of sexual attenuation would break forth. He had ridiculed as a young man the search of the Saint-Simonians for 'la femme libre,' which had indeed sent them to a quest among the harems of Constantinople. But in his old age, his final tribute to the Saint-Simonians was to 'the boldness and freedom from prejudice with which they treated the subject of the family, the most important of any, and needing more fundamental alterations than remain to be made in any other great social institution, but on which scarcely any reformer has the courage to touch,'[54] He responded with dislike to the statue of the Venus de Medici, saying 'the expression of the face is complete old maidism.'[55] Above all, in his unconscious he argued, though not very successfully, with Harriet's high discipline. He wrote her an account of a memorable dream: 'I was seated at a table like a table d'hôte, with a woman at my left hand & a young man opposite—the young man said, quoting somebody for the saying, "there are two excellent & rare things to find in a woman, a sincere friend and a sincere Magdalen." I answered "the best would be to find both in one"—on which the woman said "no, that would be *too* vain"— whereupon I broke out "do you suppose when one speaks of what is good in itself, one must be thinking of one's own paltry self-interest? no, I spoke of what is abstractedly good and admirable." '[56] Mill finally recognized explicitly that his father James together with his fellow Associationists had obliterated sexuality and the biological basis of emotions from psychology. James Mill had dismissed the whole subject delicately with one sentence: 'The affection of the husband and wife is, in its origin, that of two persons of different sex, and need not be further analysed.'[57] By contrast, John wrote: 'there is evidently in all our emotions an animal part . . . which these philosophers have passed without any attempt at explanation.'[58] The phenomenon of 'intense bodily suffering,' the 'screams, groans, contortions, etc.,' the experience and manifestations of Fear, were systematically omitted by the Associationist.

Mill's own experience, borne too stoically, had left him imprinted with a permanent dislike for human society. One wonders finally whether Mill at the last would have interceded with any enthusiasm with his Limited God to spare the human race. He could scarcely engage in the kind of panoramic mythologizing and ideologizing to which Karl Marx could surrender himself; he lacked Marx's capacity to disregard doubts, and to rely on dialectical paste-paper to underwrite the transitions between the stages of society as well as to collapsing arguments.

There are signs that the father James Mill felt at times a restiveness with his own life and was rebellious with the austere Calvinist asceticism which still underlied his Benthamite associations of pleasure. He was strangely drawn to the personality of the South American revolutionary general, Francisco de Miranda. As the French Saint-Simonians later fed the exotic longings of his son, so General Miranda delighted the Calvinist calculating utilitarian with his effervescence and spontaneity. The general was famed for exploits which had likened him to Casanova; the lover of the admired Delphine de Custine, who had escaped the guillotine which annihilated her family, he had been characterized by the Emperor Napoleon as 'a Don Quixote except that he is quite sane.' Under Miranda's influence, James Mill felt himself fortified to reject his traditional religion. James worked together with the general on an article 'Emancipation of Spanish America,' published in 1809 in the *Edinburgh Review*. He rendered tribute to the general as one in whose 'breast the scheme of emancipation, if not first conceived, seems at least to have been first matured.'[59] Betrayed in the revolutionary war, Miranda died in 1816 in a Spanish dungeon. Meanwhile, James Mill had found in vicarious revolutionary participation an alleviation of the strains of a grey doctrine, a harsh Presbyterian upbringing, and an unhappy marriage. The son John repeated the pattern. The misfortune of James Mill's life was that he too had known a Harriet Taylor, but in his case their separation was inevitable. She was the daughter of the Scottish baron, Sir John Stuart, after whom James had named his son. James wrote about her as his son did of Harriet: 'besides being a beautiful woman, [she] was in point of intellect and disposition one of the most perfect human beings I have ever known.' They grew up and studied together from childhood on, 'and were about the best friends that either of us ever had.' When she was dying, she spoke of James 'with almost her last breath.'[60] James Mill could never forget that his wife was not the one he loved. His utilitarianism was as much the outcome of a personal quarrel with society as his son's *On Liberty*.

Mill did try periodically to alleviate the pessimistic tendencies in his character and his sociological theory by partaking in the enthusiasm and fellowship of socialist movements. Nevertheless, in the final consideration, he always refrained from any overt commitment to socialism, and at the time of his death was writing what even in its fragmentary form was the most powerful critique of socialism written in the nineteenth century. As a young man, Mill had long, eager discussions with the young French Saint-Simonian Jew, Gustave d'Eichthal, whom he admired almost as much as his father had venerated the Dutch-English Jew David Ricardo. He shared the revolutionary zeal of the Saint-Simonians, and almost became one of them; tactical considerations alone, he said, kept him from enlistment in their ranks. Meanwhile, he awaited the revolutionary hurricane:

> And if the hour were yet to come for England . . . I know not that I should not renounce everything and become, not one of you, but as you.
> But our 10 août, our 20 juin, and perhaps our 18 Brumaire, are yet to come. And which of us will be left standing when the hurricane has blown over, Heaven only knows.[61]

This period of high revolutionary zeal was also that of John's growing involvement with Mrs. Harriet Taylor, the period of his own most intense rebellion against the authority of his father James Mill. He wrote in 1831 to his close friend Sterling with the swathes of the revolutionary intellectual: 'until the whole of the existing institutions of society are levelled with the ground, there will be nothing for a wise man to do . . .' and he looked forward with pleasure to what later would be called the 'physical liquidation' of the English middle class and its intellectual advocates. It will be recalled that the father James was an unashamed admirer of the middle class.[62] The son, however, wrote that if only 'a few dozens of persons' were salvaged to be 'missionaries of the great truths' for posterity, he would not care though a revolution were to exterminate every person in Great Britain and Ireland who has £500 a year. Many very amiable persons would perish, but what is the world the better for such amiable persons?'[63] Mill's ideological, Messianic mood coincided with his peaks of aggressive resentment. But with his remarkable strength of character, he always regained his rationality and independence of judgment. He finally estimated the Saint-Simonians as having a 'narrow & bigoted understanding, & a sordid & contracted disposition,' and being so fixated on the notion that history was governed by a unidirectional progressive law that they forgot it could go backwards.[64] He welcomed the advent of the International Workingmen's Association in 1864 when Karl Marx, to please the British labour leaders, allowed himself to write like a Millite rather than a Marxist. But when the secretary of its Nottingham branch sent Mill its literature, he responded that, though he 'warmly approved' of its principles: 'What advantage is there in designating the doctrines of the Association by such a title as "the principles of the political and social Revolution?" '[65] Mill in his *Autobiography* classed himself and Harriet 'under the general designation of Socialists.' During the intervening years, moreover, he kept himself well-informed through reading the journals of the Association.[66] His misgivings concerning socialism deepened. He had always been troubled by what he saw as 'the social problem of the future'—how to preserve 'the greatest individual liberty' within the framework of a socialist society.[67] He now however observed those whom he called 'the revolutionary socialists,' and concluded that the motives which impelled them were malevolent. Those 'who play this game,' he wrote, 'unconfirmed as yet by any experimental verification . . . and [who] would brave the frightful bloodshed . . . must have a serene confidence in their own wisdom on the one hand and a recklessness of other people's sufferings on the other, which Robespierre and St. Just, hitherto the typical instances— . . . scarcely came up to.'

For Mill stated clearly what he took to be the tenets of 'the revolutionary socialists'; they were exactly those of Marx. They proposed on the economic side, wrote Mill, 'the management of the whole productive resources of the country by one central authority,' while on the political side 'their purpose' was 'that the working classes, or somebody in their behalf, should take possession of all the property of the country, and administer it for the general benefit.' Thus, the three elements of the revolutionary socialist programme were total central planning, the alleged ownership of the national property by the working class, and its achievement by a revolutionary act. Their avowed aim was 'to substitute the new rule for the old at a single stroke, and to exchange the amount of good realized under the present system, and its large possibilities of improvement, for a plunge without any preparation into the most extreme form.' The programme had 'great elements of popularity,' Mill noted, not shared by 'the more cautious and reasonable form of Socialism.' For it made its appeal above all to the emotions of hatred. This, then, was the ethological middle principle of revolutionary socialism: 'for if appearances can be trusted, the animating principle of too many of the revolutionary Socialists is hate.' One might undertake to excuse their absorption in hatred by invoking their dialectical belief, in Mill's words, 'that out of chaos would arise a better Kosmos.' But this entirely overlooked, wrote Mill, 'that chaos is the very most unfavourable position for setting out in the construction of a Kosmos, and that many ages of conflict, violence, and tyrannical oppression of the weak by the strong must intervene . . .' In the character of the dialectical revolutionists Mill perceived a readiness to destroy the world if it could not be changed: 'its apostles could have only the consolation that the order of society as it exists now would have perished first, and all who benefit by it would be involved in the common ruin . . .'[68]

The leaders of the English workingmen alone, Mill observed with some pride, refused to be drawn into the destructivism of the European revolutionary socialists. 'The leaders of the English workingmen—whose delegates at the congresses of Geneva and Bâle contributed much the greatest part of such practical common sense as was shown there—are not likely to begin deliberately by anarchy . . .'[69] The congresses of the International Workingmen's Association in September 1866 and 1869 had been the scene of a struggle between Marx's faction and the followers of Proudhon. Marx's initial programme for the International, adopted at Geneva, was rather moderate, placing the empasis on achieving the eight-hour day and a system of public education; it stressed, however, the importance of trade unions as agencies for 'complete emancipation' whereas the Proudhonists advocated cooperative associations and denigrated the recourse to strikes. At Basel, however, the legendary figure of Mikhail Bakunin emerged to lead the most formidable challenge to Marx's 'centralized Communism' on behalf of anarchy; Bakunin indeed won majorities against Marx in the actual voting on resolutions. The English trade-union leaders supported Marx's proposals as against Bakuninist anarchy, yet they too were unhappy with the underlying note of Marx's 'centralized Communism.' Mill stood with such trade unionists as Odger, Cremer, and Applegarth, against both Marx and Bakunin; the English trade unionists soon dissociated themselves from the International. Marx at the last Congress of the International in 1872 insulted all the English labour leaders by saying publicly that 'almost every recognised leader of English working men was sold to Gladstone, Morley, Dilke and others.'[70] The English labourites never forgave him for that slur. The enfranchisement of 1867 had opened up to the British working class the prospect of using Parliament to achieve social reforms; thus seeds which were being planted for the Labour party were far more Millite than Marxist.[70a] Mill's last chapters on socialism must be regarded as essentially an evaluation founded on the added experience of the English trade unionists with the International. He knew that they had been at odds with all the ideologists, with Marx, Bakunin, and Proudhon. The two generous hopes of Mill in 1848 for the principle of association were now basically amended in the light of this new phenomenon of the revolutionary ideologists—this permeation of socialism with a ferocity and hatred which Mill had not previously foreseen.

No doubt Karl Marx was informed through their various common associates, the labour leaders and the positivist intellectuals, of Mill's opinion of 'revolutionary socialists.' Nothing else would explain Marx's outburst against John Stuart Mill in the closing days of the International, on 16 April 1872, months after Mill had protested the sanguinary suppression of the Paris Communards.[71] Mill still continued to place some hope in a moderate, experimental, piecemeal socialism which could 'be brought into operation progressively, and ... prove its capabilities by trial.' With sympathy toward communitarians (exemplified in later years by Martin Buber and William Morris), Mill looked to modest efforts 'on the scale of a village community.' Aware, no doubt, that experiment had often shown that such associations deteriorated into village tyrannies, Mill as a pluralist noted the aim for 'the multiplication of such self-acting units.'[72] Perhaps individual liberty might be safeguarded in a society of competing communal associations. Yet, in 1849 Mill had affirmed his belief that co-operative associations would not prove themselves as efficient as competitive societies.[73]

The notion of central planning itself, 'the very idea of conducting the whole industry of a country by direction from a single centre," was in Mill's view 'obviously chimerical.'[74] 'Communistic management' would be 'less favourable than private management to that striking out of new paths'; it would lack the venturesomeness 'generally indispensable to great improvements in the economic condition of mankind . . .'[75] It would of necessity decide 'in a more or less arbitrary manner' those questions which 'on the present system' for all its imperfections decide themselves spontaneously—that is, in predominant accord with rational criteria.[76] Thus in a few sentences Mill raised those problems concerning the irrationality of socialist economy which have haunted the halls of socialist proponents of later generations, and whose phantom still attends the sessions of the Soviet Planning Council. Curiously, the corollary latent in Mill's analysis was that for a communist society taken in isolation, the optimal equilibrium would be that of a stationary state. Mill indeed had written Harriet Taylor in 1849 that the objection to communism that it would make life 'a kind of dead level' could never be taken away.[77] Nevertheless, he always clung to elements of a socialistic ethic even though his analysis dissociated socialist society from any law of progress with which it had been joined so facilely by socialists.

Meanwhile, however, Mill finally rejected the notion that the misery of the working classes was increasing under capitalism. This was one of the 'Socialist exaggerations,' he wrote. 'The present system is not, as many Socialists believe, hurrying us into a state of general indigence and slavery from which only Socialism can save us. The evils and injustices suffered under the present system are great, but they are not increasing; on the contrary, the general tendency is towards their slow diminution.'[78] He still looked for guidance to the moral principles of socialism, but doubted whether in the trials of experiment the power of the 'higher' motives would be sufficient even in a large minority to sustain such a society.[79] It had become clear to

Mill that for the revolutionary socialists the experience of the revolutionary leap and the opportunities it afforded for the release of aggression was the primary purpose; they disliked an experimental approach toward socialist ventures precisely because it deprived them of the chance for revolutionary experience. In 1849 Mill had charged that the fear of socialism drove the bourgeoisie into an 'insane terror.'[80] At the end of his life he might have amended this statement to make central instead the fear not so much of socialism as of socialists, such as Blanqui, whose aim was the savouring of revolutionary dictatorship far more than the improvement of the human lot. What the sources of this compulsion for revolutionary experience might be was a problem that was left unsolved for a more developed science of ethology.

Mill was, it must be remembered, part of that corps of British sociologists whose ideas were largely shaped by their experience in the administration of India. Though Mill's work was confined to the London headquarters of the East India Company, he was dealing with the same problems that Sir Henry Maine and James Fitzjames Stephen met in India directly.[81] The administrative experience provided Mill's judgment with a ballast which Marx and Comte never had. He knew the inefficiencies and arbitrariness in the bureaucratic administration in a way the bookmen-revolutionists never did. With respect to problems where he lacked that ballast, Mill too tended to veer toward radical proposals otherwise foreign to him. He, for instance, like his friend John Cairnes, demanded in Abolitionist fashion that all Negro ex-slaves be given the right to vote at the end of the American Civil War; he would have repudiated such a demand for the people of India whom he regarded as still in their 'nonage.' In the former case, he was moved by pure ethics; in the latter, he weighted ethics with sociological experience. Ethical zeal, at times, could obliterate Mill's social judgment. For instance, he finally opposed the secret ballot because he felt that all men should be fearless in character, and not afraid to say openly what they thought. But the Cooperative leader, George Jacob Holyoake, thought that Mill had forgotten that a secret ballot gave the average man his only chance to be independent—that Mill expected the average citizen to have the character of a John Stuart Mill.[82]

What then is Mill's place in the development of sociological science? His enduring greatness is as the conscience of science. Virtually alone he sought steadfastly to keep his sociology free from ideology,. He articulated the problems which others repressed by a variety of devices. Where Marx and Engels projected a dialectical leap which would take human society from the kingdom of necessity to that of freedom, Mill, too honest to succumb to the allure of sociological metaphor, enquired into the threats which socialism would pose for individual liberty. While Weber built a whole theory of the origins of capitalism on a presumed psychological law that guilt-ridden people will embark more frequently upon technological innovation, Mill, with his analysis of the Calvinist character, would scarcely acquiesce to such an assumed law of mind. Where Durkheim as a Radical Republican exalted society and claimed that religion was society's worship of itself, Mill would observe that religious ideals provided a basis for the critique of every existing society; if religion was society's self-worship, then the tension between religion and society was unintelligible. Mill could never write a complete system of sociology because he alone would not confuse the subjective completeness of a narrowed mind with the objective completeness of sociological explanation.

That is why when we look for intellectual guidance today with respect to the problems of freedom in communist societies, the exhaustion of the environment, and the pressure of population, Mill is the only sociologist of the nineteenth century whose pages are not discoloured with the acid of fantasy. Others claiming to prefigure the law of history, were obsessed by the demon of making history; Mill held to the more modest ethic of acting as circumstances allowed on behalf of human happiness. In so doing, the 'saint of rationalism' held to a conception of scientific truth with an integrity which the prophet-ideologists never approached.

Notes

1. Alexander Bain, *John Stuart Mill: A Criticism With Personal Recollections* (London 1882), 78–9, 84.
2. Quetelet, it should be observed, created the science of social statistics during the early 1830s when he and the young Belgian intellectuals were, like young Mill in England, under the influence of Saint-Simonian ideas. See Lambert A.J. Quetelet, *A Treatise on Man and the Development of his Faculties,* facsimile of 1842 translation, introduction by Solomon Diamond (Gainesville 1969), vi, vii.
3. *Autobiography*, ed. J. Stillinger (Boston 1969), 16–17.
4. *The History of British India*, 5th ed. (1858; reprinted New York 1968), I, 466, 469, 480, 481, 371, 385, 347.
5. *Auguste Comte and Positivism, Collected Works*, x (Toronto 1969), 317, 319, 322–3.
6. *On Liberty*, 4th ed. (London 1869), 119.

7. Ibid., 118–19.
8. Letter to Harriet Mill, 15 Jan. 1855, cw, XIV, 294.
9. *Considerations on Representative Government*, 3rd ed. (London 1865), 38–40; *Auguste Comte and Positivism*, cw, x, 351.
10. 'Nature,' cw, x, 398.
11. Auguste Comte, *System of Positive Polity*, tr. E.S. Beesly et al. (London 1875–7), III, 47.
12. Cw, x, 393, 398.
13. 'Guizot's Essays and Lectures on History,' *Dissertations and Discussions: Political, Philosophical, and Historical*, II (London 1859), 268–70.
14. J.S. Mill, 'The Contest in America,' *Dissertations and Discussions*, III (London 1867), 191.
15. J.S. Mill, 'A Few Observations on the French Revolution,' *Dissertations and Discussions*, I (London 1859), 56.
16. *Principles of Political Economy*, cw, II, 186–7.
17. *On Liberty*, 112.
18. *Principles*, cw, III, 944.
19. *Logic*, cw, VIII, 926 (VI, x, 7).
20. Cw, VIII, 919 (VI, x 5)
21. *Dissertations and Discussions*, II, 286.
22. John Morley, *Recollections* (London 1917), I, 66.
23. 'The social organization of nations affects their vitality as much as their political constitution affects their power and fortunes.' George Finlay, *A History of Greece, from its conquest by the Romans to the present time*, B.C. 146 to A.D. 1864, new ed. (Oxford 1877), I, 89–90.
24. *Principles*, cw, II, 3.
25. *Autobiography*, 87, 101.
26. Freud in 1880 translated a volume of Mill which included his essays on the labour question, Harriet Taylor Mill's 'The Enfranchisement of Women.' Grote's *Plato*, and the posthumous writings on socialism. He did so at the request of Mill's Viennese friend and editor, Theodor Gomperz. Did Freud read the other translated volumes, and imbibe Mill's usage of the 'fatalism of the Oedipus'? In any case, Freud regarded Mill 'as perhaps the man of the century who best managed to free himself from the domination of customary prejudices.' (Ernest Jones, *The Life and Work of Sigmund Freud* [New York 1953], I, 55, 176) Apart from the essays on socialism, the contents of the volume Freud translated were drawn from Volume II of *Dissertations and Discussions*. Mill's views on the hatred-vector in revolutionary socialism clearly became the basis for Freud's analysis of communism. Compare Adelaide Weinberg, *Theodor Gomperz and John Stuart Mill* (Geneva 1963), 60.
27. *Principles*, cw, II, 199–200.
28. 'Mr. Maine on Village Communities,' *Fortnightly Review*, ns ix, 1871, 554–5.
29. *Representative Government*, 12. The action of the British people in abolishing the slave trade and in emancipating the slaves was, Mill wrote, 'a cause in which we not only had no interest, but which was contrary to our pecuniary interest...' (*Dissertations and Discussions*, III, 180).
30. J.S. Mill, *Representative Government*, 15. James Mill, *Analysis of the Phenomena of the Human Mind*, 2nd ed. (London 1869), II, 275n.
31. *On Liberty*, 16.
32. 'Vindication of the French Revolution of February 1848,' *Dissertations and Discussions*, II, 337.

380   Varieties of Scientific Experience

32a. John Viscount Morley, *Recollections*, Vol. I, New York, 1917, p. 106.
33. Letter to Harriet Mill, 30 Dec. 1854, cw, XIV, 272.
34. Letter to Harriet Mill, 2 Jan. 1855, cw, XIV, 277.
35. 'Theism,' cw, X, 459.
36. Ibid., 488–9.
37. *Autobiography*, 27–8. Bain, however, writes that 'John as a little boy, went to church; his maiden aunt remembered taking him, and hearing him say in his enthusiastic way "that the two greatest books were Homer and the Bible." ' (*James Mill*, 90).
38. *Autobiography*, 25–6.
39. 'The Contest in America,' *Dissertations and Discussions*, III, 181.
40. *Principles*, cw, III, 756.
41. Ibid., 752.
42. Ibid., 754.
43. Henry Trimen, 'His Botanical Studies,' H. Fox Bourne, ed., *John Stuart Mill— His Life and Works* (Boston 1873), 44. Mill's botanizing was throughout his life the avocation in which he found the most complete pleasure. He first became a plant-collector during his sojourn in Southern France, 1820–1, "the happiest six months of my youth." He would pursue it intensely day after day in arduous climbs and hikes whenever he could. When the "battle of the Barricades" was raging in 1832 during the crisis of the Reform bill, Mill was off botanizing quietly at Highgate. He told Herbert Spencer his "murderous propensities" were "confined to the vegetable world." (Anna Jean Mill, ed., *John Mill's Boyhood Visit to France* [Toronto 1960], 65).
44. *Principles*, cw, III, 752–7.
45. Ibid.,756.
46. See Carlo M. Cipolla, *The Economic History of World Population*, rev. ed. (Middlesex 1964), 74.
47. Leslie Stephen felt that Mill's theory of the stationary state was rather a temporary protest than a settled conviction. (*The English Utilitarians* [London 1900], III, 200).
48. When young Oliver Wendell Holmes, Jr., visited Mill in 1866, Mill took him to a meeting of the Political Economy Club where the subject for discussion was 'whether the financial policy of England should be governed by the prospective exhaustion of coal in 11 years as predicted by Jevons . . . ' (Mark DeWolfe Howe, *Justice Oliver Wendell Holmes: The Shaping Years* [Cambridge, Mass. 1957], 226–7).
49. *The Subjection of Women* (London 1869), 180–1.
50. Bain, *John Stuart Mill*, 149.
51. Letter to Lord Amberley, 2 Feb. 1870, cw, XVII, 1693; Diary entry, 26 March, in Hugh Elliot, *Letters of John Stuart Mill* (London 1910), II, 382.
52. Bain, John Stuart Mill, 78, 86. J.S. Mill, 'Michelet's History of France.' *Dissertations and Discussions*, II, 159.
53. Ernest Jones, *The Life and Work of Sigmund Freud* (New York 1953), I, 176.
54. Letter to Carlyle, 11 & 12 April 1833, cw, XII, 150; *Autobiography*, 101.
55. Letter to Harriet Mill, 8 June 1855, cw, XIV, 480.
56. Letter to Harriet Mill, 17 Feb. 1857, cw, XV, 523–4; *John Stuart Mill and Harriet Taylor*, 254.
57. James Mill, *Analysis*, II, 218–19.
58. 'Bain's Psychology,' *Dissertations and Discussions*, III, 132–3.
59. Bain, *James Mill*, 89. William Spence Robertson, *The Life of Miranda* (Chapel

Hill 1929), II, 48-50; I, 57. Joseph F. Thorning, *Miranda: World Citizen* (Gainesville 1952), 126. George F. Kennan, *The Marquis de Custine and his Russia in 1839* (Princeton 1971), 3–4.
60. Graham Wallas, *The Life of Francis Place: 1771–1854* (4th ed., London 1925), 70–1.
61. Letter to Gustave d'Eichthal, 30 Nov. 1831, cw, XII, 89.
62. Letter to John Sterling, 20–22 Oct. 1831, cw, XII, 78. The 'class which is universally described as both the most wise and the most virtuous part of the community, the middle rank ... which gives to science, to art, and to legislation itself their most distinguished ornaments ... ' (James Mill, *An Essay on Government* [reprinted, Cambridge 1937], 71–2).
63. Letter to Sterling, 20–2 Oct. 1831, cw, XII, 84.
64. Letter to Gustave d'Eichthal, 8 Oct. 1829, cw, XII, 37.
65. Edward S. Mason, *The Paris Commune: An Episode in the History of the Socialist Movement* (New York 1930), 50.
66. Letter to Georg Brandes, 4 March 1872, cw, XVII, 1874–5. A subscription on Mill's behalf to the journal *Progress* was made by Cowell Stepney, an eccentric blind millionaire who was a member of the International's General Council. (See James Guillaume, *L'Internationale: Documents et Souvenirs* [Paris 1905], I, 139.).
67. *Autobiography*, 138.
68. Chapters on Socialism, cw, V, 737, 749.
69. Ibid., 709. George Odger, the secretary of the London Trades Council, supported Mill increasingly while he denounced the intransigence of the General Council of the International, 'the most unfit persons I have ever come in contact with to represent the working classes.' George Howell, who in 1871 became secretary to the Parliamentary committee of the Trades Union Congress, advised workers to 'get Mill on Liberty and Political Economy. There are many other works, but go to the fountain head at once.' Marx fought Mill's influence in the labour movement by supporting the Land and Labour League against Mill's Land Tenure Reform Association. Mill in turn deplored 'the furious and declamatory virtue' of the former's resolutions. (Compare Royden Harrison, *Before the Socialists: Studies in Labour and Politics, 1861–1881* [London 1965], 234, 143, 223).

Mill always exerted his logical powers to dissuade the English labour leaders from being captivated by revolutionary rhetoric and violent postures. In March 1867 he wrote to William Randall Cremer in sharp disapproval of the calls for force which the Reform League was making to secure enactment of the bill for enfranchisement. Ultimate success in Britain, Mill said, could only be obtained 'by a succession of steps.' No justification for 'revolutionary expedients' existed in the British case (cw, XVI, 1247–8). He argued in September 1865 with George Howell, who placed an inordinate faith in the strike weapon; where strikes were successful in the generality of trades, said Mill, there would be a general rise in prices, which would be of 'no benefit to the labouring classes' (cw, XVI, 1102). But he subscribed in 1869 to the Labour Representation League on whose executive Howell was active (cw, XVII, 1673). He vigorously supported Odger in his efforts to win election to Parliament: 'No one has taken a warmer interest than I have in the candidatures of working men in general & Mr. Odger in particular, & I believe Mr. O. is well aware of this,' Mill wrote in 1870 (ibid., 1688). And when Odger lost again in February 1870 Mill still wrote him a letter of congratulation (ibid., 1697). He wrote in terms of admiration of the English labour leaders, and in December 1867 advised Thomas Hare, the exponent of proportional representation: 'if you could make a convert of even one such man as Odger, or Cremer, or Howell—the gain would be immense' (cw, XVI, 1342). He introduced to Odger his 'very old friend,' the former Saint-Simonian, Gustave d'Eichthal, when the latter was visiting Britain in May 1871 (cw, XVII, 1816). It was Odger who rose to commend Mill at the celebrated meeting when Mill, a candidate for Parliament, acknowledged that he had written that the English working class were 'generally liars.' When the 'vehement applause' subsided, Odger said that the working class 'wanted friends, not flatterers,' and were grateful to one who spoke sincerely of their faults (Mill, *Autobiography*, 168). To John Hales, the secretary of the General Council of the International, Mill wrote on 28 May 1871 saying that if any demonstration 'could arrest or mitigate the horrors now being perpetrated at Paris' there was hardly anything he would not do; but he saw no such hope (cw, XVII, 1821–2). He gave his support to preventing the dismissal from University College of Professor E.S. Beesly, the courageous Comtist professor of classics who in 1867 had spoken in defence of the trade unions (cw, XVI, 1297).

70. Henry Collins and Chimen Abramsky, *Karl Marx and the British Labour Movement: Years of the First International*, London, 1965, p. 260. Socialists and trade unionists generally admired the record of John Stuart Mill as a member of Parliament. William Morris said Mill had been 'a real success in Parliament.' (See James Mavor, *My Windows on the Street of the World* [London, 1922], I, 209.) J. George Eccarius, Marx's collaborator on the General Council, wrote: 'As a member of Parliament Mill conducted himself in exemplary fashion, and showed he had the courage to come forward in the interest of the working class to oppose both the aristocracy as well as the gelfackelbourgeoisie. His political behaviour is in contradiction with his economic philosophy.' (*Eines Arbeiters widerlegung der National-okonomischen Lehren John Stuart Mill's* [Berlin 1869], iv). On the reaction of the British labour leaders to Marx's insult, compare Boris Nicolaievsky and Otto Maenchen-Helfen, *Karl Marx: Man and Fighter*, tr. Gwenda David and Eric Mosbacher (Philadelphia 1936), 360. John Hales moved a vote of censure against Marx in the British Federal Council; it was adopted. An amendment charging Marx with political deception got an equally divided vote. (See Franz Mehring, *Karl Marx: The Story of His Life*, tr. Edward Fitzgerald [New York 1935], 518.)

70a. "The economic influence most potent among the Socialist radicals is still that of John Stuart Mill." Sidney Webb, *Socialism in England*, Sec. Ed., London, 1893, p. 185.

71. Marx ridiculed what he called 'John Stuart Mill's compendium of political economy,' and the proposal, for instance, that the state should lend capital to cooperative societies, a proposal made by Ferdinand Lassalle and in keeping with Mill's idea. (*The General Council of the First International, 1871–1872: Minutes* [Moscow nd], 160) Marx's ire was especially directed against Henry Fawcett, professor of Political Economy at Cambridge, and Mill's disciple. Fawcett, after studying the documents of the International and after 'frequent conversations with many of its members,' gave a series of lectures in 1872 against its principles. Though himself, like Mill, an advocate of co-operation on the Rochdale model, Fawcett argued that the effort to realize the programme of the International would make for a 'weakening and lessening of individual responsibility' and a decline of the 'industrial virtues.' (Henry Fawcett, 'The Nationalisation of

the Land,' *Fortnightly Review*, ns LXXII, [1872], 627, 637, 638. Leslie Stephen, *Life of Henry Fawcett*, 5th ed. [London 1886], 158–66, 470).
72. *Chapters on Socialism*, cw, V, 737.
73. 'The French Revolution of 1848 and its Assailants,' *Dissertations and Discussions*, II, 394.
74. *Chapters on Socialism*, cw, V, 748.
75. Ibid., 742.
76. Ibid., 743.
77. Letter to Harriet Taylor, 21 Feb. 1849, cw, XIV, 11. Mill's concern for Harriet Taylor's approbation led him to mute his criticism of communism. In revising his *Political Economy* in 1849 he had stated the objections to communism that he thought were valid, but he wrote in conciliatory fashion: 'if you do not think so, I certainly will not print it.' (Ibid.) Mill argued with her that she 'greatly' overrated 'the ease of making people unselfish,' and that even if they had absolute power, 'all our plans would fail from the impossibility of finding fit instruments.' (21 March 1849, ibid., 19). This passage in my opinion is not given sufficient weight by H.O. Pappé in his *John Stuart Mill and the Harriet Taylor Myth* (Melbourne 1960), 40. Not until Harriet was long dead was Mill able to bring all his critical acumen to bear on the socialist proposals.
78. *Chapters on Socialism*, cw, V, 736.
79. Ibid., 736, 741–2.
80. *Dissertations and Discussions*, II, 394.
81. For twenty-three years Mill wrote almost every 'political' despatch of the East India Company; two huge volumes, five or six inches thick, were each year written by Mill alone. (See W.T. Thornton, 'His Career in the India House,' *John Stuart Mill*, ed. Fox Bourne [Boston 1873], 32.).
82. George Jacob Holyoake, *Bygones Worth Remembering* (London 1905), I, 276.

16

The Sociobiological Theory of Jewish Intellectual Achievement: A Sociological Critique

I

The noted mathematician Norbert Wiener, writing his autobiography, meditated upon the significance of the Jewish family for the Jewish contribution to scholarship and science. He set down briefly what we might call the sociobiological theory of the Jewish intellectual endowment. Wiener wrote:

> At all times, the young learned man, and especially the rabbi, whether or not he had an ounce of practical judgment and was able to make a good career for himself in life, was always a match for the daughter of the rich merchant. Biologically this led to a situation in sharp contrast to that of the Christians of earlier times. The Western Christian learned man was absorbed in the church, and whether he had children or not, he was certainly not supposed to have them, and actually tended to be less fertile than the community around him. On the other hand, the Jewish scholar was very often in a position to have a large family. Thus the biological habits of the Christians tended to breed out of the race whatever hereditary qualities make for learning whereas the biological habits of the Jew tended to breed these qualities in. To what extent this genetic difference supplemented the cultural trend for learning among the Jews is difficult to say. But there is no reason to believe that the genetic factor was negligible. I have talked this matter over with my friend, Professor J. B. S. Haldane, and he certainly is of the same opinion. Indeed, it is quite possible that in giving this opinion I am merely presenting an idea which I have borrowed from Professor Haldane.[1]

Other writers have independently proposed the same theory as that stated by Norbert Wiener and J. B. S. Haldane. Thus, Mr. Nathaniel

Weyl in his book *The Creative Elite in America* has argued vigorously "that Jewish intellectual eminence can be regarded as the end-result of seventeen centuries of selective breeding for scholars."[2]

The Wiener-Haldane hypothesis is certainly consistent with mathematical genetics. Haldane was not only a most distinguished mathematical geneticist (the editor of the *Journal of Genetics*) but he was also the leading English Marxist scientist (an editor of the London *Daily Worker*). Curiously, the notion of selective breeding for intelligence was also quite compatible with historical materialism. For one could argue that according to the sociobiological hypothesis, the genetic traits of a population were themselves the outcome, in large measure, of cultural circumstances, which in turn, in the present case, were founded on the place of Jews in the European economy. There was, however, no evidence adduced that as a matter of sociological fact European history had seen a large-scale experiment in cultural selective breeding for intelligence on the part of the Jewish population. The facts, though scattered, sparse, and fragmentary, scarcely sustain some of the major propositions of the Wiener-Haldane hypothesis. While they corroborate part of the hypothesis, they suggest the operation of cultural factors quite other than those of selective breeding.

II

Selective breeding, to be effective, usually depends on three conditions: (1) the mating of certain selected members of the population is encouraged while others are discouraged; (2) the favored members of the population reproduce themselves in greater numbers than the unfavored; and (3) the favored members survive more than proportionately under the exigencies of existence. To what extent did the family practices of the Jewish population during its long history conform to these conditions?

From Talmudical times onward, the Jewish ethic and social practice placed a tremendous pressure upon everyone to get married. Marriage was a social imperative upon every individual. In Lithuania, for instance, at the beginning of the twentieth century, as the sociologist Arthur Ruppin observed, "the old Jewish tradition prevails that every Jew and every Jewess should marry." "In Carpatho-Russia hardly any Jewish girl remains single."[3] "In Galicia for instance, there is only an infinitesimal number of Jewish bachelors, while an old maid would be looked upon as a monstrosity."[4] The *shtetl* (the small Jewish community) did not provide a place for an "alteh moyd" (old maid): "To be a spinster is a dreadful fate which fortunately occurs far more in the anxious forebodings of girl and parents than in fact."[5]

The pressure on young men to marry was equally unremitting. The Talmudical rabbis had warned against any waste of human semen.[6] And Ben Sira before them had warned against celibacy: "Where no hedge is, the vineyard will be derelict, and where no wife is, a man is waif and stray." The Talmud reiterated: "If you are twenty and still celibate, you will be thinking of sin all your days." The precepts indeed expressed what was the practice of the Jews.[7] "The fraternity of bachelors was not popular with Jews, the Talmud speaking of the wifeless man as deficient in humanity, whilst Ben Sira stigmatises him as a vagabond, wandering up and down," wrote the learned scholar Solomon Schechter.[8] Bachelors were consequently rare in Jewish communities.[9] Early marriage was the norm insisted upon by the rabbis. The *Mishnah*, the core Talmudical text, regarded the age of eighteen years as the normal one for marriage, but Moses Maimonides, the celebrated medieval philosopher and sage, effectively reduced the norm to seventeen years by interpreting the age of eighteen to mean a man's eighteenth year.[10] In Jerusalem, it was said, no man was allowed to reside who, when over twenty and under sixty, was without a wife.[11]

The transmitters of Jewish values, the teachers, were especially under the social mandate to be married; in turn, they inculcated a similar mandate in their pupils. Medieval Christendom entrusted its teaching of children to celibate priests and monks; the latter conveyed values and attitudes to their pupils which were in keeping with St. Paul's denigration of the married status. Not so with the Jewish teachers. "An unmarried man was forbidden to teach, since he would have frequent contact with the mothers who brought their children to school."[12] In Moravia, the rabbis were actually forbidden to confer the first academic rank of *Haber* (Friend, or Comrade) on a bachelor; the degree-receiver was required to have been married for at least two years. For the higher title of *Morenu,* the candidate was required to have been married for at least five years. If he had been a part-time, post-yeshivah student because he had been engaged in trade or business, he was required to have been married for at least fifteen years.[13]

Since there was a system of virtually universal education for boys, the high valuation of marriage was reinforced as a basic tenet in the entire Jewish community. Flavius Josephus, in his polemic with Apion on the relative merits of the Hebraic as compared to the Roman-Hellenic civilization, boasted: "Above all we pride ourselves on the education of our children, and regard as the most essential task in life the observance of our laws and of the pious practices, based thereupon, which we have inherited." "It orders that they shall be taught to read, and shall learn both the laws and the deeds of their forefathers. . . ."[14] This tradition was preserved right into modern times. "Even small communities consisting of ten householders were expected to appoint a teacher and to support poor pupils if they could afford to do so."[15] Besides private schools, the communities supported public schools, taxing their members for this purpose, and using the threat of excommunication to collect these revenues. A community without a school, it was said, deserved to be excommunicated as a whole.[16]

Thus, almost every Jewish man and woman was guided toward marriage. No mechanism of social selection operated to exclude any particular class, section or stratum from the married status. Were scholars, however, the preferred choices as husbands for the ablest and healthiest women? Did they have a relatively higher number of children? Were learning and intelligence thus the preferred traits in a sexual selection which made for a selective breeding?

III

The self-taught Jewish philosopher Solomon Maimon, who at the end of the eighteenth century made the transition from the Talmud to Kant's philosophy, observed in his *Autobiography*: "The study of the Talmud is the chief object of higher education among our people. . . . A wealthy merchant, leaseholder or professional man with a marriageable daughter does everything in his power to acquire a good Talmudist as son-in-law. In other respects the scholar may be deformed, diseased, and ignorant: he will still have the advantage over rivals."[17] Such indeed is the customary account of the privileged status of the scholar as a marital prospect throughout Jewish history. The novelist Isaac Bashevis Singer recalls that in Poland at the beginning of the twentieth century women accepted it as their "lot to bear children, cook, run the household, and earn a living—while the man studied Torah. Rather than complain, our grandmothers praised God for providing them with husbands who were scholars."[18] Many Jewish parents indeed took pride in securing a young yeshivah student as their daughter's husband, and supported him in his studies for years afterward.[19] The more talented the student was, it is said, the greater the effort to have him marry young, even at fourteen or fifteen. Fathers interrogated prospective sons-in-law on their Talmudical erudition, though they realized that the wives of these "perennial students" would have to support them.[20]

Far more parents, however, were evidently loath to marry their daughters to the perennial Talmudists. Jewish fathers took a lot more into account in judging prospective sons-in-law than their book-learning. Thus, as Salo Baron writes, in the seventeenth and eighteenth centuries, educational leaders complained frequently "that the communal plutocracy no longer cared to give its children training in rabbinic law or, via marriage, to attract young talent from among the students of the yeshivot..."[21] The change in attitude in the eighteenth century toward the young scholars was drastic, as the latter declined in general high esteem. "Rabbi Jonathan Eybeschütz, one of the most prominent yeshiva heads of his day, deplored the fact that only the poor married off their daughters to scholars,..."[22] In Germany especially wealthy men were reluctant to have their daughters marry the Talmudical students. "Moses Samson Bacharach accused them of even going so far as to mock these students for devoting their whole time to the study of the Torah."[23] In the later Middle Ages, fathers saw their daughters deserted by wandering Jewish student sons-in-law who, more than emulating the goliards, left their wives behind as they went seeking the Lilith of learning in one center after another.[24] Curiously, the classical authorities, the Talmud and Ben Sira, had warned against marriage in which the wife supported the husband as unworthy and degrading. Ben Sira wrote: "There is anger and impudence and great reproach if a woman maintains her husband."[25] The Talmud had called on every Jew to learn a trade, and not to live by his learning. But European Jewry did not always follow classical wisdom.

Moreover, it must be borne in mind that the ordinary teacher in a Jewish community suffered an extremely low status. As far back as the Palestinian academies, the scholars learned in the oral law had looked down on the ordinary school teacher, whose learning was encompassed in a reading knowledge of the Bible. The latter was regarded as only a little higher than the ignoramuses, the "am ha-aretz."

The rift between the upper and lower intellectuals grew deeper with time. "Centuries later the gap between the Bible-centered teachers and the scholars who championed the Oral Law widened considerably."[26] The Jewish teacher became an object of ridicule, and the *heder,* his schoolroom, the cynosure of pupil and parent cynicism. The rector-rabbi who presided over a yeshivah retained something akin to the status of a university president in the United States before 1960, but the social distance between him and the ordinary teacher (the *melamed*) was immense. The latter, usually poverty-stricken, was often regarded as trying desperately to make as much money from his charges as possible. To eke out his living, he frequently combined his teaching with some trade or business. A pamphlet published in Lublin in 1635 ascribed the low state of the Jewish schoolroom to the low fees for instruction, the difficulty in collecting them, and the tendency of many parents to change teachers every half-year; the teacher was said to promote pupils simply because he feared to lose them.[27] In rural districts, especially, the *melamed*'s status was low. Parents, moreover "were more concerned with training their children for some occupation than providing them with religious instruction. They therefore employed any 'ignoramus' to teach their children, and in some cases, even men-servants were expected to devote some of their time to teaching the children of their employers."[28] Teachers who furthermore migrated from their native localities bore the opprobrium for having abandoned their wives.[29] Learning was indeed widespread. As Solomon Maimon wrote: "the majority of the Polish Jews consist of scholars, that is, men devoted to an inactive and contemplative life; for every Polish Jew is destined from his birth to be a rabbi, and only the greatest incapacity can exclude him from that rank."[30] When the massacres of 1648 took place, a contemporary chronicler wrote that out of fifty adult male Jews in an average community, twenty had the title *Morenu* (Our Teacher). "This is doubtless an exaggeration," writes Salo Baron, "but there is no question that a considerable number of Jews, whether or not possessing the title *Morenu,* were very learned."[31] But the actual rabbis and the governing officials of communities, synagogues and *yeshivot,* were an extremely small percentage of the people. Maimon wrote with some bitterness that the Jewish people were ruled by "a perpetual aristocracy under the appearance of a theocracy. The learned men, who formed the nobility, have for many centuries been able to maintain their position as the legislative body with so much authority among the common people that they can do with them whatever they please."[32] There was thus a high intellectual establishment, small in numbers, surrounded by a mass of literate citizenry. How small the establishment was can be seen from an official report in 1842 in Russia that found in the whole country 604 synagogues, 2,340 prayer-houses, 3,944 schools, but only 954 rabbis.[33] The latter were the professional, salaried rabbis, not the larger number of those who, by virtue of their learning and respected character, would usually in their later years be accorded the honorary title. The total Jewish population was then approximately two million out of a European Russian population of over fifty million.[34]

Under such circumstances, one can understand how a Jewish father would weigh carefully the prospects of the young scholar who was seeking his daughter's hand. What chance did the scholar have of rising into the genuine administrative rabbinate? Or would he evolve into a despised teacher of children in a dark and dank pretense of a schoolroom? Or would both father and daughter have to toil to maintain the academic ne'er-do-well as a perpetual bookworm? One can understand why the disaffection with the scholars became widespread, and that it was the poor, evidently seeking a spiritual upward mobility where the material kind was not to be had, who continued to seek young scholars as their daughters' husbands. Only occasionally did wealthy fathers select impoverished scholars as their sons-in-law.[35] That a selective mating of wealthy and healthy daughters with the most intelligent young men was taking place can hardly be shown.

Rich families, moreover, were a highly transient phenomenon in the annals of Jewish communities. A wealthy family almost never retained its wealth long enough to become a genetic center for the breeding of intelligence. "It is difficult to trace any wealthy dynasties, of the type which exist today; the children of Aaron of Lincoln or Benedict of New York enjoyed at the best a modest competence."[36] Among the great gentile banking families, such as the Fugger and the Medici, there was indeed a certain stability of fortune from generation to generation. But among the Jews of wealth in the Middle Ages, "it seldom happened that this wealth passed from father to son. Generally, it was strictly personal; and the confiscatory measures which followed the death of a wealthy Jew (the whole of whose fortune legally passed to the crown) prevented his son from enjoying it."[37] Moreover, the number of rich Jews was a small fraction of the Jewish population. The lists of contributors to the Jewish taxes in Angevin England, for instance, "show that there were many more poor Jews than rich," and they all lost most of their wealth when they were expelled from England in 1290.[38] When conditions of greater stability evolved in the eighteenth century, they were coeval however with a growing rejection of pure scholars as sons-in-law by wealthy families.

In any case, the eminent Jewish scholars, contrary to the Wiener-Haldane hypothesis. did not generally have large families. As the distinguished historian Cecil Roth wrote: "Few of the eminent Jewish scholars of the Middle Ages, Rashi, Rambam, and so on, are known to have had large families which would have been considered of more than average number even today."[39] Indeed, "the families of some outstanding personalities were very small: for example the great financier of the twelfth century, Aaron of Lincoln . . . had only two sons. . . ." The Jewish families on the average were small in size; when the Jews of Munich were massacred, the records showed that the modal family had two children.[40] In early modern times, the Jewish family became much larger, but then the trend reversed, and in Prussia, for instance, by the end of the nineteenth century, families with only two children were dominant.[41]

Legal documents have always been a valuable source on the structure and relations of families. The case histories which Irving Agus has assembled in the second volume of his *Urban Civilization in Pre-Crusade Europe* provide information as to the size of Jewish families in that medieval period. Of the 33 families involved in the described legal proceedings, the largest group was childless; 11 families had no children. Seven families had one child each; three families had two children; seven families had three children each; only five families had four children each. In the last three groups, ambiguities in the data involve variations of one in the statement of the family sizes. The impressive fact is that the average family in these proceedings, generally involving merchants, had only 1.6 children. Usually these cases, arising from such circumstances as the father's death, pertained to completed families, not those low in offspring because the family itself was still young. The high frequency of childless marriages in this middle-class group suggests still another difficulty in the way of a theory of selective breeding.

IV

It seems doubtful that the intellectual class among Jews enjoyed a higher rate of survival than their more ordinary fellow Jews. The likelihood is that in the massacres, pogroms, and legal measures directed against Jews, it was the intellectual class which sustained a disproportionate loss of numbers. The evidence here is indirect, but it was the case, for instance, that after the summer of 1096 when the chief communities of German Jews were destroyed by lynching bands of crusaders, scholarship was utterly interrupted. Rabbi Gershom, in the first quarter of the twelfth century, acquired a great reputation mainly because "the students of the students of the great teachers of Mayence did not survive the holocaust of the First Crusade."[42] "A good many Jews" were hanged in England in 1278–79 for alleged coin-clipping—three in Bedford, five at Canterbury, four at Alorwich. They were all houseowners; evidently poor, property-less Jews were exempt from this risk.[43] Polish Jewry subsequently became the most populous center for Jewish culture. Their numbers were estimated as having increased from 50,000 in 1501 to 500,000 in 1648.[44] But in the spring of 1648 their death through massacres began under the direction of the Cossack commander, Bogdan Khmelnitzki, and his "semibarbarous" followers from southern Russia. The catastrophe brought the Jews losses that exceeded those they had borne during the Crusades and the Black Death. Chroniclers reported that anywhere between 100,000 and 500,000 Jews were killed during the next decade. The Jewish communities on the left bank of the Dnieper virtually disappeared, while on the right bank, "only about one-tenth of the Jewish population survived." Seven hundred Jewish communities in Poland suffered massacre and pillage. Can we venture any plausible inference as to how the Jewish intellectual class fared relative to their fellows in these catastrophes?

In the similar calamities of our own time, the survivors, it has been observed, have tended not to be intellectuals or scholars, but rather simple people, mistrustful of all authorities; unencumbered by possessions, they were ready to leave and hide with little more than the miserable garments on their backs. When the Nazi army on its first day in Odessa executed a mass slaughter, "thousands of persons, most of them Jews including many intellectuals, were murdered in the first 24 hours."[45] Intellectuals, "rationalizers," as Raul Hilberg has said, were more apt, for instance, to invent plausible, relatively benevolent explanations as to why the Germans would want the Jews to assemble at a certain place at a certain time.[46] The survivors, on the other hand, acted instinctively at short notice and ran away. They had the will to live and the readiness to live on near nothing. They could dissemble, and in a peasant's hut, pass themselves off as domestics. The faces of the scholar-intellectuals, lit by the lamp of learning, were far more recognizable. Probably the scholars of the Polish Jewish community in 1648 perished similarly in disproportionate numbers. The community as a whole soon revived, numbering in 1788 in Poland and Lithuania 617,032 out of a total population of 8,790,000.[47]

A traumatic period of massacre was generally followed by a lowering of the permissible age of marriage. Thus, in the fourteenth century, after the Crusades had reduced the number of Jews, the prohibition against child marriage was relaxed.[48] Similarly, after 1648, the permissible age of marriage declined three to four years below the Talmudic and Maimonidean norm of eighteen and seventeen. The institution of the marriage broker likewise first emerged during the Crusading era, and was strengthened after 1648, because Jewish society was so disintegrated by the impact of massacres and expulsions that go-betweens were required to mediate between the families in isolated villages and hamlets.[49] Until modern times the matchmakers were almost always rabbis and scholars; the fees were lucrative, varying between two and three percent of the dowries.[50] The marriage brokers are said to have operated on the principle of mating like with like, and therefore of not mating tall with short, young with old, or learned with ignorant.[51] The pattern of mating defective or sickly persons might have led through reinforcement to their genetic extinction, while the healthy would have grown relatively more numerous. On the other hand the matchmakers became famous for their loquacious over-depiction of the eligible clients, and a whole department of Jewish jokes evolved on this theme. The marriage broker became preeminently a functionary among those who for one reason or another seemed to be failing to get married through the normal social channels. Apart from maximizing the number of married persons, it is doubtful that their intervention contributed to a higher degree of selective breeding than the available autonomous choices of the families and their young would have secured.

When the great waves of migration of Jews to America began after the pogroms of 1881, the rabbis and scholars tended to remain behind in disproportionate numbers. America was regarded as irreligious, as a place where Jews would fall away from their belief and ritual. Many immigrants departed for America as if by stealth, in a sort of shame at leaving the fold. Young lads, making the decision to migrate, shed Talmudic erudition for the lore of the Singer sewing machine: "Whoever recollects the beginnings of the Jewish emigration from Russia knows that almost every Jew who contemplated emigrating learned to operate a Singer sewing machine. What else could they do? 'To learn a trade' was the only way, and the easiest and quickest thing to learn was the ways of the Singer machine."[52] Thus recalled one who had emigrated. The traditionalist rabbis and scholars who remained behind were during the next decades decimated by wars, Polish, Ukrainian, and Russian pogroms, the Nazi holocaust, and Bolshevik intolerance. Survivors in the traditional Jewish communities were more likely the materially resourceful and adaptable rather than those adept in the scholarly discipline.

V

Perhaps the most potent factor, however, for the selective loss of highly endowed Jewish intelligence over the centuries was the conversion of Jews to the Christian and Moslem religions. The impact of conversion on the ranks of Jewish intellectuals has rarely been discussed but there can be no doubt that this phenomenon, especially in the latter Middle Ages, affected the intellectual Jews far more than it did their ordinary co-religionists.[53]

The number of Jews in the world (as indicated by the evidence) declined from the time of the Roman Empire in the third century A.D. to that of the papacy in the thirteenth by approximately 66 to 85 percent.[54] According to a reasonable estimate by Dora Askowith, there were at the time of Nero 4.5 million Jews within the Roman Empire, which then had a total population of 54 million.[55] Salo Baron, on the other hand, reckoned the Jewish population in Claudius' time at almost 7 million.[56] By the middle of the seventeenth century, however, the Jewish population was diminished tremendously both in absolute and relative terms; the Jews are estimated approximately to have numbered not more than 900,000 out of a total European population reckoned at 100 million. They had evidently declined from between nearly seven to ten percent of the European population to less than one percent.[57] What were the causes of this decline of the Jewish population? Probably the Jews shared in the general decline of the birth rate which commenced in Europe during the latter days of the Roman Empire. Physical destruction, too, had its part in the depopulation of Jews. But there was also, writes Cecil Roth, "the gradual seepage of individuals, not many at a time, but in the long run very numerous." There were such instances, as for example in Portugal at the end of the fifteenth century, when a mass conversion of a community as a whole took place; the threat of massacre was a catalytic agent for this type of social process. More significant, however, than massacre and mass conversion, writes Cecil Roth, was the continuous, "constant pressure of environment, of social advantage, of conviction, of petty annoyances, of Ghetto pressure, of conversionist sermons and the rest of an elaborate system organized mainly with a view to breaking down the resistance of the Jew."[58]

One sociological generalization emerges from the facts concerning Jewish conversions throughout the centuries: to the extent that the surrounding society was one which offered liberal intellectual, political, economic, and social opportunities, to that extent the rate of Jewish conversion to Christianity was high. The Jews of Italy, for instance, were not massacred, nor did they emigrate; nevertheless, their actual number tended to remain stationary, or to diminish; whereas in 1638 they constituted about two in a thousand of Italian population, in 1938 they had declined to one in a thousand. What had occurred was a steady assimilation of Jews into the Christian population: "There was a perpetual procession from the Synagogue to the Church. Between 1634 and 1700 no less than 1,195 Jews were baptized in Rome alone."[59] The Jews in Britain found a liberal society with a high civilization; therefore, the founders of the Anglo-Jewish community were soon assimilated into English society. Often they intermarried with aristocratic English families, such as those of the Duke of Norfolk and the Marquess of Salisbury; then too there were families whose calling was for service in the British Army—the Pereira, Aguilar, and Barrow.[60] "[T]here can be little doubt that the majority of them [the descendants of the founders of the Anglo-Jewish community] are outside," writes Cecil Roth.[61]

Montagu Frank Modder has described the assimilation of the British Jewish elite into the English Christendom: "When Sampson Gideon, the head of the Jewish community in the latter part of the 18th century, determined to bring up his children as Christians, he set an example which was followed by many of the chief Jewish families during the remainder of the 18th century and the early years of the 19th. The wealthy classes found that as Jews they could not satisfy their social and political aspirations. They saw, too, that conversion to Christianity would remove the obstructions which made their position intolerable. So a number of prominent families, the Bernals, the Lopezes, the Riccardos, the D'Israelis, the Aguilars, the Besavis, and the Samudas, for instance, severed their connection with the Synagogue, and allowed their children to grow up without any religion, or in the Established Church. In the main, these secessions arose not from religious convictions, but from social and personal causes."[62]

Apart from the effects of sheer physical force, in threat or actuality, two types of motives swayed the individuals who converted to Christianity. The highly ambitious and able often felt the wish to enter, remain, or rise in the governing elite.[63] Then, too, the rebellious in temperament often found the religion of their fathers stale, parochial, stifling, uninspiring. Both these kinds of persons tended to be extremely intelligent, although the second variety often exhibited neurotic traits in their revolt against their fathers. By contrast, the ordinary plebeian, impoverished Jew had no such compelling call to convert to Christianity. Ignorant of the gentile culture, he was scarcely a candidate for political or social advancement; no reading or immersion in gentile works had awakened in him a discontent with his father's religion. If some rebellious impulse moved him, it was more likely to express itself in allegiance to a Jewish sectarian movement, as Hasidism was, or fealty to another rabbi. The consequence was that conversion probably operated to drain a disproportionately large percentage of Jewish genetic intelligence into the Christian (or Moslem) ranks.

The converts from Judaism thus often achieved a high degree of eminence. Many Jewish doctors were converted to either Islam or Christianity; sometimes they then composed literary works devoted to attacking Judaism, and defending their new faiths.[64] Among the Spanish Jews, two eminent medieval scientists were converted to Christianity; Moses ha-Sephardi was baptized at the age of forty-four as Petrus Alfonsi; then, appointed as physician to Henry I, king of England, he introduced Arabic mathematical-astronomical science to that country;[65] Joseph Vecinho, physician to the king of Portugal, was consulted in 1484 on the practicability of Christopher Columbus' proposal. Vecinho was the translator of Zacuto's nautical tables, and had made improvements in the astrolabe; he accepted conversion in 1487.[66]

Many of the converts not only became priests, but achieved renown as the most zealous proselytizers and champions in disputation with the Jews. Some rose or had descendants who rose to high station in the hierarchy of the church.[67] The apostates led the ideological onslaught against the Jewish culture and religion in every center, moved often by zeal against their forebears.[68] In Rome, Pope Gregory XIII, prompted by a converted Jew, Joseph Tzarfati, instituted in 1584 a requirement that Jews were to attend weekly conversionist sermons.[69] The practice persisted in an attenuated form until 1847, when it was finally terminated. The conversionary discourses were usually given by apostates, who were not only zealous, but skillful in argument. How different were they in their intellectual character and desire for assimilation from the Jewish section of the Soviet Communist party, which directed the Marxizing, antireligious campaign in 1921–22? "The campaign was conducted almost exclusively by Jews against other Jews. . . . In fact, the *Evsektsiia* jealously guarded its monopoly over the persecution of the Jewish religion," writes a historian of the Jews in the Soviet Communist party.[70]

VI

Thus, diverse types of apostasy have recurred in the history of the Jews. If there were those who, in Yitzhak Baer's words, "had entered their new faith as penitents, become monks, and appeared chiefly as persecutors of and missionaries to their former co-religionists," a larger class was moved rather by "political considerations"; conversion provided "an 'admission ticket' to a world that was wholly secular and to a career in the civil and political bureaucracy."[71] Philosophical influences in every historical period have tended to reduce the tenacity of the specifically Jewish religious culture. The powerful influence of the Averroist philosophy, which was felt by Moslem, Jewish, and Christian thinkers alike, thus tended to devaluate the significance of any particular religious forms and organizations. The Averroist philosopher placed all religions on an equal footing; hence the Averroist enlightenment mitigated the trauma of religious conversion; for the religious myths and symbols, in its view, were matter for the masses, not for the rational philosophers. The Averroist influence was not unlike that of Renaissance humanism in Italy, Kantianism in the early nineteenth century. in Germany, and Marxism in Eastern Europe in the twentieth century.

Jewish teachers who explicated the Hebrew language to Pico della Mirandola and his fellow humanists became converts to Christianity; thus Flavius Mithridates. a translator of Hebrew works, was such a Renaissance convert.[72] Not essentially different in kind were the philosophical conversions to Christianity that took place in Germany beginning in the early nineteenth century. Karl Marx's father, Heinrich Marx, when his livelihood as a lawyer was threatened by an anti-Jewish decree, found his path to conversion facilitated by his Kantian leanings.[73] The philosopher Solomon Maimon, in the Kantian era, proposed to convert from Judaism to Christianity, because, in his words, "in practical use the latter has an advantage over the former; and since morality which consists not in opinions but in actions, is the aim of all religion in general, clearly the latter comes nearer than the former to this aim."[74] During the Romantic period, the children and grandchildren of Moses Mendelssohn, the famed intellectual leader of German Jewry, were converted to Christianity; the daughters embraced Roman Catholicism, whereas the sons were drawn to Protestantism. When Abraham Mendelssohn, Moses' eldest son, hesitated about having his children baptized, his brother-in-law Bartholdy wrote to him. "You say you owe it to your father's memory (not to abandon Judaism). Do you think that you are committing a wrong in giving your children a religion which you and they consider the better? In fact, you would be paying a tribute to your father's efforts in behalf of true enlightenment, and he would have acted for your children as you have acted for them. . . ."[75] Thus the world of liberal European culture, with its opportunities for advancement, beckoned to the ablest and most ambitious of the Jews. Theodor Herzl, the founder of organized political Zionism, had as a youth been restive with "the dull compulsion of the Ghetto." After a party he attended, he wrote to his parents: "Thirty or forty ugly little Jews and Jewesses. Not a very refreshing sight." As late as 1893, he indulged in a grandiose fantasy of a mass conversion to Catholicism. He would negotiate with the pope: "Help us against anti-Semitism, and I in return will lead a great movement amongst the Jews for voluntary and honorable conversion to Christianity." Then, "in the broad light of day, at noon on a Sunday, a solemn and festive procession accompanied by the pealing of bells shall proceed to the St. Stephen Cathedral in Vienna. There shall be no furtiveness and no shamefacedness, as hitherto; it shall be done proudly and with a gesture of dignity."[76] The realities of the Dreyfus case abrogated such fantasies on Herzl's part; they illustrate, however, precisely the cultural and social forces that moved many of the ablest Jews to conversion to Christianity. Among the Polish Jews as well, it was the "well-to-do Jewish neophyte" who sought to enter the Roman Catholic Church, not the handicraftsman or proletarian.[77] And in Russia, in the nineteenth century, notes Salo Baron, "quite a few Jews, whether in the armed forces, among the more easily assimilated intelligentsia, or in the group of straight careerists, joined the dominant faith."[78] The number of Russian Jews thus converted is estimated to have been 84,536.

Apart from the philosophical intellectuals and the politically ambitious, there were a certain number of Jews to whom a good business was worth a mass. "Arrant scoundrels," Cecil Roth calls them, yet the picaresque insouciance of these characters, stands in refreshing contrast to the zeal alike of the monks, penitents, and disputationists, and the careerism of the bureaucrats. Moses Israel of Salonica, for instance, made a kind of "racket" of baptism; he enjoyed the experience several times, and earned the temporal reward of a license to manufacture arquebus powder.[79] There were forced converts who managed to quit Portugal not because they wished to renew their Judaic religion but because they were attracted by commercial opportunities outside the Iberian peninsula. Whether they remained Christians because of religious indifference, or whether they felt that when in Rome, one must do as the Romans do, or whether they were mindful that a commercial traveller might as a Christian be better received in some lands, the seepage of talented Jewish men to the Gentiles was significant.[80] A man of brilliant abilities in finance might decline to renounce his life's calling in order to adhere to what he regarded as the meaningless dogmas of a religion. Thus, for instance, Don Samuel Abravanel, an "expert at figures and familiar with the royal revenues in the reigns of Henry II and John I," converted to Christianity around 1391, and was appointed chief accountant to the king of Castile.[81]

As we have seen, where the host culture is perceived by an alien minority as equal or superior, and nonrepressive of its own, the rate of voluntary conversion is high, especially among persons of intellectual abilities. Where the host culture has been perceived as inferior to one's own, the rate of conversion has been low. Thus, "in Czarist Russia up to 1917 . . . to a certain degree the Jewish cultural level was superior to that of the surrounding population."[82] The consequence was a low rate of conversion. Vienna, from 1901 to 1905, had a conversion rate of 39 Jews per 10,000 yearly; Russia, on the other hand, from 1891 to 1897 had an annual average of only 2. Germany from 1896 to 1900, and in 1904 had an annual average of 8.[83] The figures mirror the situations in which Viennese, German, and Russian Jews found themselves with respect to their host cultures. Earlier in the nineteenth century, however, many of the most talented Jews in Germany had adopted Christianity as a way of obtaining their passports to universal culture: Heinrich Heine, Ludwig Börne, joined later by the great mathematicians Leopold Kronecker and Georg Cantor; the ranks of the converted or their sons had genius not inferior to that of their greatest contemporary loyalists.

A marginal type of apostate has been rather ignored in our histories that have stressed economic, social, and philosophical considerations. In the Middle Ages, Jewish men and women evidently on occasion wished to marry Christians and therefore converted to Christianity. The converse process, of Christians converting to Judaism, was virtually out of the question. "Converted Jews were more numerous in the twelfth century than has sometimes been allowed," writes H. G. Richardson in *The English Jewry under Angevin Kings*.[84] Among the women who were converted, there were such as Constance who married the tailor Gerin.[85] The reverse process was so dangerous that when a deacon at Oxford fell in love in 1222 with a Jewish girl, he was degraded, and executed by the king's bailiffs.[86] If one can judge from the later romances of young Jews from the eighteenth to the twentieth centuries, love choices of Gentiles by Jews were far more frequent occurrences among the intellectuals and well-to-do than among the poor. Probably erotic motives therefore augmented a continuing disproportionate seepage of Jewish genetic intelligence into the Christian ranks through the centuries.

In modern times, conscious apostasy, in response to conversionist propaganda, has been a negligible phenomenon among the Jews. The largest organization with this aim in the United States, the American Board of Missions to the Jews, founded in 1894, claimed that by 1966 they had converted 2,500 Jews.[87] Far more important than conversionary

402   Varieties of Scientific Experience

efforts were the purely social and impersonal forces making for the assimilation of the Jews into American society. For many persons the Jewish religion became a vestigial set of symbols; one then chose that religious community most congenial with one's social preferences.

VII

It is often asserted that a large percentage of the eminent Jewish men of science are the descendants of rabbis; such a finding would naturally tend to confirm the hypothesis that a selective breeding for intelligence in rabbinical families made for their subsequent relatively higher contribution to science. Curiously, an empirical test of this hypothesis largely contradicts this view.

Let us, for instance, examine the class of the most recognized eminent Jewish scientists, the winners of Nobel Prizes. Forty-nine such names are listed in the *Encyclopaedia Judaica*. This work, the most recent and authoritative, might well be expected in its biographical articles to emphasize any rabbinical parentage or ancestry in the scientist's background. Only one scientist out of forty-nine is described as having had such a rabbinical background. Those for whom no rabbinical antecedents were indicated were the following: Adolf von Baeyer (mother Jewish), Robert Barany, Felix Bloch, Konrad Bloch, Niels Bohr (mother Jewish), Max Born, Sir Ernest Boris Chain, Melvin Calvin, Paul Ehrlich, Albert Einstein, Joseph Erlanger, Richard P. Feynman, Fritz Haber, Gustav Hertz (Jewish father), Herbert Spencer Gasser, George Charles de Hevesy, Murray Gell-Mann, Donald A. Glaser, Robert Hofstadter, François Jacob, Sir Bernard Katz, Arthur Kornberg, Sir Hans Krebs, Lev Landau, Karl Landsteiner, Gabriel Lippmann, Otto Loewi, Salvador Luria, André Lwoff, Elie Metchnikoff (mother Jewish), Otto Meyerhof, A. A. Michelson, Henri Moissan (mother Jewish), Hermann J. Muller, Marshall W. Nirenberg, Max Perutz, I. I. Rabi, Tadeus Reichstein, Emilio Segré, Julian S. Schwinger, Igor Tamm, Selman A. Waksman, George Wald, Otto Wallach, Otto Warburg and Richard Wilstaetter. No facts were provided concerning one laureate, Julius Axelrod. The only Nobel laureate who was identified as the scion of a rabbinical family was the geneticist Joshua Lederberg.

Thus, only slightly more than two percent of the Nobel laureates, only one out of a class of forty-nine, are identifiable as of rabbinical antecedents.[88] Acknowledging the possible loss in transmission of information from generation to generation, there is still no basis for the claim that a large percentage of Jewish scientists are the descendants of rabbis.

By contrast, the eminent rabbis were generally the sons of rabbis, or descended from rabbinical families. *Men of the Spirit,* a volume by Rabbi Leo Jung, is devoted to the biographies of twenty-eight famed rabbis of modern times.[89] Let us review the relevant entries: Akiba Joseph Schlesinger, "of a family renowned for its rabbis"; Levi Itzhak, "descended from a long line of rabbis" through "twenty-six consecutive generations"; Samuel Mohilever, whose family had bred rabbis for twenty-two generations; Israel of Sklov, his father a rabbi, his grandfather a Gaon; Edward Biberfeld, a scion of rabbis on both sides; Naphtali Berlin, of "a family rich in Rabbinic traditions"; Jonathan Eibeschuetz, on both sides; Michael Kahn, whose mother's rabbinical family traced back to the sixteenth century; Nathan Birnbaum and Shmuel Margulies, both of rabbinical descent. Those whose father was a rabbi: A. I. Karelitz, Y. E. H. Herzog, Y. Z. Soloweitchik, H. Ehrentreu, I. Salanter, Z. W. Gold, B. Z. Safran, M. S. Glasner, Yosef Hayyim, and Hayyim Medini. M. Friedland's mother was the daughter of a rabbi; M. Bar-Ilan's grandfather was a well-known rabbi. No mention of rabbinical descent was made concerning M. Gaster, S. Mendlowitz, H. Zeitlin, J. Duenner, J. Aszod, or M. Schick. With twenty-two of the twenty-eight eminent rabbis descended from or the sons of rabbis, it seems clear that a strong dynastic principle prevailed in the rabbinate. Joseph Duenner's father was a poor tavernkeeper, Moshe Schick's father a merchant, and Judah Aszod's father a tailor. Such cases, however, were exceptional.

Possibly selective breeding heightened the acumen of rabbinical intelligence; it may also have extinguished qualities of venturesomeness and originality. The rabbinical abilities, in any case, scarcely played any part in the scientific movement among the Jews. The Jewish scientists emerged rather from the mercantile and artisan classes. Less encumbered by traditionalism, their children moved quickly into the frontiers of scientific research, into the exploration of the unknown. Rabbinical law indeed posed intellectual obstacles to some branches of science. The injunction upon "Torah-true priests (Kohanim)" not to defile themselves by contact with any dead body led some to avoid a medical career, involving, as it did, the dissection of bodies.[90] Jews are said to have "apparently played no part in the renaissance of modern biology."[91] Perhaps a residual influence of Jewish traditional culture may have assisted in this cultural selective retardation.

In Germany and Austria, before the advent of Nazi rule, the Jews had achieved a foremost scientific place. Very few, however, of the scientific community had sprung from rabbinical roots. The names of 245 scientists were set forth in a volume by Sidney Osborne entitled *Germany and Her Jews*.[92] Unfortunately, in only 66 cases were facts concerning religious influences indicated. Of those 66, only five were described as having had a father who was either a rabbi or a Jewish scholar, while seven had some relative other than one's father, characterized by such associations. Only eight were described as having had received a Jewish education. Probably the vast majority of the less documented 179 scientists had no rabbinical linkages whatsoever, for such facts have generally been prized by the writers on the contributions of Jews, and were scarcely repressed in their encyclopedias and biographical dictionaries. The eighteen percent of the 66 cases who had some rabbinical or scholarly relative would probably be little more than five percent if more ample data were available.

VIII

We have thus seen how the universality of marriage among Jews of all classes; the seepage of a large percentage of the ablest stratum through conversion and assimilation into the gentile population; the ambivalent attitudes of Jewish communities toward their rabbis, students, and teachers; and the frequent nondescent of Jewish scientists from rabbinical families, all tend to indicate that social selective breeding was of small moment to the later scientific achievement. The Jews probably only held their own with a certain constancy, as far as the genetic endowment of intelligence was concerned. Yet it may well have been true that at the outset of the Middle Ages the small nucleus of surviving Jewish people was the most intellectually gifted in Europe. It is not generally realized how small were the numbers of European Jewry in the Middle Ages. Irving Agus estimates cautiously that "in the year 800 . . . the Jews of Northern Italy, Germany and France, must have numbered no more than eight to ten thousand souls." Thus, "the ancestors of Ashkenazic Jewry, in the year 800 C.E., numbered but about ten thousand."[93] The most reliable document on the population of the Jews among later medieval sources is the remarkable *Book of Travels* of Benjamin of Tudela. Probably for about 14 years, from 1159 to 1173, Benjamin journeyed through towns of Spain, southern France, Italy, Greece, the Aegean archipelago, and on to Middle Eastern and North African centers. He faithfully recorded the numbers of the Jewish communities, their occupation, and their state of scholarship. One gets a sense of how sparse the Jewish numbers were: Marseilles with 300 Jews, Genoa with only 3, Pisa with 20, and Rome notably with only 200. In Europe, only Constantinople and Thebes had sizeable Jewish settlements, each with 2,000 persons.[94] The recorded number of Jews in all the places he visited in Europe, including those of the Aegean archipelago, added up to 14,574. Benjamin seems, however, to have counted only the adult Jewish men in the communities; hence his figures should be multiplied by about 3.5 to include women and the average number of children. Thus the total number of Jews in the communities he visited was about 51,009. Excluded from Benjamin's travels and enumeration were 14 German towns where Jewish communities existed, but these as Agus indicates, could not have totalled more than twenty thousand Jews. In England there seem to have been no Jews prior to the Norman Conquest in 1066; then a small migration of French Jews began. By the 1270s, the total population of the English Jewish communities was not more than 3,000.[95]

Thus, European Jewry at the outset of the Middle Ages may have consisted, indeed, of only about 10,000 persons. St. Jerome had remarked on the decline of the Jewish population toward the close of the fourth century when he wrote that "in comparison to their previous multitude, there hardly remained a tenth part of them."[96] The spread of Christianity, conversions, and persecutions all operated to diminish their number. Yet the handful who chose to survive as Jews at the beginning of the Middle Ages was probably among the most highly endowed intellectually in Europe. They were a group of "exceptionalists," intellectually, economically, and politically.[97] Intellectually, they had remained relatively immune to the waves of irrationalism that grew pronounced in the later Roman era. They remained in aloof detachment as the Christian theologians of diverse orthodoxies and heterodoxies argued over the trinity and virgin birth. Their monotheism, despite their own superstitions, tended by contrast to be reinforced as a simple rationalistic doctrine. Joinville, in the crusading thirteenth century, described a typical response to argument with the Jews. The Jew, asked by a knight whether he believed in the virgin birth, "replied that of all this he believed nothing." Whereupon the knight smote him. "And so ended the disputation." The knight warned against any further disputations, "for there were a great many good Christians there who, before the disputation came to an end, would have gone away misbelievers...."[98] Those who intellectually could not yield to the complexities of Christian theology, who found it contravened their principle of simplicity in philosophizing, were among Europe's most scientific in spirit. As Europe receded into the decentralized separatism of feudal societies, the surviving Jews were among those who chose to remain outside the feudal economic and political structure.[99] They became merchants and artisans even as trade everywhere was diminishing, and the lines of commerce between countries and continents were being largely severed. The Jews continued to undertake voyages to distant places, selling their wares in the towns.[100] Every small Jewish community was an outpost, a wayfaring station, where the commercial traveller felt at home—a surviving international of commerce. A background in Talmudic learning was economically helpful to the merchant, for it guaranteed a warm reception from fellow Jews on one's travels; thereby Talmudic learning was "a strong factor in providing security of life and property to these travellers."[101] The Jews, pioneers of capitalism, stood altogether outside the system of lords, vassals, and serfs. No Christian oath of fealty bound them; inevitably they were partisans of the towns and the slowly emerging national kings. In medieval England and Normandy, for instance, the Jews were exempt from local customs and tolls, and "could sue and be sued only in the king's court" or the courts of the keepers of the king's castles; "As against all other men, except the king, he was protected."[102] As artisans they were linked to the division of labor and the exchange market. At Thebes, Benjamin of Tudela found that the Jews were "the most eminent manufacturers of silk and purple in all Greece"; in Salonica, they lived "by the exercise of handicrafts"; at Brindisi, all of them were dyers; and at Constantinople, many were "manufacturers of silk cloth."[103] In medieval Sicily, where records are available, "the bulk of the Jews of the lower classes were stevedores, dock-laborers, metal workers and so on"[104]—surviving cells of workingmen in the feudalizing society. The average villager in medieval Europe was so isolated that, as G. G. Coulton estimates, he probably never saw more than a hundred people in his life.¹⁰⁵ In this structure of social isolates, the wandering Jew brought news, ideas, and portents of things to come. Those who chose to be Jews during the period in the early Middle Ages, when Europe declined intellectually, economically, and religiously, were probably persons who were, on the average, intellectually far ahead of their surrounding culture. It was not until several centuries later that the advancement of gentile culture made conversion in the absence of force a more likely alternative for rationalist intellectuals.

IX

Once the Middle Ages were under way, there is no real evidence that there existed practices of selective breeding that would have raised the average genetic endowment of Jewish intelligence. Nevertheless, though Jewish intelligence probably remained roughly constant, the genetic level of intelligence among Christian Europeans may well have tended on the average to decline during the course of centuries. Christian social institutions and practices did make for a selective, dysgenic breeding against intelligence.

A decline of intelligence had perhaps begun during an earlier era toward the close of the Hellenic period. The intelligence of the ancient Greeks, conjectured Francis Galton, was probably much superior to that of modern Europeans, while according to the founder of British genetics, William Bateson, the decline of Greek civilization might have arisen from the mixtures that were introduced into that remarkable genetic stock. Bateson wondered if the reforms of Cleisthenes (507 B.C.), which sanctioned foreign marriages and admitted aliens and freedmen to citizenship, were "the effective beginning of a series of genetic changes which in a few generations so greatly altered the character of the people."¹⁰⁶ Apart from the problematic consequences of genetic mixtures in Hellenic society, there can be little doubt that the homosexuality which became widespread among the Greek intellectual class tended to reduce their relative fertility. Hans Kelsen, the distinguished Austrian scholar, noted: "The characteristic of homosexuality must remain an exceptional one, it can and may not be the general rule if society is not to be destroyed (by becoming extinct).''¹⁰⁷ Where it specifically affects the intellectual class, homosexuality then would entail a probable differential loss in the intellectual endowment of that given society. Aristotle had been aware that there was a relation between homosexuality and population decline, for he wrote in his *Politics* that homosexuality was fostered in Crete in order to reduce overpopulation.[108] And in Plato's *Laws,* one can read the warning that homosexuality "destroys the seeds of human increase, or sows them in stony places, in which they will take no root."[109] The intellectuals rarely had children, if one can judge from the accounts in Diogenes Laertius' *Lives of the Philosophers.* Among the Jews, on the other hand, homosexuality, as Cecil Roth wrote, was "hardly to be traced" at any place or time in their many-centuried history.[110] Meanwhile, the Roman Empire in its last centuries had to face the combined impact of homosexuality, celibacy, and the will to childlessness. The Christians not only favored celibacy but were averse to large families; especially among the Roman middle classes did childless families become the fashion, so much so that a scarcity arose of persons from the propertied classes eligible for appointment to town councils.[111]

Throughout the first part of the Middle Ages, "the great era of population decline coincided with increase in the strength of the Roman Catholic Church." as J. C. Russell has noted.[112] Although this trend was reversed in the latter part of the Middle Ages, the heritage of a depopulation-culture persisted, "so that many of its forms of expression partake of ideas derived from the preceding age."[113] The number of celibates in the laity was probably not high; a volume of the hearth taxes in a Belgian county in the early sixteenth century shows that the celibates numbered only 2 percent of the household population.[114] But the impact of the monasteries throughout Europe was toward reducing a large part of its superior genetic intelligence. One-third of the population of Spain is said to have been at one time in monastic orders,[115] and allowing for exaggeration, with an undoubted adverse consequence for the average intelligence of Spaniards. In England, as J. C. Russell observed: "Quantitatively, the effect of celibacy was probably not great. The English poll tax of 1377 showed about 30,000 clergy in an aggregate of 1,400,000 or about 2.3 percent. Qualitatively, the story may be quite different. A study of the medieval German nobility tends to show that the entrance of so many noble families into the Church hastened their extinction...." Monasticism appealed to the most energetic and ambitious as well as to the more contemplative of educated persons at that time.[116] And precisely the latter class of intellectuals did not transmit its proportionate genetic heritage to Christian Europe.[117]

The triumph of celibacy as an ideal in the era from 200 to 900 A. D. is a fascinating episode in the oscillations of human history. Christianity spread most rapidly in urban areas where depopulation and depression were having their greatest effects. A sense of exhaustion came upon the intellectuals.[118] A longing grew for supernatural intervention. The sentiment for celibacy increased so strongly that the Christian emperors abolished the penalties against it, despite the fact that larger families were needed. The laity elected celibate monks as their bishops rather than secular married clergy.[119] "Great numbers of Christians entered monasteries voluntarily, renouncing those ties of family which the Romans had once prized."[120] By the fifth century, the Christian church was supreme in the religious domain, "rooting out religious dissent except among the Jews."[121]

As late as the beginning of the twentieth century, Spanish liberals, seeking to bring their country into the mainstream of European history, sensed that clerical celibacy was a principal enemy. "If we could only get rid of our monks as easily as our colonies!," a Spaniard is represented as declaring in one of the comic papers of Madrid during the World War an aspiration voiced by many Spaniards of all classes.[122] As Charles Darwin wrote, the decline of the Spanish nation was probably the consequence of its institution of celibacy. In Darwin's words: "Almost all the men of a gentle nature those given to meditation or culture of the mind, had no refuge except in the bosom of a church which demanded celibacy; and this could hardly fail to have had a deteriorating influence on each successive generation." Added to this, "some of the best men," the doubters, the questioners, "were eliminated during three centuries at the rate of a thousand a year." The "evil" wrought was "incalculable."[123] In 1646, Fray Luis de Miranda complained how the plethora of monks and their celibacy was draining Spain of its best seed: "The towns are almost all depopulated and deserted. . . . Our Spanish monarchy is being consumed away hour by hour and moment by moment. . . . [The religious orders are carrying off] the bravest men, the healthiest, the most upstanding, those with the best faces, the most talented and skillful. There is not among them a cripple, nor hardly a dwarf, nor one that is ugly, or dull, or ignorant. . . . In the world [remain] only the dregs and dross of men."[124]

Castile's population had commenced in the sixteenth century to decline sharply. In the fifteenth century, it had numbered between 7,900,000 to 9,500,000 persons, but by 1590 it had fallen to

6,250,000.[125] The religious vocation meanwhile continued to absorb a considerable number: "The Cortes of 1626 stated that there were then 9,088 monasteries, not counting the convents. Thus the fields were unpeopled and the factories empty."[126] In the last years of the eighteenth century, when industry and the entry of foreign immigrants had been much encouraged, the population of Spain which reached 10,541,221 still included, according to the historian Altamira, 168,248 ecclesiastics.[127] Thus, a class of more than 1.5 percent of the male population, including doubtless a disproportionate number of highly intelligent, was being subtracted from the Spanish heritage.

In Britain, during the latter part of the Middle Ages, monasticism grew at a rate that far exceeded that of the population. "A wave of enthusiasm for monasticism followed the Conquest and set up a system which included probably twenty times as many members by 1300 as in 1066. During the same period the population increased about 3.5 times. Probably the peak of the numbers was reached about 1275–1300 although the population of England continued to increase until the Black Death. . . . The decline was in about the same proportion as the loss in general population between 1346 and 1377." Subsequently, the monastic population tended to be stable until the dissolution of their institution, while the population increased about fifty percent.[128] In 1377, the combined numbers of the religious and secular clergy were about 35,000, or 1.6 percent of the total population of 2,200,000.[129]

Long after the Protestant Reformation had withdrawn its universities from the Roman Catholic fold, the ethic of celibacy continued to exert its influence on English scientists. As J. B. S. Haldane remarked: "Fellows of Colleges at Oxford and Cambridge lost their jobs on marriage up till the late nineteenth century, and the tradition of celibacy still persists. One result of this has been a lesser fertility of the educated."[130] In 1882, there were evidently not more than sixty families, and hardly ten young women in all Cambridge. Then the University of Cambridge removed the ban on marriage for their fellows. At once, a "stampede" for marriage ensued.[131] Several centuries of genetic disendowment, however, had been wrought against British scientific capacity.

Is there any way of estimating, even roughly, the decline of intelligence that may have taken place in the Christian community as a result of clerical celibacy? J. B. S. Haldane wrote in 1938 that "if the existing differences in fertility of social classes continue, [we may] expect a slow decline perhaps of 1 or 2 percent per generation in the mean intelligence quotient of the country. That is, on the whole, deplorable."[132] Haldane was particularly mindful of the lower fertility of the professional classes—the "doctors, clergymen, teachers, and so on," because apart from that stratum "the well-to-do do not have children appreciably more intelligent, as judged by these tests, than the remainder of the population."[133] Of course European countries have varied considerably as far as their relative numbers of clergy are concerned; moreover, one must bear in mind the degree of the clergy's commitment to celibacy as well as the date of the country's reception of the Christian religion. If we assume that the average western and central European country adhered to a celibacy-practicing Christianity on an average of ten centuries, and if each century is taken as having consisted of four generations, and lastly, if we assume that the differential fertility of their celibate clergy was at least less than that of the Western professional class (a very modest assumption), then the negative impact on the mean intelligence quotient must have been very large indeed. To be sure, once the most intelligent had been eliminated, the dispersion of its remaining groups with respect to intelligence would have been lessened. The differences in the intelligence, in other words, between priest and parishioner would have become smaller. The law of diminishing intelligence under a negative differential fertility would have had a milder effect as the population lost its most talented. The damage inflicted would have grown proportionately less. Meanwhile, however, European intelligence would have sustained a traumatic injury especially in the region of the highest ability.

This whole domain of the rise and fall of civilizations and their genetic status continues to be enveloped in obscurity. If some geneticists have deplored the consequences of certain racial intermixtures which took place in antiquity, the eminent R. A. Fisher, on the other hand, has declared "The fact of the decline of past civilizations is the most patent in history, and since brilliant periods have frequently been inaugurated, in the great centres of civilization, by the invasion of alien rulers, it is recognized that the immediate cause of decay must be the degeneration or depletion of the ruling classes."[134] To what extent alien infusions have been like external sources of energy counteracting an entropic trend is unclear. The historian Tenney Frank felt that the Roman decline was coeval with racial mixture.[135] What, might one ask, was the effect on Arab civilization in North Africa of admixtures with such peoples as the Berbers and the Sudanese Negroes? If sheer conquest has a revitalizing potential, one might perversely argue that the pogroms and rapine committed against the Jews had an intermittent eugenic effect through the agency of ethnic or racial mixture. The areas of ignorance here are immense: which racial mixtures are deleterious, which favorable, and in what respects and in what degrees, remains virtually *terra incognita*. Many, moreover, would prefer that these problems remain unexplored.[136]

X

Mark Twain once wrote that the Jew's "contributions to the world's list of great names in literature, science, art, music, finance, medicine, and abstruse learning are . . . way out of proportion to the weakness of his numbers." He felt that "nine-tenths of the hostility to the Jews comes from the average Christian's inability to compete successfully with the average Jew in business. . . ."[137] Justice Oliver Wendell Holmes, Jr. in 1926 told his friend the British jurist Sir Frederick Pollock, that it was "queer" to see the widespread prejudice against the Jews. To which Pollock replied that the others were "jealous."[138] Hitherto, explanations for this high attainment have been either racial or environmental. It has been felt that the choice was between these two hypotheses; the racial hypothesis proposed that the Jews as a race were primordially biologically endowed with a higher intelligence; the environmental hypothesis held rather that the higher achievements were due to such social influences as the tradition for scholarship, the closeness of the Jewish family, the persecution of the Jews which led them to prefer the learned professions, and the pressures on the Jewish child to work harder in order to survive and prosper in a hostile world.

The modified sociobiological theory, on the other hand, is neither racial nor environmental. No hypothesis is proposed that the Jews as a race *ab initio* were endowed with a genetic basis for intelligence superior to that of other races. In their origins the races may have been equal or unequal in intelligence; no assumption is made concerning the original state. The modified sociobiological theory rather affirms: whatever the primordial genetic endowments of races, certain values and practices in sexual relations will maintain the constancy of that endowment, whereas other practices, especially a differential adherence of the intellectual class to celibacy and homosexuality, will tend to diminish that endowment.[139] The biological basis of intellectual achievement is thus not a constant among the various races and peoples; the social practices with regard to sexuality can elevate the average genetic endowment for intelligence of a given race, or lower it.

The sociobiological theory of intellectual achievement does not, however, reduce to a case of sociological determinism. Standing, for instance, outside the deterministic model was the decision of ten thousand or so Jews, a small minority of them, to resist at the outset of the Middle Ages the sociological processes of assimilation and conversion. Their decision, exceptional as it was, arose from no requirement of the late Roman or early medieval mode of production, nor was it rational in terms of helping one's chances to survive. Such "exceptional" decisions can, however, be crucial in the evolution of a culture and a people.

Notes

1. Norbert Wiener, *Ex-Prodigy: My Childhood and Youth* (New York, 1953) pp. 11–12. Also cf. J. B. S. Haldane, *Heredity and Politics* (New York, 1938), p. 162.
2. Nathaniel Weyl, *The Creative Elite in America* (Washington, 1966) p. 92. Also cf. Lewis S. Feuer, *The Scientific Intellectual: The Psychological and Sociological Origins of Modern Science* (New York, 1963) p. 308. Raphael Patai, *The Jewish Mind* (New York, 1977) pp. 306, 334.
3. Arthur Ruppin, *The Jews in the Modern World* (London, 1934) p. 79.
4. Arthur Ruppin, *The Jews of To-day,* tr. Margery Bentwich (London, 1913), p. 73.
5. Mark Zborowski and Elizabeth Herzog, *Life Is with People: The Jewish Little-Town of Eastern Europe* (New York, 1953), p. 129.
6. Salo W. Baron, *A Social and Religious History of the Jews.* Vol. II (New York, 1952), p. 210.
7. Cf., the proverbs on "celibacy" in Reuven Alcalay, ed., *Words of the Wise* (Jerusalem, 1910), p. 62.
8. Solomon Schechter, *Studies in Judaism; Second Series* (Philadelphia, 1908), p. 95.
9. Salo Wittmayer Baron, *The Jewish Community: Its History and Structure to the American Revolution,* Vol. II (Philadelphia, 1942), p. 308.
10. Israel Abrahams, *Jewish Life in the Middle Ages,* New Ed. (London, 1932),
11. Ibid., p. 106.
12. Max Arzt, "The Teacher in Talmud and Midrash," in Jewish Theological Seminary of America, *Mordecai M. Kaplan Jubilee Volume,* ed. Moshe Davis (New York, 1953), p. 45.
13. Isidore Fishman, *The History of Jewish Education in Central Europe from the End of the Sixteenth to the End of the Eighteenth Century* (London, 1944), pp. 30–31.

414   Varieties of Scientific Experience

14. *Josephus,* tr. H. St. J. Thackeray (London, 1926), Vol. I, pp. 187, 375.
15. Isidore Fishman, *The History of Jewish Education in Central Europe,* p. 17.
16. Ibid., p. 38.
17. Solomon Maimon, *An Autobiography,* tr. J. Clark Murray, ed. Moses Hadas (New York, 1967), p. 16.
18. Isaac Bashevis Singer, *In My Father's Court* (New York, 1966), p. 44.
19. Isidore Fishman, *The History of Jewish Education in Central Europe,* p 125.
20. Zborowski and Herzog, *Life Is with People,* pp. 131, 136, 276, 272.
21. Salo W. Baron, *The Jewish Community,* Vol. II, p. 362.
22. Chaim Wolf Reines, "Public Support of Rabbis and Scholars" in *Yivo Annual of Jewish Social Science,* Vol. VII (New York,1966), p. 44.
23. Isidore Fishman, *The History of Jewish Education in Central Europe,* p. 125.
24. Salo W Baron, *The Jewish Community,* Vol. II, p. 187.
25. Cf. Schechter, *Studies in Judaism,* Second Series, pp. 95–96.
26. Max Arzt, "The Teacher in Talmud and Midrash," p. 43.
27. Cf. Elijah Bortniker, "Education," *Encyclopaedia Judaica,* Vol. 6 (Jerusalem, 1971), p. 414.
28. Isidore Fishman, *The History of Jewish Education in Central Europe,* p. 56.
29. Ibid., pp. 62–63.
30. Solomon Maimon, *An Autobiography,* p. 52.
31. Salo Wittmayer Baron, *Steeled by Adversity: Essays and Addresses on American Jewish Life* (Philadelphia, 1971), p. 155.
32. Solomon Maimon, *An Autobiography,* p 108.
33. Salo W. Baron, *The Russian Jew under Tsars and Soviets,* New York, 1964, p. 139.
34. In 1820, the Jews numbered 1,600,000 out of a European Russian population of 46,000,000. This rose in 1851 to 2,400,000 Jews in a population of 61,000,000 in European Russia.
35. Salo W. Baron, *The Jewish Community,* Vol. II, p. 81.
36. Cecil Roth, *The Jewish Contribution to Civilization* (London, 1938), p. 229. The estate of Aaron of Lincoln, "the wealthiest Jew of his time," was evidently mostly confiscated by the king at his death in 1186, so that his two sons actually were not his heirs. Cf. H. G. Richardson, *The English Jewry under Angevin Kings* (London, 1960), p. 115.
37. Cecil Roth, *The Jewish Contribution to Civilization, loc. cit.*
38. H G. Richardson. *The English Jewry,* pp. 5, 229.
39. Cecil Roth, "The Ordinary Jew in the Middle Ages," in *Studies and Essays in Honor of Abraham A. Newman* (Philadelphia, 1962), p. 22.
40. Ibid., pp. 23–24.
41. Raphael Straus, *Regensburg and Augsburg,* tr. Felix N. Gerson (Philadelphia, 1939), p. 228. Arthur Ruppin, *The Jews in the Modern World,* London, 1934, pp. 71–72.
42. Irving A. Agus, *Urban Civilization in Pre-Crusade Europe,* Vol. I (Leiden, 1968), p. 40.
43. H. G. Richardson, *The English Jewry under Angevin Kings,* pp. 219–220.
44. S. M. Dubnow, *History of the Jews in Russia and Poland,* tr. I. Friedlaender (Philadelphia, 1916–1920), Vol. I, p. 66.
45. Dora Litani, "The Destruction of the Jews of Odessa in the Light of Rumanian Documents," in *Yad Vashem Studies on the European Jewish Catastrophe and Resistance,* ed. Nathan Eck and Aryeh Leon Kubovy (Jerusalem, 1967), p. 137.
46. Discussion with Raul Hilberg, author of *The Destruction of the European Jews,*

Burlington, Vermont, June 14, 1972. The first lists of survivors, narrated Simon Wiesenthal, an organizing spirit in the search for Nazi war criminals, were "nomads, vagabonds, beggars." Cf. Gerh Korman ed. *Hunter and Hunted: Human History of the Holocaust* (New York, 1913), p. 290. Also cf. Helen Epstein, *Children of the Holocaust: Conversations with Sons and Daughters of Survivors* (New York, 1979), pp. 161, 166. Also cf. Yehuda Bauer, *Flight and Rescue: Brichah* (New York, 1970), pp. 3–4.
47. Dubnow, *History of the Jews in Russia and Poland,* pp. 263–264.
48. Abrahams, *Jewish Life in the Middle Ages,* p 169.
49. Ibid., p. 171.
50. Editorial Staff, "Shadkhan," *Encyclopaedia Judaica,* Vol 14, pp. 1254–1255.
51. Zborowski and Herzog, *Life Is with People,* p 273. Felix Aron Theilhaber, *Die Schädigung der Rasse durch soziales und wirtschaftliches Aufsteigen, bewiesen an den Berliner Juden* (Berlin, 1914).
52. Jacob Milch, "New Movements amongst the Jewish Proletariat: VII." *The International Socialist Review,* Vol VII, No. 10 (April 1907), p. 605.
53. Moses Hadas has suggested a distinction between the terms "apostate" and "convert." "The term apostate has a more unfavorable connotation than that of convert; whereas the convert may have acted from conviction or self-protection, the apostate is felt to have changed his religion for selfish purposes." The usages are, however, vaguely differentiated, and the terms are often used interchangeably. "Apostasy" has a more voluntary connotation than "convert"; thus, although the literature often discusses "forced conversions," there is no such usage as "forced apostasy." Also, "apostasy" is more often used to refer to an individual's decisions, in deviation from his group, whereas "conversion" refers more often to a collective phenomenon. The opprobrium attached to "apostate" reflects the close in-group feeling violated by the "deviationist," or "renegade," or excommunicate. Cf. Moses Hadas, "Apostates," *The Universal Jewish Encyclopedia,* Vol. I (New York, 1939), p. 427.
54. Cecil Roth, "Are the Jews Unassimilable?" *Jewish Social Studies,* Vol. III (1941), p. 5.
55. Dora Askowith, *The Toleration of the Jews under Julius Caesar and Augustus* (New York, 1915), p. 52.
56. Cecil Roth, "Are the Jews Unassimilable?," p. 5.
57. Ibid., p. 7.
58. Ibid., p. 7.
59. Ibid., p. 9
60. Ibid., p. 10.
61. Ibid. Also cf. Cecil Roth, *A History of the Jews in England,* Third Ed. (Oxford, 1964), pp. 209, 225.
62. Montagu Frank Modder, *The Jews in the Literature of England to the End of the 19th Century* (1939, reprinted, Cleveland, 1960), p. 82. The novel *Reuben Sachs,* published by the unhappy Amy Levy in 1888, depicted vividly the emotional strains of the educated youth among upper class Jews in England.
63. The role of Jews in the medieval bureaucracy is discussed in Heinrich Graetz, *History of the Jews,* ed. Bella Lowy (London, 1892), Vol. III, pp. 299–301, 527–538.
64. Salo Wittmayer Baron, *A Social and Religious History of the Jews,* Vol. VIII, Sec. Ed. (New York, 1952–1973), pp. 246, 249. Among the list of Jewish medical writers in Arabic during the Middle Ages, the names of converts to Islam recur frequently. Cf. Harry Friedenwald, *The Jews and Medicine: Essays*

(Baltimore, 1944), Vol. I, p. 172, ff.
65. Cf. Barry Spain, "Mathematics," *Encyclopaedia Judaica,* Vol. II, p. 1122. Baron, *A Social and Religious History of the Jews,* Vol. VIII, p. 173.
66. Cecil Roth, *Gleanings: Essays in Jewish History, Letters and Art* (New York, 1967), p. 174.
67. Ibid., p. 35. Cecil Roth, *A History of the Marranos,* Sec. Ed. (Philadelphia, 1959), pp. 14, 48. Also, Haim Hillel Ben-Sasson, "Apostasy," *Encyclopaedia Judaica,* Vol. 3, pp. 202–211.
68. Cf. Cecil Roth, *The Jewish Book of Days,* Rev. Ed. (New York, 1966), pp. 45, 195, 143, 215–216.
69. David Philipson, *Old European Jewries* (Philadelphia, 1894), pp. 143–145. Louis Wirth, *The Ghetto* (Chicago, 1928), pp. 59–60. Cecil Roth. *A History of the Jews in Italy* (Philadelphia, 1946), pp. 316, 409.
70. Zvi Y. Gitelman, *Jewish Nationality and Soviet Politics: The Jewish Sections of the CPSU, 1917–1930* (Princeton, 1972), pp. 298–299.
71. Yitzhak Baer, *A History of the Jews in Christian Spain,* Vol. II (Philadelphia, 1966), p. 93. Also cf. Heinrich Graetz, *History of the Jews from the Earliest Times to the Present Day,* Vol. III, tr. Bella Löwy (London, 1891), pp. 179, 299, 301, 527, 533, 538. "[A] large number of Bolshevik leaders and active party members were of Jewish parentage; but they had been completely assimilated . . ." Solomon M. Schwarz, *The Jews in the Soviet Union* (Syracuse, 1951), p. 93. The Jewish socialist parties, on the other hand, opposed the Bolshevik Revolution. Consequently, the Commissariat for Jewish National Affairs had difficulty in finding persons who could write in Yiddish for their Communist Jewish newspaper. Ibid., pp. 94–95.
72. Cf. Haim Hillel Ben-Sasson, "Apostasy," *Encyclopaedia Judaica,* Vol. 3, p. 206.
73. Lewis S. Feuer, "The Conversion of Karl Marx's Father." *The Jewish Journal of Sociology,* Vol. XIV (1972), pp. 154–157.
74. Solomon Maimon, *An Autobiography,* p. 89.
75. Gustav Karpeles, *Jewish Literature and Other Essays* (Philadelphia, 1895), p. 308.
76. Alex Bein, *Theodore Herzl: A Biography,* tr. Maurice Samuel (Cleveland, 1962), pp. 49, 94, 35. *The Complete Diaries of Theodore Herzl,* tr. Harry Zohn (New York, 1960), Vol. I, p. 7.
77. Beatrice C. Baskerville, *The Polish Jew: His Social and Economic Values* (New York, 1906), p. 331.
78. Baron, *The Russian Jew under Tsars and Soviets,* p. 81.
79. Cecil Roth, "Forced Baptisms in Italy," *Gleanings,* p. 243.
80. A. S. Halkin, "A Contra Cristianos by a Marrano," *Mordecai M Kaplan Jubilee Volume* (New York, 1953), p. 399.
81. Baer, *A History of the Jews in Christian Spain,* Vol. I, p. 378.
82. H. H. Ben-Sasson, "Apostasy," *Encyclopaedia Judaica,* Vol. 3, p. 207.
83. Ruppin, *The Jews of To-day,* p. 185.
84. H. G. Richardson, *The English Jewry under Angevin Kings* (London, 1960), p. 28.
85. Ibid., pp. 29–30.
86. Ibid., p. 35.
87. Robert E. Blumstock, "Mission to Jews: Reduction of Inter-Group Tension," *Practical Anthropology,* Vol. 14 (1967), p. 39. Robert Blumstock, "Fundamentalism, Prejudice, and Missions to the Jews," *The Canadian Review of Sociology*

and Anthropology, Vol. 5 (1968), p. 31.
88. Some error may arise from the fact that some scientists of Jewish ancestry have been extremely recalcitrant to any effort to link them to the Jewish community. Perhaps some data were suppressed, although journalists and biographers, eager for information, would make this difficult. "There was the celebrated case of Kari Landsteiner, a Viennese Jew, the famous discoverer of various blood groups, who sued the editors of a Jewish encyclopedia for libel because they dared to mention that he was of Jewish extraction." Cornelius Lanczos, *Judaism and Science* (Leeds, 1970), p. 2.
89. Leo Jung, ed., *Men of the Spirit* (New York, 1964).
90. Ibid., p. vi.
91. Mordecai L. Gabriel, "Biology," *Encyclopaedia Judaica,* Vol. 4, p. 1030.
92. Sidney Osborne, *Germany and Her Jews* (London, 1939). My assistant, Deena Mandel, has helped me with the research on the names in this volume.
93. Irving Agus, *Urban Civilization in Pre-Crusade Europe,* Vol. I, pp. 12–13.
94. *The Itinerary of Rabbi Benjamin of Tudela,* tr. A. Asher (reprinted, New York, 1927), Vol. I, pp. 36, 37, 38, 47, 55. The appellation "rabbi" was evidently honorific. Also, cf. Cecil Roth, "Benjamin of Tudela," *Encyclopaedia Judaica,* Vol. 4, p. 538.
95. H. G. Richardson, *The English Jewry under Angevin Kings,* p. 216.
96. S. W. Baron, *A Social and Religious History of the Jews,* Vol. II, p. 210. Also, Cecil Roth, *The History of the Jews of Italy,* p. 34.
97. In the Soviet Communist party, the number of Jews was likewise disproportionately large among the "deviationists," both left and right, who were embattled with Stalin and the dominant group. Cf. Z. Y. Gitelman, *Jewish Nationality and Soviet Politics,* p. 449 ff.
98. Villchardouin and De Joinville, *Memoirs of the Crusades,* tr. Sir Frank Marzials (London, 1908), p. 148. St. Jerome much earlier had complained of the Jews' polemical powers in the disputations between church and synagogue. Cf. S. Krauss, "The Jews in the Works of the Church Fathers," *The Jewish Quarterly Review,* Vol. VI (1894), p. 239.
99. "In the midst of a social organization where the populace was attached to the land, and where everyone was dependent upon a liege lord, they (the merchants) presented the strange picture of circulating everywhere without being claimed by anyone." Henri Pirenne, *Medieval Cities,* tr. Frank D. Halsey (Princeton, 1925), p. 131.
100. L. Rabinowitz, *Jewish Merchant Adventurers: A Study of the Radanites* (London, 1948), pp. 9–10, 15–22.
101. Agus, *Urban Civilization in Pre-Crusade Europe,* Vol. I, pp. 56–57.
102. H. G. Richardson, *The English Jewry under Angevin Kings,* p. 110.
103. *The Itinerary of Rabbi Benjamin of Tudela,* pp. 47, 49, 55, 45.
104. Cecil Roth, *Gleanings,* p. 26.
105. G. G. Coulton, *The Medieval Village,* p. 393.
106. *William Bateson, Naturalist, His Essays and Addresses,* ed. Beatrice Bateson (Cambridge, 1928), p. 311. "The average ability of the Athenian race is, on the lowest possible estimate, very nearly two grades higher than our own.... It has been a severe misfortune to humanity, that the high Athenian breed decayed and disappeared...." Francis Galton, *Hereditary Genius: An Inquiry into Its Laws and Consequences* (London, 1869, Sec. Ed., 1892), pp. 330–331.
107. Hans Kelsen, "Platonic Love," *The American Imago,* Vol. III, (1942), p. 9. Flavius Josephus writes of "the contempt for marriage" which had prevailed

among the Lacedaemonians, and "the unnatural vice so rampant" among the people of Elis and Thebes. Cf. *Josephus,* Vol. I, p. 403.
108. Aristotle, *Politics,* tr. William Ellis (London, 1912), p. 58.
109. Cf. Kelsen, "Platonic Love," p. 39.
110. Cecil Roth, *The Jews in the Renaissance,* Philadelphia, 1959, p. 45.
111. Arthur E. R. Boak, *Manpower Shortage and the Fall of the Roman Empire in the West* (Ann Arbor, 1955), pp. 80, 84, 129. Tenney Frank, "Race Mixture in the Roman Empire," *The American Historical Review,* Vol. XXVI (1916), pp. 704–705.
112. Josiah Cox Russell, "Late Mediaeval Population Patterns," *Speculum,* Vol. XX (1945), p. 171.
113. Ibid.
114. J. C. Russell, "Recent Advances in Medieval Demography," *Speculum,* Vol. XI (1965), p. 97.
115. Cf. Trout Rader, *The Economics of Feudalism* (New York, 1971), p. 58.
116. Josiah Cox Russell, "Medieval Population," *Social Forces,* Vol. 15 (1937), p. 506. In England, it should be observed, the number of the religious was halved by the plague, evidently being affected more than the population generally. Cf. J. C. Russell, "Late Medieval Population Patterns," p. 170. The four great orders of mendicant friars, Dominican, Franciscan, Carmelite, and Augustinian were, however, exempted from the poll tax and records of 1377. Their numbers have been estimated respectively as 1,889; 2,219; 945; and 765. Cf. Josiah Cox Russell, "The Clerical Population of Medieval England," *Traditio,* Vol. II (1944), p. 209.
117. The sociological factors, apart from the theological, which impelled the church toward a discipline of celibacy were discussed at length by Henry Charles Lea, *History of Sacerdotal Celibacy in the Christian Church* (New York, 1907), Vol. I, pp. 60–63, 408–409.
118. E. M. Sanford, "Contrasting Views of the Roman Empire," *American Journal of Philology,* Vol. LVIII (1937), pp. 454–456.
119. J. C. Russell, "The Ecclesiastical Age: A Demographic Interpretation of the Period 200–900 A.D.," *The Review of Religion,* Vol. V (1941), pp. 142–143. H. R. Betterman, "The Beginning of the Struggle Between the Regular and the Secular Clergy," *Medieval and Historiographical Essays in Honor of J. W. Thompson,* ed. Cate and Anderson (Chicago, 1938), p. 25.
120. J. D. Russell, "The Ecclesiastical Age," p. 145.
121. Ibid.
122. "A century and a half ago there was one priest to every thirty inhabitants in Spain," wrote Havelock Ellis, with some possible exaggeration. Havelock Ellis, *The Soul of Spain,* New Ed. (London, 1937), pp. 395–396.
123. Charles Darwin, *The Descent of Man and Selection in Relation to Sex,* Sec. Ed., (New York, 1874), p. 160.
124. Américo Castro, *The Structure of Spanish History* (Princeton, 1954), pp. 645–646.
125. Harold Livermore, *A History of Spain,* Sec. Ed. (London, 1966), p. 278. Another estimate for Spain as a whole which confirms this trend is given in Massimo Livi Bacci, "Fertility and Nuptiality Changes in Spain from the 18th to the Early 20th Century," *Population Studies,* Vol. XXII (1968), p. 83.
126. Altamira, op. cit., p. 137.
127. Ibid., p. 164. According to another source, the Spanish clergy at this time numbered 182,564 out of a population (including Minorca) of about ten and a half million. Antonio Domínguez Ortíz, *La Sociedad Española en el Siglo XVIII*

(Madrid, 1955), pp. 58, 123.
128. Russell, "The Clerical Population of Medieval England," p. 212.
129. Ibid., p. 179.
130. Cf. J.B.S. Haldane, "Alfred Kinsey," *Kinsey: A Biography,* ed. Cornelia V. Christenson (Bloomington, 1971), p. 230. Charles Edward Maliet wrote that "prolonged celibacy" has tended to make the College fellows "dreary and vinose." Cf. *A History of the University of Oxford* (London, 1927), Vol. III, p. 348. "A don of thirty," wrote Leslie Stephen, "was ten years older than a rising young barrister of forty." Sheldon Rothblatt, *The Revolution of the Dons: Cambridge and Society in Victorian England* (London, 1968), p. 191.
131. George Gordon Coulton, *Fourscore Years, an Autobiography* (Cambridge, 1943). p. 94. Joseph John Thomson, *Recollections and Reflections* (Toronto, 1936), pp. 74, 90, 274. Sheldon Rothblatt, *The Revolution of the Dons,* p. 242.
132. J. B. S. Haldane, *Heredity and Politics* (New York, 1938), p. 126.
133. Ibid.
134. Ronald Aylmer Fisher, *The Genetical Theory of Natural Selection* (Oxford, 1930), p. 237. J. B. S. Haldane, *Heredity and Politics,* p. 21.
135. Tenney Frank: "Race Mixture in the Roman Empire " *American Historical Review,* Vol. XXI (1916), p. 705 ff.
136. H. J. Eysenck, *The Inequality of Man* (London, 1973), pp. 14, 24. R. J. Herrnstein, *I.Q. in the Meritocracy* (Boston, 1973), p. 45.
137. Mark Twain, "Concerning the Jews," *Literary Essays,* Vol. 24, New York, pp. 275, 286.
138. Mark De Wolfe Howe. ed., *Holmes-Pollock Letters: The Correspondence of Mr. Justice Holmes and Sir Frederick Pollock 1874–1932* (Cambridge, Mass., 1941), Vol. 2, pp. 191–192.
139. The later distinguished Nobel laureate in genetics, and earlier a Marxist and admirer of the Soviet Union, H. J. Muller, once wrote that "it is easy to show that in the course of a paltry century or two . . . it would be possible for the majority of the population to become of the innate quality of such men as Lenin, Newton, Leonardo, Pasteur, Beethoven, Omar Khayyam, Pushkin, Sun Yat Sen, Marx." The list indeed was a curious conglomerate. Muller, however, feared that the Nazis would avail themselves of genetic knowledge to breed instead for "a maximum number of Billy Sundays, Valentinos, Jack Dempseys, Babe Ruths, even Al Capones." H. J. Muller, *Out of the Night: A Biologist's View of the Future* (New York, 1935), pp. 113–114.

17

Causality in the Social Sciences

In the social sciences, man is studying himself. And his basic attitudes towards himself, his hopes, loves, fears, and hatreds, reflect themselves in what we may call his meta-sociological convictions. These, in turn, express themselves in a choice, broadly speaking, between two modes of social analysis, which I shall call the *interventionist* and the *necessitarian*. The interventionist social scientist believes that men can intervene in social situations to change conditions and determine, in significant measure, the direction of trends. The necessitarian believes, on the contrary, that social science can never be used to deflect the lines of evolution, that men's contrary decisions are perturbations in irresistible movements. From meta-sociological beliefs, there thus arise two corresponding types of laws or models, interventionist and necessitarian, which may be characterized as follows:

A law conforms to a necessitarian model if it is one according to which no decision on the part of a person or group of persons can, given the existing social state of affairs, prevent the predictable successive states from coming into existence.

A law conforms to an interventionist model if it is one according to which the decision of a person or group of persons can intervene to alter the existing state of affairs so that it will be followed by states which would not have occurred and would have been unpredictable apart from those decisions.

It is the argument of this essay that contemporary social science is increasingly giving adherence to an interventionist mode of thought. In economics, political science, anthropology, interventionist models are becoming dominant, and scientists are operating on the basis of a meta-sociological principle of interventionism. Before we undertake to justify this analysis of social science, I should like to point out that the distinction between interventionist and necessitarian laws will be familiar to the student of ordinary economic theory. A competitive market, for instance, is described as one in which there is "a large number of buyers and sellers so that the influence of any one or several in combination is negligible."[1] The forms of the laws of supply and demand for such competitive situations are determinate; market prices are unique resultants without equiprobable alternatives, and individual or group intervention cannot affect their determination. By contrast, the laws of monopoly situations allow for the contribution of interventionist decision. The negotiation of wages, for instance, between a strong trade union and a monopolist employer has no unique possible outcome; there is no single equilibrium position, but rather a neighborhood of equally possible wage-rates. The laws of bilateral monopoly are said to have a "zone of indeterminateness."[2] Under such conditions, the action of a single individual or group can materially affect the determination of price. Where domains of indeterminacy are found, sociological laws have begun to lose their necessitarian character.

1. In political science, the greatest example in the present century of a necessitarian classic is Michels' *Political Parties*. The iron law of oligarchy, according to this work, is "the fundamental sociological law of political parties."[3] Michels argued that no movement can hope to produce profound or permanent changes in the social structure. The inevitable tendency to oligarchy, he said, is inherent in all organization. Trade unions, monastic orders, corporations, and socialist parties are all subject to its workings. With masterly realism, Michels drew an impressive documentation for the law of oligarchy from the organization of revolutionary parties. Other investigators had been led to similar conclusions. Sidney and Beatrice Webb, devoted servants of the Fabian ideal, concluded, after their many years' study of British trade unionism, that there is a universal tendency for the primitive democracy of trade unions to be transformed into personal dictatorships or bureaucracies.[4]

Michels observed the drama of human revolt against the iron law of oligarchy. He counseled the wisdom of resignation. But political sociologists today are often reluctant to acquiesce to Michels' law. They argue that the tendency to oligarchy is neither universal nor necessary; they hold that oligarchic leadership arises under special social and psychological conditions. If most unions tend to become oligarchical, there is still the example of the Typographical Union with its long history of an organized two-party system. If Soviet forms are examples of bureaucratic control, there are also the collective settlements of Israel which are seeking by institutional safeguards to restrain oligarchic trends.[5] A sociology of democratic leadership tries to supplant Michels' law; it investigates the conditions of political apathy, it inquires into the extent to which a proliferation of rival oligarchies weakens their respective internal powers, it applies the psychoanalytical method to power-seeking and leader-craving personalities, it studies how oligarchical trends can be mitigated and controlled. The entirety of these efforts will not as yet measure up to the conviction and evidence behind Michels' law. But the meta-sociological standpoint which animates political sociologists today is clear. They are seeking interventionist models of sociological law which will provide the guide for effective contra-oligarchical action.

From the logical standpoint, what characterizes necessitarian laws is that their independent variables are (what I shall call) *inaccessible*. Michels' law, for instance, holds that as an organization matures, it becomes oligarchical. Oligarchical structure is thus a function of time, and nothing can be done by human intervention to arrest the passage of time. The time-variable is inaccessible. Again, in the case of the laws of the competitive market, supply and demand as independent variables cannot be affected by any individual or group. They determine price, the dependent variable, but they themselves are inaccessible to human intervention in the given context. Interventionist models of causal law, on the other hand, are characterized by the quest for independent variables which will be, in large part, *accessible*. If oligarchical trends can be shown to depend on certain specific psychological traits, then intervention on the level of basic personality structure may avail to counteract them. The independent variable, basic personality structure, is presumably accessible through the controls of infant care and child rearing.

2. The power of the meta-sociological principle of interventionism guides social thought in regions where the criterion of verification grows tenuous. A sociological theory wins the allegiance of social scientists not so much because it has more empirical evidence on its

side but because it opens up possibilities of human action, of human intervention. The system of economic thought which has become regnant in the last generation is the Keynesian. It is amazing how little verification has had to do with its reception.[6] Keynesian ideas have been accepted not because they explained more than others but because they provided a set of causal laws whose independent variables were accessible to action in the immediate present. Roy Harrod in his monumental biography of Keynes states the matter incisively:

> Keynes certainly claimed to be promoting a revolution of thought. The more comprehending critics have had some doubt, on the ground that his main work did no more than substitute one system of concepts for another. In the physical sciences some crucial test is usually available to decide between conflicting theories. If Keynes was really to be successful, he should have been able, it is argued, to refute, say Mr. D. U. Robertson, by showing a set of facts which the Keynesian doctrine would fit, while the other would not. Unhappily the state of economics is not so advanced. It is true to say that the Keynesian scheme consisted in essence of new definitions and a re-classification. . . . In a certain sense one cannot dogmatically affirm one way to be right and the other to be wrong. . . . It is by actual use and application, not by logic, that Keynes has been, and will, I am confident, continue to be triumphantly vindicated.[7]

Underlying the reception of Keynesian theory was the will to intervene in economic processes. When many men are unemployed, and when the social feelings of the economist are genuine, there is a desire to be able to do something about human misery. We don't care to say: "in the long run, the automatic, self-regulating operations of supply and demand will eliminate sub-marginal firms, and lower wages till the demand for labor rises." "In the long run," as Keynes said, "we are all dead." Not that we can really refute the non-interventionist. Herbert Hoover is convinced that if only we had followed his policy of doing nothing for another six months in 1932, then recovery would have set in. And he ascribes the unemployment of the next decade to the pursuit of Keynesian policies.[8] There is no crucial refutation of this thesis, but it runs counter to the will to intervene which is the response of unresigned men to social crisis.

A meta-economic standpoint, a philosophy of history, was the source of the Keynesian interventionist system of causal laws. Schumpeter has observed that every comprehensive economic theory consists of two complementary but distinct elements. There is first the thinker's vision, his view as to the basic features of society, "about what is and what is not important in order to understand its life at a given time."

Secondly, there is the theorist's, technique, the apparatus "by which he conceptualizes his vision."[9] Keynes had stated his vision in his early *Economic Consequences of the Peace*. His later theory was the product "of a long struggle to make that vision of our age analytically operative." What was Keynes' vision? In essence, it was a commitment to the postulate of interventionism during an era when impersonal forces might otherwise make for the collapse of capitalism. Keynes' mind was filled with images of the struggle between human ideas and an Immanent Will. In one mood, he was depressed by the necessitarian forces of history, and he wondered if a self-destructive impulse inevitably overcame ruling classes. "Perhaps it is historically true," he wrote, "that no order of society ever perishes save by its own hand. In the complexer world of Western Europe the Immanent Will may achieve its ends more subtly and bring in the revolution no less inevitably through a Klotz or a George than by the intellectualisms ... of the bloodthirsty philosophers of Russia."[10] He spoke of the spectacle of the extraordinary weakness of the great capitalist class, the terror and personal timidity of its individuals. At the last, however, Keynes held fast to the faith that human ideas and individual decision can affect the course of history. The hidden currents beneath the surface of political history, he said, are unpredictable in their outcome. "In one way only can we influence those hidden currents,—by setting in motion those forces of instruction and imagination which change *opinion*. The assertion of truth, the unveiling of illusion, the dissipation of hate, the enlargement and instruction of men's hearts and minds, must be the means."[11] The practical fruit of Keynes' interventionist faith were the proposals in his *General Theory* to apply economic wisdom in fiscal policy and public investment in order to achieve a high level of employment, and preserve the capitalist system.

Whenever the attempt is actually made to verify the efficacy of human decision and attitude in historical crisis, the blind alley of unverifiability besets us. Schumpeter, for instance, affirmed that the decline of the capitalist system does not arise from the workings of an economic pattern like vanishing investment opportunity. He held that capitalism is decaying because of the spread of an anti-capitalist mentality, because profit-seeking is under the taboo of moral disapproval.[12] How can this theory be weighed as against the Marxian or Keynesian? Prolonged economic depression will invariably be accompanied by a loss of confidence in the dominant system. To test Schumpeter's theory, we should have to exhibit a battered economy which still preserved the undaunted confidence of its citizens. But we are dealing with inseparable variables, loss of faith and economic depression, which cannot be exhibited in isolation. How shall we verify Schumpeter's theory concerning the causal role of human attitudes? In the end, it embodies like the Keynesian philosophy a conviction in the power of human attitudes to mould the workings of impersonal economic forces.

3. It might perhaps be gathered that the necessitarian mode of thought in economics is the monopoly of the Marxians. Certainly, the Marxian law of the decline of capitalism is a supreme example of the necessitarian model, and its apocalyptic finalism has given to the communist movement a kind of religious exaltation.[13] At the same time, however, it is striking that there is a widespread school which conceives of the inevitable cataclysm of socialism in the same necessitarian manner. The hasty prophets of the debacle of socialism, Hayek and Mises, have evolved a species of dialectical law akin to the Marxian. These thinkers, in Hayek's words, hold that in economic analysis there are "inherent necessities determined by the permanent nature of the constituting elements."[14] In conformity with this belief, Mises, as far back as 1920, argued that rational calculation is impossible in a socialist economy, and that the Soviet economy was bound to collapse. Soviet society, he wrote, is "in a state of entire dissolution," and a "closed peasant household economy" is replacing the disintegrating order.[15] The necessitarian thus held to a simple law: every socialist economy must inevitably founder because it cannot solve the problem of economic calculation.

As the years went on, Mises' prediction remained unfulfilled. But no necessitarian need ever abandon his hypothesis if facts delay its verification. To save his "law" after fifteen years' delay in its working, Mises proposed an unusual modification in its terms. Soviet society, he said, had endured because it was surrounded by capitalist economies from which it derived its standards of rationality. Leon Trotsky held that socialism was impossible in one country; Mises, we might say, held that it's possible only in one country.[16] Still another decade and generation passed by, and the collapse of Soviet economy was still postponed. The economic necessitarians, such as Hayek, have now tended to become sociological necessitarians. They no longer emphasize the prediction of the decline of socialist economy. What they now insist upon is the sociological causal law: that every socialist society must inevitably become totalitarian, because the power of the socialist planners must grow to a dictatorial magnitude. Hayek, therefore, predicted that individual freedom, civil rights, intellectual liberties, would all vanish in a socialist world. It is curious that such critics of socialism have been led to adopt an extreme economic determinism, the principle that the form of economy determines the political superstructure. And the defense of socialist thinkers, notably R. H. Tawney, has been that socialist culture is not bound by a necessitarian law, that human initiative and intervention can help fashion the moral and political relations which will rest on the socialist economic foundation. Socialists today are beginning to base their proposals not on a necessitarian creed, but on the postulate of interventionism.[17]

4. The quest for a new type of interventionist theory has become most pronounced in the science of anthropology during the years after the Second World War. With the development of programs for technical and educational assistance to the peoples of backward areas, programs associated with the United Nations, the Food and Agricultural Organization, the World Health Organization, and the American Point-Four, the pre-war necessitarian anthropological theory has been receding into a rapid obsolescence. Functionalism was the dominant standpoint in the pre-war years. Some may be surprised to hear it described as an example of the necessitarian mode. But the accuracy of this description can be briefly shown.

Functionalism, in the form which Malinowski gave it, affirms that culture is an "organic unity"; it is the principle that in every culture, each custom, belief, and behavorial form "represents an indispensable part within a working whole."[18] According to this scheme of thought, deliberate intervention to modify or reconstruct a culture is foredoomed to failure. To the functionalist, there are, indeed, no accessible variables in the analysis or control of cultures. To tamper with one institution is at once to rend the fabric of the total society.[19] Each culture is like an idealistic Absolute, a totality which cannot be altered by modifying strategic segments, an "organic unity" which is disrupted by the crude hands of social planners. The policy consequence of functionalism was straightforward; its practical bearing for applied anthropology was the admonition: don't try to disturb cultures, don't try to change them. Colonial administrators were exhorted not to lay hands upon the delicate patterns of native culture. An imperial proconsul like Lord Lugard was especially praised because his method of so-called "indirect rule" exemplified functionalism in action. The practical value of his standpoint, said Malinowski, was "that it can help the white man to govern, exploit, and 'improve' the native with less pernicious results to the latter." How shall we, for instance, deal with sorcery? Shall we undermine the sorcerer's authority with medical and technological assistance? Malinowski's answer is that "since it invariably ranges itself on the side of the powerful, wealthy, and influential, sorcery remains a support of vested interest; . . . It is always a conservative force, . . . There is hardly anything more pernicious, therefore, in the many European ways of interference with savage peoples, than the bitter animosity with which Missionary, Planter, and Official alike pursue the sorcerer."[20] Those who would manage native labor and exploit effectively the resources of tropical countries would do well, says Malinowski, to study functionalist anthropology.[21] Would it contravene the evidence to assert that functionalism was an expression of the wisdom of a managerial imperialism?

Positivist critics have sometimes suggested that functionalism is nothing more than a tautology. If that were the case, we should be much less concerned with its consequences. It is characteristic, however, of some thinkers that they desire to prove that nobody is saying anything about anything, an aspect of the self-destructive drive which seems to underlie much contemporary philosophizing. The functionalist concept in anthropology, however, is used in much the same way as the notion of equilibrium in economics.[22] Where a culture has persisted virtually without change for the lifetime of several generations or more, one can assume that the institutions of its social system have achieved an equilibrium with respect to each other. The religious institutions, for instance, have come to sustain the tribal economy, they raise no voice of criticism, the mythology and folk tales which are prevalent support the political authority, the economy subsidizes the religious leaders. Social institutions are, in the last analysis, founded on human needs, but what is important to observe is that each cultural equilibrium imposes a kind of schedule on needs. It decides which needs are more and less important, it determines the extent to which given needs will be expressed or repressed. Every culture has its unique *repression-expression* schedule, the resultant line of equilibrium located by the interactions of institutions upon men's emotional needs. That every culture is an "organic unity" is a vague way of trying to say something similar to what the economist says when he holds that all the agents of production and consumption in a system are in general equilibrium.

What functionalism, however, has failed to emphasize is that there are cultural equilibria at higher and lower levels of human satisfaction. Keynes dwelled on the fact that there are alternative states of equilibrium possible, each with its respective level of unemployment. An economy can be in equilibrium, despite a high measure of unemployment with the consequent suffering of citizens. Similarly, a culture can be at functional equilibrium with a repression-expression schedule which imposes much hardship upon its members. And such a society, although it's "integrated" and a "living whole," is exceedingly vulnerable to external stimuli; its equilibrium is unstable. The anthropological philosophy associated with Point Four recognizes the vulnerability of cultures in unstable equilibrium, and proposes that social science help intervene in the domain of those accessible variables which control the proportions of repression.[23]

Hitherto, the anthropologist in the field has tried hard to be inobtrusive, not to affect the social system which he was observing. The skilled anthropologist has been something like a painter, and it is not surprising that his categories have been aesthetic. When directed change is, however, introduced within segments of a culture, latent dissatisfactions are brought to the surface, unvoiced aspirations for different ways of living are articulated, the degree of intensity of resistances to specific reforms can be ascertained. There are in India, for instance, religious resistances to a program of birth control education. But an experiment in cultural reconstruction has been undertaken by the Government together with the World Health Organization. Does the culture of the Indian peasantry fall apart? Rather the peasants welcome the new program, and urge its extension as a way of dealing, at least in part, with poverty and starvation.[24] The functionalist was concerned to strengthen those agencies which would resist cultural change. The interventionist seeks out the loci of discontent, helps people to express their needs, and renders assistance to those persons and groups who have felt especially the impact of repressive forces, and who would lead their societies in cultural innovation.[25]

5. There is a tendency to assume that a necessitarian standpoint, if adopted, applies to all social systems, and the interventionist is likewise apt to extend his model to all history. It may be the case, how-

ever, that human decision counts for more in certain social systems than in others. Not even a hero might have prevailed against the forces making for the decline of the Western Roman Empire, and much less than heroic intervention might change hopefully the course of events in our time. A recent necessitarian historian insists that "there was no way out for the Roman Empire."[26] He builds up a kind of immanent dialectical law for its decline: the Empire was based on slavery, the internal market flagged because of the extremes of wealth and poverty, the Roman peace entailed a drying-up of the main source for the supply of slaves. But the necessitarian schemas are never quite convincing. Manumission of slaves became extensive in Rome, and to add to the confusion, if slaves had been needed, their breeding could have been attempted. Moreover, there is always the embarrassing fact of the continued survival of the Eastern Roman Empire for almost a thousand years despite the operation of the same factors that are alleged to have brought Western Rome to its end. Barbarian invasions, the dominance of large estates, the corporative economy, were as much a part of the history of the Byzantine Empire as of the West, but it maintained its existence, nonetheless.[27] The historical pattern is made of unruly material, recalcitrant to designs of inevitability. But at the same time, we find ourselves reluctant to accept the view of J. B. Bury that it was a conflux of coincidences which brought the Roman Empire to its downfall. General causes, Bury held, do not usually explain the great events of history; the chance conjunction of unforeseeable Asian irruptions and a series of incompetent emperors, in his opinion, brought Rome to its decline.[28] Impersonal social forces of a powerful kind do, however, impress one as having been at work in the Western Empire; perhaps the social determinants were not sufficient to constitute a closed necessitarian system, but the margin for human intervention, the degree of freedom which the Roman social system allowed, was small as compared to later societies. Individual decision was overwhelmed in a relatively necessitarian environment. Anything like direct experimental evidence is beyond us with respect to this question, but we can argue that just as economic laws are sometimes necessitarian and sometimes, more or less, interventionist, so the same diversity may likewise characterize the modes of social systems.

The old formulations in the philosophy of social science have become obsolete. The social scientist today is neither historical idealist nor historical materialist; he tends to be what we might call an historical interventionist. He may agree with the Marxian that economic variables are, in many sociological laws, the accessible ones. The importance which sociologists assign to land reform in the Middle and Far East is one such acknowledgment of economic primacy. At the same time, the historical interventionist denies that the fact of human decision can itself be always subsumed in a scheme in which it is dependent on economic independent variables.

Take, for instance, the choice which confronted the Jews of Spain in 1492. On March 30th of that year, the Spanish monarchs decreed that Jews must become Catholics or otherwise leave the country. A few months' grace was allowed for their decision; the Jews furthermore were forbidden to take with them any gold or silver. An inventory of all the facts of the Spanish mode of production, a knowledge of the role of the Jews in the Spanish economy,—none of these economic facts would have enabled us to predict the decision of the greater part of the Jews. "Stunned at first by the blow, as soon as they rallied from the shock, they commenced preparations for departure. . . . The sacrifices entailed on the exiles were enormous. . . . There were comparatively few renegades. . . . There was boundless mutual helpfulness; the rich aided the poor and they made ready as best they could to face the perils of the unknown future."[29] Nor was the decision of the Jews the outcome of the influence of some single outstanding individual or individuals. Without a hero to guide them, it was the group which made its decision as to how to act in a great crisis.

6. At every great historical crisis, a situation obtains which we may describe as one of "existential indeterminacy." When the laws of a social system which is breaking down provide no clear basis for predicting the structure of the successor social system, an interval of existential indeterminacy ensues.[30] Let us imagine, for instance, that some primitive society has been making its living through some form of hoe culture. Over a period of years, its population has come to exceed its output of subsistence. A social crisis then confronts the group. What shall be its response? There are, indeed, a number of possible responses to the problem. The society may revise its values to allow for infanticide, or it may decide to destroy its old, or it may embark upon cannibalism. Through chance or inventive genius, it may perhaps discover and adopt a more intensive mode of agriculture; on the other hand, perhaps some warlike leaders will gain the ascendancy, and lead the people into a war for their neighbor's lands.[31] Or perhaps, like the ancient Hellenes on an arid soil, they will take to the

sea, to become fishermen or traders. A society in crisis is in unstable equilibrium; some fortuitous occurrence may then be decisive in determining which of several possible outcomes will be realized. But an exhaustive knowledge of the structure and functioning of the system in crisis will not enable us to foretell by which alternative order it will be followed.[32]

The postulate of interventionism is the assertion that in critical social situations, men can shape their own history in some partial sense; the postulate affirms that human decision can be crucial in determining which of several alternative solutions to a crisis will be realized. To put it in more analytic terms, the postulate of interventionism affirms that, in critical situations, the unknown governing laws are laws nonetheless in which one or more of the independent variables are accessible, and which we discover experimentally. Now there is a temptation to set up an interventionist metaphysics just as there is a necessitarian one. Existentialism seems to me precisely a metaphysics of intervention in the sense that it holds that free choice is inherent in every situation. But the social scientist, we might say, is not a metaphysical existentialist, but rather a logical or methodological one. He finds situations where human freedom is reduced to nullity, where man far from being the captain of his soul is more like a private on fatigue. He finds fortunate conditions under which men are sometimes free. The postulate of interventionism cannot be proved. It is a working conviction of human beings who know at any rate that no necessitarian law of history can be demonstrated. Perhaps, unbeknown to us, the ultimate meta-historical truth may be that man has been lured by a malevolent God to seek for invention, that the inevitable culmination of human history is the self-extinction of the human race through cobalt bombs. The rise and decline of the human race would then be an underlying necessitarian law of history. And the Luddites, the occasional machine-wreckers, would have been the illiterate, unsung tragic heroes who tried to save mankind from itself. Bertrand Russell, indeed, has frequently wondered if this were not the truth of things; he felt, as his hope in Soviet communism dwindled, that "all politics are inspired by a grinning devil, " and, as he watched Europe on its way to a Second World War, he speculated that perhaps the death-wish was the best explanation of human history.[33] But if such be the underlying law of the macro-social cosmos, we shall never know it.

A directional law of evolution is not logically impossible. The law of entropy in physical science asserts that for closed systems there is a definite direction that energy transformations will take. A directional law of social change cannot therefore be excluded on purely logical grounds. The law of entropy, however, holds only for closed systems; if there were an influx of energy from other parts of an unlimited universe, a given region might not actually evolve toward thermodynamical equilibrium. And laws of history are beyond our grasp precisely because human initiative, human intervention, upsets any approximation toward a closed social universe. The factor of human creativity corresponds to the influx of energy into the physical system. It makes, moreover, for the unpredictable aspect of historical causation. The discovery of atomic fission, for instance, was not something foreseeable from our laws of social science; it was an unpredictable event, dependent on what happened to be the case with sub-atomic events and on the abilities and interests of a small number of persons.

We are once more confronted by the indeterminate aspect of history, the contribution of human individuals, in all its elusive, unresolvable vagueness. A necessitarian law of history may exist, but if so, it does not fall within the domain of humanly constructible laws of social science.[34]

And on this slender foundation, the postulate of interventionism exists. It is a hope of human beings groping in darkness. The necessitarian may hold that all our interventions have been and will be, in principle, predictable, and he might argue that as human beings we are under an illusion of our effective intervention in history, that perhaps an unkind god watches us in our futile efforts to fend off Necessity. It may be that the limits of human intervention will gradually become evident. The countless frustrated efforts to contrive perpetual machines finally provided an empirical basis for the law of entropy. Perhaps a repeated failure of schemes for world peace will similarly persuade social scientists that Freud's theory of the innate aggressive drive in man is sound. But, for the present, there is the rare chance to dare to use the interventionist philosophy. The necessitarian has amended Heraclitus and, instead of saying that "man's character is his fate," declares as a principle of social science that "man's social system is his fate, but that that fate is uncertain." The interventionist works on the hope that men can, under certain conditions, choose their social world. We act in blindness but in the hope that our action takes us toward light.

Notes

1. Edward Chamberlain, *The Theory of Monopolistic Competition,* Harvard University Press, 1933, p. 7.
2. Joseph A. Schumpeter, *Business Cycles,* Vol. I, New York, 1939, p. 59.
3. Robert Michels, *Political Parties,* translated by E. and C. Paul, London, 1915, pp. 401–402.
4. Sidney and Beatrice, *Industrial Democracy,* New York, 1902, p. 36.
5. S. M. Lipset, "The Two Party System in the International Typographical Union," *Labor and Nation,* Vol. VI (1950), p. 34. Lewis S. Feuer, "Leadership and Democracy in the Collective Settlements of Israel," in *Studies in Leadership,* ed. by Alvin W. Gouldner, New York, 1950, pp. 363–385.
6. Cf. Lawrence R. Klein, *The Keynesian Revolution,* New York, 1947, p. 107. Also, Lord Beveridge, *Power and Influence,* London, 1953, p. 253.
7. R. F. Harrod, *The Life of John Maynard Keynes,* New York, 1951, pp. 462–463.
8. *The Memoirs of Herbert Hoover, The Great Depression 1929–1941,* New York, 1952, pp. 267, 351, 475, 482.
9. Joseph A. Schumpeter, *Ten Great Economists,* New York, 1951, p. 268. "He was childless and his philosophy of life was essentially a short-run philosophy. So he turned resolutely to the only 'parameter of action' that seemed left to him, both as an Englishman and as the kind of Englishman he was—monetary management." Ibid., p. 275.
10. John Maynard Keynes, *The Economic Consequences of the Peace,* New York, 1920, pp. 237–238.
11. Ibid., pp. 296–297.
12. Joseph A. Schumpeter, "Capitalism in the Postwar World," in *Postwar Economic Problems,* ed. by Seymour E. Harris, New York, 1943, pp. 119–121.
13. It is noteworthy that according to Stalin, bureaucratic interventionism has terminated the Marxian law of dialectical necessity. The "law of transformation of quantity into quality," he wrote, "is not at all compulsory for a society which has no hostile classes." The transition to collective farms was peaceable and gradual, he declared. "And we succeeded in doing this because it was a revolution from above, because the revolution was accomplished on the initiative of the existing power . . ." (Joseph Stalin, *Marxism and Linguistics,* New York, 1951, pp. 27–28). The political use of doctrines is capable of strange permutations. A movement of social liberation may espouse a necessitarian philosophy, while interventionism, in its bureaucratic version becomes the apologetic for tyranny.
14. *Collectivist Economic Planning,* ed. by F. A. von Hayek, London, 1935, p. 12.
15. Ibid., p. 124.
16. "The attempt of the Russian Bolsheviks to transfer Socialism from a party programme into real life has not encountered the problem of economic calculation under Socialism, for the Soviet Republics exist within a world which forms money prices for all means of production. . . . Without the basis for calculation which Capitalism places at the disposal of Socialism, in the shape of market prices, socialist enterprises would never be carried on,. . . ." Ludwig von Mises, *Socialism,* translated by J. Kahane, London, 1936, p. 136.
17. A form of necessitarian doctrine meanwhile has been adopted by sociologists like W. Lloyd Warner who argue that "classless societies are impossible" because the division of labor requires supervisory functions. If a group of supervisors is a class, then the existence of classes is a tautological consequence of the division of

labor. A necessitarian argument is valid in this sense, but this tautology is scarcely the meaning intended by those who have argued for social and economic equality. Cf. W. Lloyd Warner, Marchia Meeker, Kenneth Eels, *Social Class in America,* Chicago, 1949, pp. 3–32.
18. Bronislaw Malinowski, "Social Anthropology," *Encyclopaedia Britannica,* Fourteenth Edition, New York, 1929, Vol. 20, p. 864.
19. As S. H. Roberts stated: "An attack on one part of a closely interrelated native structure usually meant the collapse of the whole" ("Native Policy," *Encyclopaedia of the Social Sciences,* Vol. XI, 1933, p. 274).
20. Bronislaw Malinowski, *Crime and Custom in Savage Society,* London, 1926, p. 93.
21. Ibid., p. 1. "Social Anthropology," loc. cit., p. 864. Malinowski adds: "Since conservatism is the most important trend in a primitive society, sorcery on the whole is a beneficent agency, of enormous value for every culture" (*Crime and Custom in Savage Society,* p. 94).
22. Cf. A. R. Radcliffe-Brown, "A Note on Functional Anthropology," *Man,* Vol. XLVI, 1946, p. 40. This valuable essay traces the various stages in Malinowski's thought, and records the fact that the name "functionalism" was derived from the functional jurisprudence of Roscoe Pound.
23. As the English psychologist, F. C. Bartlett, has written: "Every culture has its 'hard' or its 'soft' points. If change is first sought at the former, it will provoke resistance and very likely open discord, while the latter are yielding and it is from them that reformation will spread." Cf. "Psychological Methods for the Study of 'Hard' and 'Soft' Features of Culture," *Africa,* Vol. XVI (1946), p. 145.
24. S. Chandrasekhar, "The Prospect for Planned Parenthood in India," *Pacific Affairs,* Vol. XXVI (1953), pp. 324–325.
25. See especially the essays of Ralph Linton, Melville J. Herskovits, and Morris E. Opler, in *The Progress of Underdeveloped Areas,* ed. by Bert F. Hoselitz, Chicago, 1952, pp. 109, 126, 90–91. Also, cf. Lewis S. Feuer, "End of Coolie Labor in New Caledonia," *Far Eastern Survey,* Vol. XV (1946), pp. 264–267. Kennard and MacGregor state: "The great opportunity for the anthropologist lies in the problem of the extent to which purposive change can be introduced, how rapidly, and with what consequence to different segments of the Society." ("Applied Anthropology in Government: United States," in *Anthropology Today,* "Anthropology and the Problems of Indian Administration," *The Southwestern Social Science Quarterly,* Vol. XVII (1937), pp. 1–10.
26. F. W. Walbank, *The Decline of the Roman Empire in the West,* London, 1946, p. 73.
27. A. V. Vasiliev, *History of the Byzantine Empire,* Madison, Wisconsin, 1928, Vol. I, pp. 191–197, 417–423. Peter Charanis, "Economic Factors in the Decline of the Byzantine Empire," *The Journal of Economic History,* Vol. XVIII (1953), p. 421.
28. Cf. J. B. Bury, *Selected Essays,* ed. by Harold Temperley, Cambridge University Press, p. 66, xxvi–xxvii.
29. Cf. Henry Charles Lea, *A History of the Inquisition in Spain,* New York, 1906, Vol. 1, pp. 137–142.
30. For the significance of indeterminacy in the analysis of economic change, cf. Lewis S. Feuer, "Indeterminacy and Economic Development," *Philosophy of Science,* Vol. 15 (1948), pp. 225–241.
31. The historian today finds the outburst of aggressive energy among the Mongols in the thirteenth century as much a "psychological riddle" as it was to the medieval chronicler who wrote: "God alone knows who they were and whence they came."

Cf. George Vernadsky, *The Mongols and Russia,* New Haven, 1953, p. 5; N. K. Gudzy, *History of Early Russian Literature,* translated by S. W. Jones, New York, 1949, p. 201.
32. Why, for example, did the Romans fail to develop the intensive agriculture, with the use of excrement as manure, which became prevalent in China and Japan? The historian answers: "We do not know," Cf. Vladimir G. Simkhovitch, *Toward the Understanding of Jesus,* New York, 1927, p. 111.
33. Bertrand Russell, *The Problem of China,* London, 1922, p. 19. "Freud's 'death wish' may be more or less mythological, but on the whole it affords a better explanation of the present behaviour of Europe than is possible on a purely economic view" ("Two Prophets," *The New Statesman and Nation,* Vol. XIII, 1937, p. 416).

Index

Aaron of Lincoln, 391–92
Aboab, Rabbi Isaac de Fonseca, 259–61, 266
Abravanel, Don Samuel, 400
The Acquisitive Society (Tawney), 345
Acton, Lord, 261
Adler, Alfred, 167
Adler, Friedrich, 186
Adler, Victor, 186
Aggression, 86, 167, 188, 432–33
Agus, Irving, 392, 404
Albo, Joseph, 174
Alexander, Samuel, 19, 189, 348–49
Altamira, 410
American Board of Missions to the Jews, 401
American Revolution, 325
Ampere, Andre, 6
Anselm (Saint), 74–80, 83, 87–88
Anthropology, 427
Anthropology (Kant), 212, 214, 223, 230–31, 238
Anti-Semitism, 165–78, 181–89
Aragon, 343
Aristotelian metaphysics, 106–7
Aristotle, 62, 343, 408
Armenian massacre, 166–67, 185
Arrowsmith (Lewis), 47
Askowith, Dora, 395
Athens, 362
Atomic bomb, 185
Ausonius, Decimus Magnus, 279–82
Autobiography (Maimon), 388
Autobiography (Mill), 370, 373

Auto-eroticism, 275
Averroes, 106–7, 398–97

Bacharach, Moses Samson, 389
Bacon, Francis, 185, 337
Baer, Yitzhak, 398
Baillet, Adrian, 269, 274–75, 292, 295–96, 299, 301–2
Bain, Alexander, 355
Bakunin, Mikhail, 375
Balling, Pieter, 254
Barnes, E.W., 6
Baron, Salo, 389–90, 395, 400
Barth, Karl, 75, 77–78, 83–84
Bateson, William, 407
Bauer, Otto, 186
Beeckman, Isaac, 314
Behaviorism, 199–200
Benedict of New York, 391
Ben-Israel, Menasseh, 259
Benjamin of Tudela, 405–6
Ben Sira, 387, 389
Bentham, Jeremy, 137–38, 304
Bergson, Henri, 188, 203, 347–48
Berkeley, George, 90, 345
Bernstein, Eduard, 186
Besso, Michele Angelo, 5
Bialik, Hayyim N., 182
Big Bang, 30–31, 55
Biped posture, 213
Black Death, 410
Black hole, 28
Blanqui, 377
Boas, Franz, 104–5, 112–13, 156

Boas, George, 211
Bohr, Niels, 366
Boltzmann, 10, 14
Bondi, Hermann, 7, 50–52
"Book of Fallacies" (Bentham), 138
The Book of Job as a Greek Tragedy (Kallen), 340
Book of Travels (Benjamin of Tudela), 405
Boolean algebra, 130
Born, Max, 5, 7–8, 10, 28, 42, 49, 366
Boyle, Robert, 155
Boyle's law, 125
Bradley, F.H., 173
Bradley, J., 58
Bridgman, P.W., 9, 49, 58–61
Brook Farm, 330
Brouillard, Henri, 83
Buber, Martin, 4, 305
Buffon, Georges, 123, 226
Bury, J.B., 430
Butler, Samuel, 239
Byzantine Empire, 430

Calvinism, 361, 371
Cambridge Apostles, 146, 149, 158
Camus, Albert, 309
Capitalism, 345–46, 361, 376–77, 426
Capital (Marx), 159
Carnap, Rudolf, 200–202
Catastrophic gravitational collapse, theory of, 56
Celibacy, 408–11
Chandrasekhar, Subrahayam, 28
Chaos theory, 1–2, 14–15, 19, 30
Chariots of Fire (film), 182
Charlesworth, M.J., 77
Checks and balances, 344
China, 369
Chinese language, 108–10
Chmielnicki Massacre, 171, 182, 328
Christianity, 186–87, 387
   and celibacy in Middle Ages, 408–11
Churchill, Winston, 184
Civilization, 197
Civil War, U.S., 332, 360, 367
Cleisthenes, 407
Cohen, Morris, 118, 142, 336–42
Collingwood, Robin, 92
Common sense, 192–93
Communist Party (Soviet), 398

Comte, Auguste, 244, 355–56, 358–59
Condorcet, Marquis de, 172, 368
Consciousness, 31–34
Conservation of energy, law of, 10–11
Constitution, U.S., 344
Copernican system, 131
Cornford, F.M., 105–6
Corpus Omnium Veterum Poetarium, 280–82
Cossacks, 171, 182
Coulton, G.G., 406–7
Court, Pieter de la, 343
The Creative Elite in America (Weyl), 386
Crescas, Don Hasdai, 21
Critique of Pure Reason (Kant), 211–12, 216, 219, 225, 242
Critique of Teleological Judgment (Kant), 227
Cromwell, Oliver, 327–28
Crusades, 77, 170, 393–94
Cultural pluralism, 339–40
Cultural relativism, 198

Dalton, John, 18
Dana, Charles A., 330
Darwin, Charles, 21–22, 25, 27, 66, 148, 188, 409
Death-wish, 167, 188, 432
Declaration of Independence, 327, 341
Democracy, 343–46
Democracy in America (Tocqueville), 344–45
De Polignac, Cardinal, 294
Descartes, Francine, 302
Descartes, Rene, 90, 118, 149, 218, 346
   animal automatism of, 290–94
   dreams of, 269–86
   four principal tenets of, 289
   maternal deprivation as source of anxiety in, 294–303
   ontological argument, 303–12
   psychological existence-argument, 312–17
   struggle in unconscious of, 289–90
De Scudery, Mlle., 291
Determinism, 323–24, 348, 363
Devil, disproof of existence of, 88–89
Dewey, John, 109, 143, 148, 153, 188
   Lovejoy on, 193–208
De Witt, John, 322

D'Holbach, 325
The Dial, 330–31
Diaz, Henrique, 258
Dickinson, Emily, 333
Dickinson, G. Lowes, 146
Dictatorship, 359
Diderot, Denis, 325
Diminishing returns, law of, 364
Dingle, Herbert, 55, 61–62
Diogenes Laertius, 408
Dirac, Paul, 9, 44, 51, 56–57, 59, 63
Discourse on Method (Descartes), 302, 309
DNA, 27
Double negation, 102–3
Dramatic present, 103
Dreams of a Spirit-Seer (Kant), 90–91, 215, 238–39, 241
Dreyfus, Alfred, 188
Durant, Will, 338
Durkheim, Emile, 24–25, 358, 378
Dutch West India Company, 257

Economic Consequences of the Peace (Keynes), 425
Eddington, Sir Arthur Stanley, 5, 9, 28, 57, 177
Edelman, Marek, 184
Eichmann, Adolf, 89–90
Einstein, Albert, 129, 131, 156, 188–89
 argument for existence of God, 1–35, 347
 determinism of, 159–60, 340
 mysticism of, 41, 54, 58–59, 63–64
 on simplicity in mathematics, 122–25, 366–67
 theory of relativity, 5–6, 8–9, 34, 59, 148, 153, 165
Eliot, George, 337
Emergent evolution, 32–34
Emerson, Ralph Waldo, 329–31
Encyclopedia Judaica, 402
Engels, Friedrich, 24, 367, 378
England, 396–97, 408, 410
The English Jewry under Angevin Kings (Richardson), 401
Enlightenment, 240
Entropy, law of, 14, 175, 432–33
Equilibrium, 429
Ergodic Theorem, 14
Eskimo, 112

Essay on Man (Pope), 170
Ethics (Spinoza), 20, 322–23, 334–35
Ethology, 355
Euphues (Lyly), 133
Evil, 171–73, 187
Evolutionary hypothesis, 226–29
Evsektsiia, 398
Ewing, A.C., 234
Existential indeterminacy, 431
Expanding universe, theory of, 53–57
Eybeschutz, Rabbi Jonathan, 389

February Revolution, 365
Feudalism, 360
The Feud of Oakfield Creek (Royce), 81–82
Feuerbach, Ludwig, 331
Finlay, George, 362
Fire symbolism, 277–78
First Principles (Spencer), 333
Fisher, R.A., 2, 6–27, 411
Fourier, Jean, 6
France, Anatole, 33
Frank, Philipp, 5
Frank, Tenney, 411
Franklin, Benjamin, 172, 189
Freedom of speech and thought, 326, 346
Frequency theory of probability, 12–13
Freud, Sigmund, 24, 78, 128, 141, 151, 159, 311
 on aggression, 167, 188, 432–33
 on child's concept of God, 304–5
 on Descartes dreams, 284
 on dreams, 232–33
 on fire, 277–78
 on gambling, 275
 on homosexual tendencies, 285
 on J.S. Mill, 370
 on lies childhood lies, 243–44
 on passion, 342
 on senses and sexuality, 212–14
 on slips of the tongue, 348
Friedmann, Alexander, 54
Frothingham, Octavius Roy, 333
Fuller, Margaret, 330–31
Functionalism, 427–29
Fung-Yu-Lan, 109

Galileo Galilei, 283
Galton, Francis, 407
Gambling, 275

440 Varieties of Scientific Experience

Gamow, George, 54, 60–61
Gassendi, Pierre, 290, 307
Gaunilo, 78–80
Gell-Mann, Murray, 42, 62
Gender, 101
Genesis, 15
Genetic analysis, 196, 206
Genetic fallacy
    divergence concerning psychoanalysis, 141–45
    genetic privilege and handicap, 145–50
    incompleteness of genetic analyses, 150–54
    modes of genetic analysis, 154–60
Germany, 408
Germany and Her Jews (Osborne), 404
Gershom, Rabbi, 393
Gilson, Etienne, 74, 83
Gitlow, Benjamin, 335
Gladston, Iago, 285
Glorious Revolution, 327
God, 166, 172
    concept of, 206, 208
    Descartes on, 303–12
    Einstein's argument for existence of, 1–35
    Freud on, 149, 151
    ontological argument for existence of, 73–93
    Spinoza on, 264–65, 321–23, 347
    and transcendentalists, 329–33
Godel, Kurt, 34–35, 174
The God that Failed, 176
Gold, Thomas, 50–52
Good, 175
Goropius, 121
Graetz, Heinrich, 171
Grant, C.K., 88
Greeks, 153, 280, 407
Greene, Graham, 313
Gregory XIII (Pope), 398
Grote, George, 370
Guilt-consciousness, 85–87
Guizot, Francois, 360

Haldane, Elizabeth, 295, 299, 301, 315
Haldane, J.B.S., 147, 385–86, 410–11
Haldeman-Julius Blue Books, 338
Hall, G.S., 246
Hamilton, Alexander, 344

Hamilton, Sir William Rowan, 10
Harrison, Jane, 106
Harrod, Roy, 424
Hawthorne, Nathaniel, 194
Hayek, F.A. von, 426–27
Hearing, sense of, 216
Hebrew, 107, 113, 156
Hedge, Frederic H., 332
Hegel, Georg, 91–92
Heidegger, Martin, 188
Heines, 211
Heisenberg, Werner, 28, 42, 49, 366
Hellenic period, 407
Henderson, Lawrence J., 29–30
Herzl, Theodor, 399–400
Higginson, Thomas Wentworth, 333
Hilberg, Raul, 393–94
Historicism, 342
History of British India (Mill), 357–58
Hitler, Adolf, 60, 169–71, 187, 328
Holland, 256
Holmes, Oliver Wendell Jr., 167, 324, 333–35, 412
Holocaust, 165–78, 181–89
Holyoake, George Jacob, 377
Homosexuality, 146–47, 149, 158, 407–8
Hook, Sidney, 142–43, 194
Hoover, Herbert, 424
Hoxie, Robert, 155–56
Hoyle, Fred, 50–52, 54
Hubble, Edwin P., 21, 53–54
Hudibras (Butler), 239
Human needs, 428
Hume, David, 16–20, 33, 84, 242, 245
Huxley, Thomas H., 20, 177

Inalienable rights, 325–26
Inaugural Dissertation (Kant), 242–43
Incest, 245
India, 364, 377, 429
Infeld, Leopold, 7, 64
Insanity, 222–24
International Index, 89
International Workingmen's Association, 373–75
Interventionist model, 421–33
Inverse deductive method, 356, 358
Israel, 166, 168, 171
Italy, 396

James, William, 9, 53–54, 80, 110–11,

154–55, 167, 175, 187, 237
Jeans, Sir James, 5, 9
Jefferson, Thomas, 152, 325–27, 341
Jerome (Saint), 283, 405
Jespersen, 101, 103, 115, 117
Jevons, William, 124, 158
Jewish Daily Forward, 337
The Jewish Tribune, 337
Jews
    backgrounds of Jewish scientists, 402–4
    conversions of, 395–402, 431
    in early Middle Ages, 404–7
    and Holocaust, 168–78, 181–89
    influence of Spinoza on, 336–42, 346–49
    intellectual achievement of, 385–413; and education, 388–92; and marriage practices, 386–88
    massacres of, 393–95
    in Middle Ages, 407–11
    and revolution, 328
    in sixteenth century Holland, 260–63
Jones, Ernest, 128
Josephus, Flavius, 388
Jung, Rabbi Leo, 403

Kallen, Horace M., 324, 339–42
Kant, Immanuel, 35, 84, 90–91, 99, 149, 172, 177, 311, 321
    antipsychological revolution in life of, 237–46
    and *Critique of Pure Reason*, 211–12, 216, 219, 225, 242
    dreams of, 230–33
    antagonism of to evolutionary hypothesis, 226–29
    analysis of insanity, 222–24
    lawless element in psyche of, 224–26
    attack on psychological method, 234–37
    estrangement from sensation, 212–17
    and transcendental deduction, 217–22
Kaufmann, Walter, 91
Kelsen, Hans, 407
Kelvin, Lord (William Thomson), 124, 358
Kepler, Johannes, 66, 130–31
Keynes, John Maynard, 158, 424–26, 429
Keyserling, Contess, 238

Khmelnitzki, Bogdan, 393
Khruschchev, Nikita, 343
Kierkegaard, Soren, 85
Kung-sun Lung, 108

La Fontaine, 291
Laird, John, 64
Lamarck, Chevalier de, 26
La Naissance de la Paix (Descartes), 299–300
Language, 99–100
    common, universal categories, 115–19
    linguistic relativity, 111–15
    social sources of metaphors, 108–11
    syntactic variation and philosophic ideas, 101–7
Lankester, E. Ray, 189
La Piana, George, 261
Lardner, Ring, 102
Larmor, Sir Joseph, 62
Lassalle, Ferdinand, 186
Latin, 116–18
Lavoiser, Antoine, 128–29
Laws (Plato), 408
Least action, principle of, 10
Le Chatelier's Law, 31
Leeuwenhoek, Anton van, 256
Leibniz, Gottfried, 3, 9–10, 28, 42–44, 62–63, 90, 121–22, 178
    industrial interests of, 153–55
    on innate ideas, 144–45, 151
Lemaitre, Abbe George, 55
Lenin, V.I., 135, 157, 346
Leroy, Maxime, 296
Leuba, James H., 23
Lewes, George Henry, 337
Lewis, Sinclair, 47
The Light on the Candlestick (Balling), 254
Linguo-genetic theory, 156
Lippmann, Walter, 324, 342
Lisbon Earthquake, 170, 172
Lives of the Philosophers (Diogenes Laertius), 408
Locke, John, 73, 144–45, 149, 151, 154, 291, 324, 327
Loeb, Jacques, 42, 47–49
Loewenberg, Jacob, 205
Logan, Thomas, 325
Logic (Clauberg), 133

Logic (Mill), 152
The Logic of Modern Physics (Bridgman), 59
Longfellow, Henry Wadsworth, 333
Lorentz transformations, 6
Lovejoy, Arthur O., 42, 151, 191–208
  on behaviorism, 199–200
  on Dewey's pragmatism, 193–98
  on Kant, 226–27
  on positivism, 200–202
Luddites, 432
Lugard, Lord, 427
Lying, 243–44
Lyly, John, 133

Mach, Ernst, 42, 57–58, 65, 132–33, 149, 153
Machajski, Waclaw, 157
Machiavelli, Niccolo, 343, 345
Mach's Principle, 57–58
Madison, James, 344
Maimon, Solomon, 388, 390, 399
Maimonides (Moses Ben Maimon), 170, 174, 387
Malcolm, Norman, 74–75, 85–87
Malinowski, Bronislaw, 115, 128, 427–28
Malthus, Thomas, 24, 367, 369
Manichaeanism, 366–69
Maritain, Jacques, 284
Marsh, James, 331
Marx, Heinrich, 399
Marx, Karl, 25, 152, 159, 356, 367–68, 371, 378
  on Descartes, 291, 311
  on greed and power, 359–60
  and International Workingmen's Association, 373–75
  on religion, 265–66
Marxism, 143, 157–59, 426–27
Mason, Gabriel R., 338–39
Massaniello (Tomas Aniello), 263–64
Maternal deprivation, 294–303
Maupertuis, 10, 122
Maurice, Prince of Orange, 296, 298, 300
Maxwell, Clerk, 8
Maxwell, James C., 31
Mayall, H.U., 53
McTaggart, John, 35, 146
Mead, George Herbert, 198
Mediocrity, 358–59

Meditations (Descartes), 302, 309
Mencken, H.L., 103
Mendel, Gregor, 156
Mendeleev, Dmitri, 43–46, 62
Mendelssohn, Moses, 212, 399
Mennonites, 254
Men of the Spirit (Jung), 403
Messiah, 173–74
Metaphysics, 239–40, 242, 312
Meyerson, Emile, 4, 31, 347
Michels, Robert, 25, 157, 422–23
Middle Ages, 404–11
Mill, James, 357–58, 365, 367, 371–72
Mill, John Stuart, 24–25, 33, 124, 132, 150, 152, 157, 324
  failure as sociologist, 355–78
  Manichaeanism of, 366–69
  on revolutionary socialism, 374–78
  stationary state theory of, 368–69
Milton, 222
Miranda, Francisco de, 371–72
Miranda, Luis de, 409
Mirandola, Pico della, 399
Mises, Ludwig von, 141–42, 426
Mishnah, 387
Mithridates, Flavius, 399
Modder, Montagu Frank, 396–97
Moliere (Jean Baptiste Poquelin), 133, 291
Monash, Sir John, 184
Monasticism, 408–11
Moore, G.E., 146–47, 149
Morley, John, 366
Moscati, 213
Moses ha-Sephardi (Petrus Alfonsi), 397
Moses Israel of Salonica, 400
Moslems, 387, 398–99
Mo Ti, 108–9
Muller, Max, 109
Mydorge, Claude, 283

Naples, 263–64
Natural Theology (Paley), 21
Nature, 66, 122–25, 227–28, 347
"Nature" (Mill), 359
Nazism, 5, 89, 126, 148, 156, 167, 185, 202, 285, 393
Necessitarian model, 421–33
Ne'eman, Y., 43
Newton, Sir Isaac, 15, 122, 124, 130–31
New York school (psychoanalysis), 143

Niebuhr, Barthold, 207
Nietzschean Recurrence, 14
Norton, Rev. Andrews, 330
Noumenalism, 4–35

Oberndorf, C.P., 278
Observations on the Feeling of the Beautiful and Sublime (Kant), 225
Occam's Razor, 121–37, 150
October the First is Too Late (Hoyle), 52
Oedipus Complex, 128, 363–64
Old News, 194
Oligarchy, 422–23
On Liberty (Mill), 365, 369
"On the Logic of the Moral Sciences" (Mill), 355
Ontological argument, 303–12
Operationism, 59–61
Oppenheimer, J. Robert, 55–56
The Organism as a Whole (Loeb), 48
The Origin of Species (Darwin), 21–22
Osborne, Sidney, 404
Ovid, 296

Pantheism, 329–32
Pareto, Vilfredo, 25, 367
Parker, Theodore, 331
Parricide, 294, 301, 311
Pauli, Wolfgang, 43
Paul (Saint), 186–87, 387
Paulsen, F., 242–43
Pedagogy (Kant), 225
Peirce, Charles, 19–22, 82, 187, 192
Perez, Antonio, 343
Periodic Table of the Elements, 43–46
Pernambuco, Brazil, 257–59
Pfister, O., 246
Phenomenalism, 6
Phenomenology (Hegel), 91
Phillips, Wendell, 333
Philosophes, 18
Philosophical Commentaries (Berkeley), 90
Piaget, Jean, 147, 151–52
Pinson, K.S., 243
Planck, Max, 10
Plato, 62, 343, 408
Platonism, 105–6
Plenitude, principle of, 42–43
Poem on the Disaster of Lisbon (Voltaire), 170

Pogroms, 182
Political Parties (Michels), 422–23
Political Treatise (Spinoza), 343
Politics (Aristotle), 408
Pollock, Sir Frederick, 334, 412
Pope, Alexander, 170
Popper, Karl, 50
Portugal, 258, 396
Positivism, 152–53, 156, 200–202
Positron, 43–44
Pragmatism, 110–11, 188
of Dewey, 193–98
Primal Hypothesis, 64–66
Primitive languages, 104–5
Principia Mathematica (Newton), 131
The Principles of Chemistry (Mendeleev), 45–46
The Principles of Descartes' Philosophy (Spinoza), 254
Principles of Political Economy (Mill), 355, 364–69
Prolegomena (Kant), 222
Protestant Ethic, 133
Protestant Reformation, 410
Protophilosophy, 211
Proudhon, Pierre, 374–75
Psychoanalytic theories, 141–44
Psychological-existence-argument, 312–17
Psychological method, 234–37, 245
Ptolemaic system, 131

Quakers (English), 254
Quetelet, Lambert, 356

Radical Evil, 171–73
Rashdall, Hastings, 151
Rationalism, 303
Realism, psychoanalytical, 205–6
Red shift, 54
Reichenbach, Hans, 141
Relativity, 5–6, 8–9, 34, 59, 148, 153, 165, 347
Repression-expression schedule, 428
Rescher, Nicholas, 88
Revolution, 327–29
Revolutionary socialism, 374–78
Richardson, H.G., 401
Ripley, George, 330–31
Rolland, Romain, 188
Roman Catholic Church, 408

Roman Empire, 186–87, 362, 395, 408, 430
Roscoe-Bunsen light energy law, 48
Rosenfield, Leonora Cohen, 291
Rosicrucians, 300–301
Roth, Cecil, 392, 396, 400
Rousseau, Jean Jacques, 245–46
Royce, Josiah, 74–82, 87, 110–11
Ruja, Harry, 92
Runyon, Damon, 103
Ruppin, Arthur, 386
Russell, Bertrand, 9, 24, 29, 62, 66–67, 93, 148, 167, 177, 188, 366, 432
  on analysis of language, 99–100, 116, 156, 317
  on principle of simplicity, 134–37
Russell, J.C., 408

Sagan, Carl, 29
Saint-Simonians, 370, 372–73
Santayana, George, 147–48, 342
Sayce, 100, 102
Schechter, Solomon, 387
Schiller, F.C.S., 133
Schiller, Johann, 222
Schonberger, Stephen, 285–86
Schopenhauer, Arthur, 136, 212
Schrodinger, Erwin, 9, 12, 58, 132
Schumpeter, Joseph, 346, 424–26
Scientific plenitude, principle of, 43–46
Scientific Revolution, 285
Scottish Renaissance, 18
Self-aggression, 312, 315
Semitic languages, 106–7
Sensation, 212–17
Sexuality, 220–22, 228–29
Shaftesbury, Earl of (Anthony Ashley Cooper), 239, 241, 243, 245
Shaw, Bernard, 25, 176
Sidgwick, Henry, 141, 149
Simplicity
  biological basis of, 132–34
  critique of mathematical, 129–32
  meta-scientific principle of, 121–25
  Occam's Razor as case of verifiability, 125–29
  pseudophilosophical principle of, 134–38
Singer, Isaac B., 183–84, 388
Slavery, 330, 360
Slave trade, 256–58

Smell, sense of, 212–13
Smith, Norman Kemp, 280, 282, 290
Socialism, 346
Social science, 421–33
Sociology, 356
Solzhenitsyn, Aleksandr, 328
Sommerfeld, Arnold, 5
Soviet Union, 426
Spain, 408–10, 431
Spencer, Herbert, 24, 154, 188, 336, 356
Spinoza, Benedict de, 1–3, 20, 35, 88–90, 153, 189, 310, 321–22
  on democracy, 343–46
  determinism of, 323–24
  dreams of, 253–56
  excommunication of, 260–63
  on God, 264–65
  influence on first generation Jewish immigrants, 336–42
  influence on Oliver Wendell Holmes, Jr., 333–35
  influence on transcendentalists, 329–33
  influence on twentieth century scientists and philosophers, 346–49
  on natural rights, 325–27
  on revolution, 327–29
Spinoza: His Life and Philosophy (Pollock), 334
Stalin, Joseph, 166, 183–84, 187, 343, 346
Stationary state theory, 368–69
Steady-state theory, 50–52
Steiner, George, 181, 184, 186–87
The Story of Philosophy (Durant), 338
Strachey, Lytton, 146
Subjunctive mood, 101–2
Suicide, 357
Swedenborg, Emanuel, 240–41
System of Logic (Mill), 355–63

Tachyons, 41
Talmud, 106, 387–89
Tawney, R.H., 152, 345, 427
Taylor, Harriet, 370, 372
Teleological principles, 41–67
Thermodynamics, second law of, 14
Tocqueville, Alexis de, 329, 331–32, 344–45, 356
Totalitarian society, 168

Toynbee, Arnold, 50
Tractatus Theologico-Politicus (Spinoza), 325
Transcendental deduction, 217–22, 234–38
Transcendentalism, 329–33
Trotsky, Leon, 426
Twain, Mark, 412
Tyutchev, Fyodor Ivanovich, 44–45
Tzarfati, Joseph, 398

Unified field theories, 4, 7–9
Unity of Science Congress, 177
Urban Civilization in Pre-Crusade Europe (Agus), 392
Urban II (Pope), 77
Uriel Acosta (play), 261–62, 337

Vaihinger, Hans, 218
Van der Waals' law, 125
Veblen, Thorstein, 48, 118, 152–53, 155–56
Vecinho, Joseph, 397–98
Venice, 343, 345
Verifiability, principle of, 125–29

Vernacular language, 117–18
Vichy France, 176
Vision, sense of, 213–15
Voltaire (Francois Marie Arouet), 170

Warsaw Ghetto, 183–84
Wasianski, 230, 232
Watson, J.B., 317
Webb, Sidney and Beatrice, 422
Weber, Max, 24–25, 361, 367
Wells, H.G., 25
Weyl, Nathaniel, 385–86
What Is to Be Done? (Lenin), 157
Whitehead, Alfred N., 62
Whittaker, Edmund, 368
Whorf, B.L., 111–12, 114
Wiener, Norbert, 385–86
Wisdom, J.O., 285–86
Wittgenstein, Ludwig, 84–86
Wolfson, Harry A., 74, 308
World Health Organization, 429
World War I, 32, 143–44, 167, 185
World War II, 165–78, 332

Zionism, 341